Radiation Effects of Advanced Electronic Devices and Circuits

Radiation Effects of Advanced Electronic Devices and Circuits

Editors

Yaqing Chi
Li Cai
Chang Cai

Basel • Beijing • Wuhan • Barcelona • Belgrade • Novi Sad • Cluj • Manchester

Editors

Yaqing Chi
College of Computer
National University of
Defense Technology
Changsha
China

Li Cai
Institute of Modern Physics
Chinese Academy of Sciences
Lanzhou
China

Chang Cai
State Key Laboratory of ASIC
and System
Fudan University
Shanghai
China

Editorial Office
MDPI
St. Alban-Anlage 66
4052 Basel, Switzerland

This is a reprint of articles from the Special Issue published online in the open access journal *Electronics* (ISSN 2079-9292) (available at: https://www.mdpi.com/journal/electronics/special_issues/K198F4P50R).

For citation purposes, cite each article independently as indicated on the article page online and as indicated below:

Lastname, A.A.; Lastname, B.B. Article Title. *Journal Name* **Year**, *Volume Number*, Page Range.

ISBN 978-3-7258-1481-7 (Hbk)
ISBN 978-3-7258-1482-4 (PDF)
doi.org/10.3390/books978-3-7258-1482-4

Cover image courtesy of Yaqing Chi

© 2024 by the authors. Articles in this book are Open Access and distributed under the Creative Commons Attribution (CC BY) license. The book as a whole is distributed by MDPI under the terms and conditions of the Creative Commons Attribution-NonCommercial-NoDerivs (CC BY-NC-ND) license.

Contents

About the Editors vii

Yaqing Chi, Chang Cai and Li Cai
Radiation Effects of Advanced Electronic Devices and Circuits
Reprinted from: *Electronics* **2024**, *13*, 1073, doi:10.3390/electronics13061073 1

Mingzhu Xun, Yudong Li, Jie Feng, Chengfa He, Mingyu Liu and Qi Guo
Effect of Proton Irradiation on Complementary Metal Oxide Semiconductor (CMOS) Single-Photon Avalanche Diodes
Reprinted from: *Electronics* **2024**, *13*, 224, doi:10.3390/electronics13010224 7

Hualiang Zhou, Hao Yu, Zhiyang Zou, Zhantao Su, Qianyun Zhao, Weitao Yang and Chaohui He
Evaluation of Single Event Upset on a Relay Protection Device
Reprinted from: *Electronics* **2024**, *13*, 64, doi:10.3390/electronics13010064 19

Ruiqiang Song, Jinjin Shao, Yaqing Chi, Bin Liang, Jianjun Chen and Zhenyu Wu
Machine Learning-Based Soft-Error-Rate Evaluation for Large-Scale Integrated Circuits
Reprinted from: *Electronics* **2023**, *12*, 4978, doi:10.3390/electronics12244978 33

Yutang Xiang, Xiaowen Liang, Jie Feng, Haonan Feng, Dan Zhang, Ying Wei, et al.
Refined Analysis of Leakage Current in SiC Power Metal Oxide Semiconductor Field Effect Transistors after Heavy Ion Irradiation
Reprinted from: *Electronics* **2023**, *12*, 4349, doi:10.3390/electronics12204349. 48

Lan Lin, Zhongchao Cong and Chunlei Jia
Recovery Effect of Hot-Carrier Stress on γ-ray-Irradiated 0.13 μm Partially Depleted SOI n-MOSFETs
Reprinted from: *Electronics* **2023**, *12*, 4233, doi:10.3390/electronics12204233 60

Mingyu Liu, Chengfa He, Jie Feng, Mingzhu Xun, Jing Sun, Yudong Li and Qi Guo
Analysis of Difference in Areal Density Aluminum Equivalent Method in Ionizing Total Dose Shielding Analysis of Semiconductor Devices
Reprinted from: *Electronics* **2023**, *12*, 4181, doi:10.3390/electronics12194181 74

Tongde Li, Jingshuang Yuan, Yang Bai, Chunqing Yu, Chunliang Gou, Lei Shu, et al.
Research on High-Dose-Rate Transient Ionizing Radiation Effect in Nano-Scale FDSOI Flip-Flops
Reprinted from: *Electronics* **2023**, *12*, 3149, doi:10.3390/electronics12143149 87

Xiaowen Liang, Haonan Feng, Yutang Xiang, Jing Sun, Ying Wei, Dan Zhang, et al.
Oxide Electric Field-Induced Degradation of SiC MOSFET for Heavy-Ion Irradiation
Reprinted from: *Electronics* **2023**, *12*, 2886, doi:10.3390/electronics12132886 95

Zhikang Yang, Lin Wen, Yudong Li, Jie Feng, Dong Zhou, Bingkai Liu, et al.
Heavy Ion Single Event Effects in CMOS Image Sensors: SET and SEU
Reprinted from: *Electronics* **2023**, *12*, 2833, doi:10.3390/electronics12132833 107

Jie Feng, Hai-Chuan Wang, Yu-Dong Li, Lin Wen and Qi Guo
Mechanism of Total Ionizing Dose Effects of CMOS Image Sensors on Camera Resolution
Reprinted from: *Electronics* **2023**, *12*, 2667, doi:10.3390/electronics12122667 118

Haonan Feng, Xiaowen Liang, Xiaojuan Pu, Yutang Xiang, Teng Zhang, Ying Wei, et al.
Total Ionizing Dose Effects of ^{60}Co γ-Ray Radiation on Split-Gate SiC MOSFETs
Reprinted from: *Electronics* 2023, *12*, 2398, doi:10.3390/ electronics12112398 129

Weitao Yang, Yonghong Li, Yang Li, Zhiliang Hu, Jiale Cai, Chaohui He, et al.
Neutron Irradiation Testing and Monte Carlo Simulation of a Xilinx Zynq-7000 System on Chip
Reprinted from: *Electronics* 2023, *12*, 2057, doi:10.3390/electronics12092057 139

Yihao Cui, Jie Feng, Yudong Li, Lin Wen and Qi Guo
Proton Radiation Effects of CMOS Image Sensors on Different Star Map Recognition Algorithms for Star Sensors
Reprinted from: *Electronics* 2023, *12*, 1629, doi:10.3390/electronics12071629 149

Xiao Li, Jiangwei Cui, Qiwen Zheng, Pengwei Li, Xu Cui, Yudong Li and Qi Guo
Study of the Within-Batch TID Response Variability on Silicon-Based VDMOS Devices
Reprinted from: *Electronics* 2023, *12*, 1403, doi:10.3390/electronics12061403 164

Zheng Zhang, Gang Guo, Futang Li, Haohan Sun, Qiming Chen, Shuyong Zhao, et al.
Effects of Different Factors on Single Event Effects Introduced by Heavy Ions in SiGe Heterojunction Bipolar Transistor: A TCAD Simulation
Reprinted from: *Electronics* 2023, *12*, 1008, doi:10.3390/electronics12041008 177

Hongyu Ding, Jiangwei Cui, Qiwen Zheng, Haitao Xu, Ningfei Gao, Mingzhu Xun, et al.
Effect of Trapped Charge Induced by Total Ionizing Dose Radiation on the Top-Gate Carbon Nanotube Field Effect Transistors
Reprinted from: *Electronics* 2023, *12*, 1000, doi:10.3390/electronics12041000 196

Xiaoyu Pan, Hongxia Guo, Chao Lu, Hong Zhang and Yinong Liu
The Inflection Point of Single Event Transient in SiGe HBT at a Cryogenic Temperature
Reprinted from: *Electronics* 2023, *12*, 648, doi:10.3390/electronics12030648 210

Jindou Xin, Xiang Zhu, Yingqi Ma and Jianwei Han
Study of Single Event Latch-Up Hardness for CMOS Devices with a Resistor in Front of DC-DC Converter
Reprinted from: *Electronics* 2023, *12*, 550, doi:10.3390/electronics12030550 221

Peixiong Zhao, Bo Li, Hainan Liu, Jinhu Yang, Yang Jiao, Qiyu Chen, et al.
The Effects of Total Ionizing Dose on the SEU Cross-Section of SOI SRAMs
Reprinted from: *Electronics* 2022, *11*, 3188, doi:10.3390/electronics11193188 233

About the Editors

Yaqing Chi

Yaqing Chi, Ph.D., Associate Researcher, Master's Supervisor in the School of Computer, National University of Defense Technology, director of the Hunan Aerospace Society, and member of the "High Performance Microprocessor Technology" Innovation Team of the Ministry of Education of China. He received the Ph.D. degree in 2009 from the National University of Defense Technology and has worked there since. He leads a team testing radiation effects for advanced integrated circuits, and has conquered many key technologies such as single-event effect characterization and measurement for nano-integrated circuits, radiation-resistance testing, and evaluations of complex microprocessors and so on. He participated in the development and evaluation of many aerospace microprocessors, which have been widely applied in aerospace projects such as satellites, rockets, spacecrafts, space stations, and deep space explorations. His research interest includes the radiation effects in integrated circuits, radiation-hardening techniques for microprocessors based on advanced semiconductor processes, high-speed radiation-tolerant SerDes, and FPGAs. He has won an award of national scientific and technological progress and four awards of the provincial and ministerial levels, obtained more than 20 national invention patent authorizations, and published more than 40 papers in important journals and conferences in the field of integrated circuit radiation effects.

Li Cai

Li Cai, Ph.D., Researcher, was selected for the CAS Talent Program Youth Project, Doctor's Supervisor in the Institute of Modern Physics, Chinese Academy of Sciences. After graduating with a Master's degree in 2009, she worked in the Chinese Academy of Atomic Energy (CIAE), and obtained her doctoral degree from the CIAE in 2017. She began to work in the Institute of Modern Physics in 2020. The main research field is the radiation effect of semiconductor devices, especially the experimental technology based on accelerators. Dr. Cai is skillful in broadbeam and microbeam irradiation experiments, has established a sample temperature measurement and control system based on the SEE test terminal, and has evaluated the impact of temperature on SEE in SRAMs and inverter chains. During the work, she has collaborated with several universities and institutes to study the radiation effects on SRAMs, DSPs, FinFET inverter chains, CMOS image sensor, MRAM, etc. She has won the second prize for National Defense Science and Technology Progress and first prize for the China National Nuclear corporation Science and Technology Award, and she has published more than 20 papers as the first or corresponding author in important journals and conferences in the field of radiation effects on semiconductor devices. As the project leader, she has led multiple scientific research projects, such as the National Natural Science Foundation of China, national defense pre-research, talent project, etc.

Chang Cai

Chang Cai received his Ph.D. degree from the Institute of Modern Physics, Chinese Academy of Sciences, in 2021. After graduation, he joined Fudan University as a researcher, where he has continued to make significant contributions to the field of radiation effects in integrated circuits. Dr. Cai's research focuses on a wide range of topics within this field, including the radiation mechanisms for bulk, FDSOI, and FinFET devices. He investigates the single-event effects in advanced novel electronic devices, which are critical for understanding and mitigating radiation-induced failures. His work also includes the design of fault-tolerant field-programmable gate arrays (FPGA) and systems on chip (SoC), where he develops strategies to enhance radiation hardness through both layout and logical design. In addition to his theoretical and design work, Dr. Cai is actively involved in practical experiments. He conducts radiation experiments in extreme conditions to test and validate the robustness of various electronic devices. His expertise includes single-event effect characterization and evaluation techniques for VLSI, which are essential for ensuring the reliability of complex integrated circuits in radiation environments. Dr. Cai has published more than 30 peer-reviewed papers as the first or corresponding author in recent five years. His research has been widely recognized and presented at numerous international conferences, including prestigious symposiums such as the IEEE IRPS.

Editorial

Radiation Effects of Advanced Electronic Devices and Circuits

Yaqing Chi [1,2], Chang Cai [3,*] and Li Cai [4]

[1] College of Computer, National University of Defense Technology, Changsha 410073, China; yqchi@nudt.edu.cn
[2] Key Laboratory of Advanced Microprocessor Chips and Systems, National University of Defense Technology, Changsha 410073, China
[3] State Key Laboratory of ASIC and System, Fudan University, Shanghai 201203, China
[4] Institute of Modern Physics, Chinese Academy of Sciences, Lanzhou 730000, China; caili@impcas.ac.cn
* Correspondence: caichang@fudan.edu.cn

1. Introduction

Research on the effects of radiation on advanced electronic devices and integrated circuits has experienced rapid growth over the last few years, resulting in many approaches being developed for the modeling of radiation's effects and the design of advanced radiation-hardened electronic devices and integrated circuits [1–10]. With the progressive scaling of integrated circuit technologies and the growing complexity of electronic devices, their susceptibility to radiation's effects has presented many exciting challenges that are expected to propel research in the coming decade [11–14]. Additionally, regarding single-event effects (SEEs), continued scaling has drastically introduced new challenges, resulting in multiple-cell upsets, multipulse propagations, and other complex effects [15–22]. These issues necessitate the development of new solutions to assess and mitigate radiation sensitivity in advanced devices and integrated circuits.

The first edition of "Radiation Effects of Advanced Electronic Devices and Circuits" features nineteen high-quality submissions that showcase emerging applications and address recent breakthroughs. One key focus is the exploration of materials and device architectures designed to enhance radiation tolerance. This Special Issue also studies the development of advanced simulation tools and modeling techniques for accurately predicting the behavior of electronic devices exposed to radiation. These efforts encompass the refinement of existing simulation methodologies and the development of new computational approaches to better capture the complex interactions between radiation particles and basic materials. Additionally, this Special Issue addresses the growing importance of testing and validation methodologies for assessing the radiation hardness of integrated circuits and electronic systems. Researchers are exploring innovative testing protocols to ensure the reliability and robustness of electronic components in radiation environments, highlighting recent advancements in the field of radiation-tolerant electronics for space applications. Overall, this Special Issue serves as a comprehensive platform for researchers to showcase their latest findings and advancements in the effects of radiation on advanced electronic devices and circuits. By addressing a wide array of topics spanning from fundamental mechanisms to practical applications, this first Special Issue aims to foster collaboration and innovation within the radiation effects community and to contribute to the ongoing advancement of radiation-hardened electronics technology.

2. Highlighting Key Contributions

The nineteen articles in this Special Issue focus on not only systematic evaluation methods such as technology computer-aided design (TCAD), geometry and tracking (GEANT4), and novel numerical computation techniques but also the basic mechanisms and hardening results regarding the radiation performance of key components or devices such as sensors,

Citation: Chi, Y.; Cai, C.; Cai, L. Radiation Effects of Advanced Electronic Devices and Circuits. *Electronics* 2024, *13*, 1073. https://doi.org/10.3390/electronics13061073

Received: 27 February 2024
Accepted: 8 March 2024
Published: 14 March 2024

Copyright: © 2024 by the authors. Licensee MDPI, Basel, Switzerland. This article is an open access article distributed under the terms and conditions of the Creative Commons Attribution (CC BY) license (https://creativecommons.org/licenses/by/4.0/).

FinFET, silicon-on-insulator (SOI), system-on-chip (SoC), direct current (DC)–DC converters, SiC, heterojunction bipolar transistors (HBTs), and carbon nanotubes (Contribution 1–19).

With the development of integrated circuit technology, radiation's effects such as total ionizing dose (TID) effects, the high-dose-rate transient ionizing radiation response, and the single-event upset (SEU) of electron devices under advanced SOI CMOS processes have attracted considerable attention. Three articles provide recent and relevant research on the effects of radiation on SOI technology. The detailed TID effects and SEU features for SOI static random-access memories (SRAMs) with different layout structures were explored by Zhao, P. et al. (Contribution 1). The experimental results indicate that the SEU cross-sections are not only influenced by TID irradiation but also closely related to the layout structure of the memory cells. Li, T. et al. (Contribution 13) conducted an experimental and simulation study on the high-dose-rate transient ionizing radiation response and factors influencing fully depleted SOI (FDSOI) D flip-flop (DFF) circuits. The results demonstrate that the number of errors in DFFs nonlinearly increases with increasing dose rate, and the increasing supply voltage leads to an increase in data errors due to increased charge collection efficiency. Lin, L. et al. (Contribution 15) investigated the effect of hot-carrier injection (HCI) on γ-ray-irradiated partially depleted (PD) SOI n-MOSFETs with a T-shaped gate structure. The results indicate that the HCI has a recovery effect on the long-term reliability of n-MOSFETs when applied to a space environment.

Bulk silicon complementary metal oxide semiconductor (CMOS) devices encounter distinct single event latch-up (SEL) problems in aerospace. The traditional method fails to release devices from the latch-up state due to the narrow resistance range. Therefore, Xin, J. et al. (Contribution 2) developed an improved design for the resistor in front of the DC–DC buck converter, which increases the resistance range according to the input characteristics of the DC–DC buck converter. The method enhances the latch-up hardness performance by expanding the resistance range in comparison with that of the conventional design.

Some studies focused on the basic radiation effects of transistors or diodes have been published in our Special Issue. Pan, X. et al. (Contribution 3) investigated the inflection point of a single-event transient in a SiGe HBT. The collector's transient inflection point is jointly determined by the transient current of the emitter, substrate, and base, and the characteristics of the transient peaks widely vary among electrodes. Additionally, the contributors proposed a method to introduce the initial ionized EHPs' distribution of the Geant4 simulation to a TCAD simulation, thereby increasing the simulation accuracy and efficiency of the heavy-ion-induced SEE. To understand the microphysical mechanism of SEEs in SiGe HBTs, the effects of the heavy-ion striking location, incident angle, LET value, projected range, ambient temperature, and bias state were investigated by Zhang, Z. et al. (Contribution 5). The results indicate that the current transient peak value increases with the LET and the projected range of the heavy ions and decreases with the ambient temperature. The SEEs of SiGe HBTs are influenced not only by heavy-ion irradiation parameters such as the incident angle, LET value, and projected range but also by the striking location, ambient temperature, and bias state. In addition, the effects of proton irradiation on CMOS single-photon avalanche diodes with and without shallow trench isolation were examined by Xun, M. et al. (Contribution 19). The I–V characteristics, dark count rate, and photon detection probability of the diodes were measured under proton irradiation, contributing to meeting the dramatically increasing demands for satellite-to-ground quantum communication and space environment detection. Furthermore, semiconductor devices have entered the post-Moore era, where new materials and new technology have emerged. The excellent performance and radiation-hardness potential of carbon nanotube field-effect transistors (CNTFETs) have widely attracted attention. Ding, H. et al. (Contribution 4) investigated the TID effect of top-gate structure CNTFETs and the influence of the substrate on top-gate during irradiation. Studies regarding the influence mechanism of trapped charge introduced by TID irradiation on the characteristics of the top-gate CNTFETs are urgently needed for the design of CNT-based devices.

SiC power devices require resistance to both SEEs and TIDs in a space radiation environment, and several articles in our Special Issue present detailed results on simulation or irradiation experiments. Li, X. et al. (Contribution 6) investigated the impact mechanism and regularity of using the split-gate-enhanced process to determine the radiation resistance and long-term reliability of SiC vertically diffused MOS (VDMOS). The split-gate-enhanced VDMOSFET process can effectively enhance the radiation resistance of SiC VDMOS but impacts on the gate oxide reliability of SiC VDMOSs. Feng, H. et al. (Contribution 9) investigated the impact mechanism and regularity of using the SGE process to determine reliability of SiC VDMOS under radiation conditions. The use of the new process leads to more defects in the oxide layer, reducing the long-term reliability of the device, but its stability recovers after accelerated high-temperature annealing. Liang, X. et al. (Contribution 12) experimentally studied heavy-ion irradiation with different particle LETs, gate biases, and drain biases. The experimental results, along with those of TCAD simulations, suggest that the latent damage induced by irradiation in gate oxide is closely related to the peak electric field in the gate oxide at the time of particle incidence. The peak electric field, determined via the potential difference between the two sides of the gate oxide, is affected by the particle LETs, gate biases, and drain biases together. The leakage current is the most critical parameter for characterizing heavy-ion radiation damage in SiC MOSFETs. Moreover, an accurate and refined analysis of the source and generation process of leakage current is the key to revealing the failure mechanism. Xiang, Y. et al. (Contribution 16) finely tested the online and postirradiation leakage changes in and leakage pathways of SiC MOSFETs caused by heavy-ion irradiation, reverse-analyzed the damaged location of the device, and discussed the mechanism of leakage generation. The experimental results further confirm that an increase in the leakage current of a device during heavy-ion irradiation is positively correlated with the applied voltage of the drain, but the leakage path is indirect from the drain to the source. This study provides a theoretical basis for the radiation resistance reinforcement of SiC power devices.

Star sensors are widely used on satellites owing to their precise pointing accuracy. However, space radiation environments ill cause cumulative effects and single-event transients (SETs) in the imaging systems of star sensors, which can affect their star map recognition success rate. In this Special Issue, three articles illustrate the radiation effects on sensors. Cui, Y. et al. (Contribution 7) individually analyzed the influence of the decrease in the number of stars to be identified caused by proton irradiation, hot pixels, and SET spots on the success rate of different star map recognition algorithms. The findings of this study provide theoretical and technical bases for the improvement in star map recognition algorithms for long-term on-orbit star sensors. In addition, Feng, J. et al. (Contribution 10) conducted gamma-ray TID radiation experiments on CMOS image sensors and camera systems, and they thoroughly analyzed the impact mechanisms of dark current, full well capacity, and quantum efficiency of CMOS image sensors on camera resolution. Yang, Z. et al. (Contribution 11) investigated the relationship between the variation in SET bright spots under different conditions by conducting heavy-ion irradiation of image sensors. The authors propose identifying and classifying SEUs using the characteristics of set bright spo.t They established a fast identification method to analyze SEU patterns and sensitive areas based on transient bright spot size, background gray value, and other parameters. These studies provide theoretical bases for the evaluation of the radiation resistance of sensors in radiation environments and the development of radiation-resistant cameras.

The reliability of nanoscale electronic systems is crucial in various applications. Current research has confirmed that atmospheric neutrons can induce single-event effects in advanced relay protection devices as well. Yang, W. et al. (Contribution 8) investigated a Xilinx Zynq-7000 SoC manufactured with 28 nm CMOS technology using two rounds of spallation neutron irradiation. They conducted spallation neutron irradiation and analyzed the results in combination with those of Monte Carlo simulation to explore the impact of atmospheric neutrons on the SEEs of the target system-on-chip. Zhou, H. et al. (Contribution 18) preliminary assessed the SEEs on relay protection devices using neutron-based analysis and provide valuable insights for evaluating the reliability of advanced technology relay protection devices.

Several novel evaluation methods for radiation effects have been developed. Based on the illustrations of Liu, M. et al. (Contribution 14), depending on the particle energy, the areal density aluminum equivalent method may over- or underestimate the absorbed dose in a shielded silicon detector, especially for the ionization total-dose shielding effect of low-energy electrons. For integrated circuits used in space applications, the soft errors caused by transient pulses must first be evaluated, and the conventional evaluation approaches are limited to the circuit scale. Additionally, Song, R. et al. (Contribution 17) developed an approach for evaluating the soft error rate using machine learning technology. A back propagation neural network is implemented in the proposed approach. The proposed approach helps with determining the probability of transient pulse propagation. Compared with the conventional soft-error-rate evaluation results, the proposed approach strong correlations in both trend and magnitude.

3. The Future

The space radiation environment strongly impacts electronic devices, thereby seriously affecting the service life of spacecraft on-orbit electronic equipment. Consequently, the need is critical to thoroughly investigate the basis of radiation effects and develop innovative strategies to enhance the radiation resistance of electronic devices. The diverse array of articles featured in this Special Issue underscore the breadth of research in the field of the effects of radiation on advanced electronic devices and circuits. These articles span from cutting-edge advancements in nuclear and solid-state physics to sophisticated device and circuit-level modeling techniques as well as innovative hardening design methodologies. Moreover, these researchers have explored the application of progressive algorithms and deep learning methodologies to optimize system performance across various radiation environments. Together, these articles represent a collective leap forward in the pursuit of understanding radiation's effects and devising efficient methods for assessing the reliability and responses of novel electronic devices under radiation conditions.

In addition to the aforementioned areas of focus, the second edition of "Radiation Effects of Advanced Electronic Devices and Circuits" will delve deeper into several key aspects of radiation effects on electronic systems. This includes exploring the impact of radiation on emerging technologies such as quantum computing, neuromorphic computing, photonic devices, etc. The second edition will feature research on the development of radiation-hardened sensors and actuators, as well as advances in fault-tolerant computing architectures designed to mitigate the effects of radiation-induced errors. Moreover, given the push toward miniaturization and the complexity of electronic systems, the second edition will highlight research on radiation's effects at the nanoscale level. This will encompass investigations into the susceptibility of advanced electronic devices, such as carbon nanotubes, graphene-based transistors, and nanostructured materials, to radiation-induced degradation and failure mechanisms. Furthermore, the second edition will address the growing importance of system-level approaches to radiation hardening, including the integration of redundant components, fault-tolerant algorithms, and adaptive error correction techniques. Additionally, the next Special Issue will include articles exploring the role of machine learning and artificial intelligence in enhancing the resilience of electronic systems to radiation's effects, particularly in autonomous spacecraft, low-orbit commercial satellites, and space station systems. Overall, the second edition aims to provide a comprehensive overview of the latest advancements in radiation effects research and their implications for the design and operation of advanced electronic devices and circuits in space applications. These insights are expected to drive innovation and development in the field, paving the way for the creation of more robust and reliable electronic systems for future space missions.

Author Contributions: Y.C., C.C. and L.C. worked together in the whole editorial process of the Special Issue, "Radiation Effects of Advanced Electronic Devices and Circuits". Y.C., C.C. and L.C. worked closely together in the overall editorial activities towards the completion of the Special

Issue. Y.C. and C.C. drafted this manuscript. Y.C., C.C. and L.C. reviewed, edited, and finalized the manuscript. All authors have read and agreed to the published version of the manuscript.

Funding: This study was jointly supported by National Natural Science Foundation of China (grant No. 62174180 and No. 12205052), CAS Talent Program Youth Project (grant No. E129193YR0), and the fund of Innovation Center for Radiation Application (grant No. KFZC2022020301).

Conflicts of Interest: The authors declares no conflicts of interest.

List of Contributions

1. Zhao, P.; Li, B.; Liu, H.; Yang, J.; Jiao, Y.; Chen, Q.; Sun, Y.; Liu, J. The Effects of Total Ionizing Dose on the SEU Cross-Section of SOI SRAMs. *Electronics* **2022**, *11*, 3188.
2. Xin, J.; Zhu, X.; Ma, Y.; Han, J. Study of Single Event Latch-Up Hardness for CMOS Devices with a Resistor in Front of DC-DC Converter. *Electronics* **2023**, *12*, 550.
3. Pan, X.; Guo, H.; Lu, C.; Zhang, H.; Liu, Y. The Inflection Point of Single Event Transient in SiGe HBT at a Cryogenic Temperature. *Electronics* **2023**, *12*, 648.
4. Ding, H.; Cui, J.; Zheng, Q.; Xu, H.; Gao, N.; Xun, M.; Yu, G.; He, C.; Li, Y.; Guo, Q. Effect of Trapped Charge Induced by Total Ionizing Dose Radiation on the Top-Gate Carbon Nanotube Field Effect Transistors. *Electronics* **2023**, *12*, 1000.
5. Zhang, Z.; Guo, G.; Li, F.; Sun, H.; Chen, Q.; Zhao, S.; Liu, J.; Ouyang, X. Effects of Different Factors on Single Event Effects Introduced by Heavy Ions in SiGe Heterojunction Bipolar Transistor: A TCAD Simulation. *Electronics* **2023**, *12*, 1008.
6. Li, X.; Cui, J.; Zheng, Q.; Li, P.; Cui, X.; Li, Y.; Guo, Q. Study of the Within-Batch TID Response Variability on Silicon-Based VDMOS Devices. *Electronics* **2023**, *12*, 1403.
7. Cui, Y.; Feng, J.; Li, Y.; Wen, L.; Guo, Q. Proton Radiation Effects of CMOS Image Sensors on Different Star Map Recognition Algorithms for Star Sensors. *Electronics* **2023**, *12*, 1629.
8. Yang, W.; Li, Y.; Li, Y.; Hu, Z.; Cai, J.; He, C.; Wang, B.; Wu, L. Neutron Irradiation Testing and Monte Carlo Simulation of a Xilinx Zynq-7000 System on Chip. *Electronics* **2023**, *12*, 2057.
9. Feng, H.; Liang, X.; Pu, X.; Xiang, Y.; Zhang, T.; Wei, Y.; Feng, J.; Sun, J.; Zhang, D.; Li, Y.; et al. Total Ionizing Dose Effects of 60Co -Ray Radiation on Split-Gate SiC MOSFETs. *Electronics* **2023**, *12*, 2398.
10. Feng, J.; Wang, H.; Li, Y.; Wen, L.; Guo, Q. Mechanism of Total Ionizing Dose Effects of CMOS Image Sensors on Camera Resolution. *Electronics* **2023**, *12*, 2667.
11. Yang, Z.; Wen, L.; Li, Y.; Feng, J.; Zhou, D.; Liu, B.; Zhao, Z.; Guo, Q. Heavy Ion Single Event Effects in CMOS Image Sensors: SET and SEU. *Electronics* **2023**, *12*, 2833.
12. Liang, X.; Feng, H.; Xiang, Y.; Sun, J.; Wei, Y.; Zhang, D.; Li, Y.; Feng, J.; Yu, X.; Guo, Q. Oxide Electric Field-Induced Degradation of SiC MOSFET for Heavy-Ion Irradiation. *Electronics* **2023**, *12*, 2886.
13. Li, T.; Yuan, J.; Bai, Y.; Yu, C.; Gou, C.; Shu, L.; Wang, L.; Zhao, Y. Research on High-Dose-Rate Transient Ionizing Radiation Effect in Nano-Scale FDSOI Flip-Flops. *Electronics* **2023**, *12*, 3149.
14. Liu, M.; He, C.; Feng, J.; Xun, M.; Sun, J.; Li, Y.; Guo, Q. Analysis of Difference in Areal Density Aluminum Equivalent Method in Ionizing Total Dose Shielding Analysis of Semiconductor Devices. *Electronics* **2023**, *12*, 4181.
15. Lin, L.; Cong, Z.; Jia, C. Recovery Effect of Hot-Carrier Stress on -ray-Irradiated 0.13 um Partially Depleted SOI n-MOSFETs. *Electronics* **2023**, *12*, 4233.
16. Xiang, Y.; Liang, X.; Feng, J.; Feng, H.; Zhang, D.; Wei, Y.; Yu, X.; Guo, Q. Refined Analysis of Leakage Current in SiC Power Metal Oxide Semiconductor Field Effect Transistors after Heavy Ion Irradiation. *Electronics* **2023**, *12*, 4349.
17. Song, R.; Shao, J.; Chi, Y.; Liang, B.; Chen, J.; Wu, Z. Machine Learning-Based Soft-Error-Rate Evaluation for Large-Scale Integrated Circuits. *Electronics* **2023**, *12*, 4978.
18. Zhou, H.; Yu, H.; Zou, Z.; Su, Z.; Zhao, Q.; Yang, W.; He, C. Evaluation of Single Event Upset on a Relay Protection Device. *Electronics* **2024**, *13*, 64.

19. Xun, M.; Li, Y.; Feng, J.; He, C.; Liu, M.; Guo, Q. Effect of Proton Irradiation on Complementary Metal Oxide Semiconductor (CMOS) Single-Photon Avalanche Diodes. *Electronics* **2024**, *13*, 224.

References

1. Allison, J.; Amako, K.; Apostolakis, J.; Araujo, H.A.A.H.; Dubois, P.A.; Asai, M.A.A.M.; Barrand, G.; Capra, R.; Chauvie, S.; Chytracek, R.; et al. Geant4 developments and applications. *IEEE Trans. Nucl. Sci.* **2006**, *53*, 270–278. [CrossRef]
2. Weller, R.A.; Mendenhall, M.H.; Reed, R.A.; Schrimpf, R.D.; Warren, K.M.; Sierawski, B.D.; Massengill, L.W. Monte Carlo simulation of single event effects. *IEEE Trans. Nucl. Sci.* **2010**, *57*, 1726–1746. [CrossRef]
3. Reed, R.A.; Weller, R.A.; Mendenhall, M.H.; Fleetwood, D.M.; Warren, K.M.; Sierawski, B.D.; King, M.P.; Schrimpf, R.D.; Auden, E.C. Physical processes and applications of the Monte Carlo radiative energy deposition (MRED) code. *IEEE Trans. Nucl. Sci.* **2015**, *62*, 1441–1461. [CrossRef]
4. Munteanu, D.; Autran, J.L. Modeling and simulation of singleevent effects in digital devices and ICs. *IEEE Trans. Nucl. Sci.* **2008**, *55*, 1854–1878. [CrossRef]
5. Artola, L.; Gaillardin, M.; Hubert, G.; Raine, M.; Paillet, P. Modeling single event transients in advanced devices and ICs. *IEEE Trans. Nucl. Sci.* **2015**, *62*, 1528–1539. [CrossRef]
6. Hughes, H.L.; Benedetto, J.M. Radiation effects and hardening of MOS technology: Devices and circuits. *IEEE Trans. Nucl. Sci.* **2003**, *50*, 500–521. [CrossRef]
7. Lacoe, R.C. Improving integrated circuit performance through the application of hardness-by-design methodology. *IEEE Trans. Nucl. Sci.* **2008**, *55*, 1903–1925. [CrossRef]
8. Lacoe, R.C.; Osborn, J.V.; Koga, R.; Brown, S.; Mayer, D. Application of hardness-by-design methodology to radiation-tolerant ASIC technologies. *IEEE Trans. Nucl. Sci.* **2000**, *47*, 2334–2341. [CrossRef]
9. Leray, J.L.; Dupont-Nivet, E.; Pere, J.F.; Coïc, Y.M.; Raffaelli, M.; Auberton-Hervé, A.J.; Bruel, M.; Giffard, B.; Margail, J. CMOS/SOI hardening at 100 Mrad(SiO$_2$). *IEEE Trans. Nucl. Sci.* **1990**, *37*, 2013–2019. [CrossRef]
10. Wang, H.; Dai, X.; Ibrahim, Y.M.Y.; Sun, H.; Nofal, I.; Cai, L.; Guo, G.; Shen, Z.; Chen, L. A Layout-Based Rad-Hard DICE Flip-Flop Design. *J. Electron. Test.* **2019**, *35*, 111–117. [CrossRef]
11. Fleetwood, D.M. Radiation effects in a post-Moore world. *IEEE Trans. Nucl. Sci.* **2021**, *68*, 509–545. [CrossRef]
12. Gaspard, N.J.; Jagannathan, S.; Diggins, Z.J.; King, M.P.; Wen, S.-J.; Wong, R.; Loveless, T.D.; Lilja, K.; Bounasser, M.; Reece, T.; et al. Technology Scaling Comparison of Flip-Flop Heavy-Ion Single-Event Upset Cross Sections. *IEEE Trans. Nucl. Sci.* **2013**, *60*, 4368–4373. [CrossRef]
13. Fleetwood, Z.E.; Lourenco, N.E.; Ildefonso, A.; Warner, J.H.; Wachter, M.T.; Hales, J.M.; Tzintzarov, G.N.; Roche, N.J.-H.; Khachatrian, A.; Buchner, S.P.; et al. Using TCAD modeling to compare heavy-ion and laser induced single event transients in SiGe HBTs. *IEEE Trans. Nucl. Sci.* **2017**, *64*, 398–405. [CrossRef]
14. Chatterjee, I.; Narasimham, B.; Mahatme, N.N.; Bhuva, B.L.; Reed, R.A.; Schrimpf, R.D.; Wang, J.K.; Vedula, N.; Bartz, B.; Monzel, C. Impact of technology scaling on SRAM soft error rates. *IEEE Trans. Nucl. Sci.* **2014**, *61*, 3512–3518. [CrossRef]
15. Sheshadri, V.B.; Bhuva, B.L.; Reed, R.A.; Weller, R.A.; Mendenhall, M.H.; Schrimpf, R.D.; Warren, K.M.; Sierawski, B.D.; Wen, S.-J.; Wong, R. Effects of multi-node charge collection in flip-flop designs at advanced technology nodes. In Proceedings of the 2010 IEEE International Reliability Physics Symposium, Anaheim, CA, USA, 2–6 May 2010; pp. 1026–1030.
16. Vogl, T.; Sripathy, K.; Sharma, A.; Reddy, P.; Sullivan, J.; Machacek, J.R.; Zhang, L.; Karouta, F.; Buchler, B.C.; Doherty, M.W.; et al. Radiation tolerance of two-dimensional material-based devices for space applications. *Nature* **2019**, *10*, 1202. [CrossRef] [PubMed]
17. Wang, P.; Kalita, H.; Krishnaprasad, A.; Dev, D.; O'Hara, A.; Jiang, R.; Zhang, E.; Fleetwood, D.M.; Schrimpf, R.D.; Pantelides, S.T.; et al. Total-ionizing-dose response of MoS$_2$ transistors with ZrO$_2$ and h-BN gate dielectrics. *IEEE Trans. Nucl. Sci.* **2019**, *66*, 1584–1591. [CrossRef]
18. Gao, S.; Li, X.; Zhao, S.; He, Z.; Ye, B.; Cai, L.; Sun, Y.; Xiao, G.; Cai, C.; Liu, J. Heavy Ion Induced MCUs in 28 nm SRAM-based FPGAs: Upset Proportions, Classifications, and Pattern Shapes. *Nucl. Sci. Tech.* **2022**, *33*, 10. [CrossRef]
19. Chi, Y.; Huang, P.; Sun, Q.; Liang, B.; Zhao, Z. Characterization of Single-Event Upsets Induced by High-LET Heavy Ions in 16-nm Bulk FinFET SRAMs. *IEEE Trans. Nucl. Sci.* **2022**, *69*, 1176–1181.
20. Chi, Y.; Wu, Z.; Huang, P.; Sun, Q.; Liang, B.; Zhao, Z. Characterization of single-event transients induced by high LET heavy ions in 16 nm bulk FinFET inverter chains. *Microelectron. Reliab.* **2022**, *130*, 114490. [CrossRef]
21. Dodds, N.A.; Martinez, M.J.; Dodd, P.E.; Shaneyfelt, M.R.; Sexton, F.W.; Black, J.D.; Lee, D.S.; Swanson, S.E.; Bhuva, B.L.; Warren, K.M.; et al. The contribution of low-energy protons to the total on-orbit SEU rate. *IEEE Trans. Nucl. Sci.* **2015**, *62*, 2440–2451. [CrossRef]
22. Sierawski, B.D.; Mendenhall, M.H.; Reed, R.A.; Clemens, M.A.; Weller, R.A.; Schrimpf, R.D.; Blackmore, E.W.; Trinczek, M.; Hitti, B.; Pellish, J.A.; et al. Muon-induced single event upsets in deep-submicron technology. *IEEE Trans. Nucl. Sci.* **2010**, *57*, 3273–3278. [CrossRef]

Disclaimer/Publisher's Note: The statements, opinions and data contained in all publications are solely those of the individual author(s) and contributor(s) and not of MDPI and/or the editor(s). MDPI and/or the editor(s) disclaim responsibility for any injury to people or property resulting from any ideas, methods, instructions or products referred to in the content.

Article

Effect of Proton Irradiation on Complementary Metal Oxide Semiconductor (CMOS) Single-Photon Avalanche Diodes

Mingzhu Xun [1,2,3], Yudong Li [1,2,*], Jie Feng [1,2], Chengfa He [1,2], Mingyu Liu [1,2,3] and Qi Guo [1,2]

1. Xinjiang Technical Institute of Physics and Chemistry, Chinese Academy of Sciences, Urumqi 830011, China; xunmz@ms.xjb.ac.cn (M.X.); fengjie@ms.xjb.ac.cn (J.F.); hecf@ms.xjb.ac.cn (C.H.); liumingyu21@mails.ucas.ac.cn (M.L.); guoqi@ms.xjb.ac.cn (Q.G.)
2. Xinjiang Key Laboratory of Electronic Information Material and Device, Urumqi 830011, China
3. University of Chinese Academy of Sciences, Beijing 100049, China
* Correspondence: lydong@ms.xjb.ac.cn

Abstract: The effects of proton irradiation on CMOS Single-Photon Avalanche Diodes (SPADs) are investigated in this article. The I–V characteristics, dark count rate (DCR), and photon detection probability (PDP) of the CMOS SPADs were measured under 30 MeV and 52 MeV proton irradiations. Two types of SPAD, with and without shallow trench isolation (STI), were designed. According to the experimental results, the leakage current, breakdown voltage, and PDP did not change after irradiation at a DDD of 2.82×10^8 MeV/g, but the DCR increased significantly at five different higher voltages. The DCR increased by 506 cps at an excess voltage of 2 V and 10,846 cps at 10 V after 30 MeV proton irradiation. A γ irradiation was conducted with a TID of 10 krad (Si). The DCR after the γ irradiation increased from 256 cps to 336 cps at an excess voltage of 10 V. The comparison of the DCR after proton and γ-ray irradiation with two structures of SPAD indicates that the major increase in the DCR was due to the depletion region defects caused by proton displacement damage rather than the Si-SiO$_2$ interface trap generated by ionization.

Keywords: CMOS SPAD; proton radiation; DCR; displacement damage

1. Introduction

A Single-Photon Avalanche Diode (SPAD) is a photodiode that operates in Geiger mode with a reverse bias voltage higher than its avalanche breakdown voltage, and it utilizes an avalanche process to achieve single-photon detection capability. When photons are absorbed in the multiplication region of SPADs, a self-sustaining avalanche may be generated, and the current increases rapidly in the realm of picoseconds. By measuring the detectable current during the avalanche process, the arrival time of photons can be recorded [1]. An external quenching circuit is used to restore a SPAD to its initial state while waiting for the next photon to enter the multiplication region. By repeating the process above, single-photon detection and counting can be achieved. It can be seen that there is a dead time after the avalanche process, which may limit the maximum counting rate of the detected photons.

Due to their performance of high sensitivity, high detection efficiency, reliability, and low jitter noise, SPADs are widely used in some applications that require a low dark count rate (DCR), a low breakdown voltage, a low leakage current, a high gain, and high photon detection efficiency. SPADs are used in various fields, such as light detection and ranging (LiDAR), Non-Line of Sight (NLOS) imaging, fluorescence spectroscopy analysis, astronomic observations, optical communication, and quantum key distribution in weak light detection. When non-visible light must be used, especially in the near-infrared spectrum, high efficiency is very important [2].

At present, many research institutions and companies are developing SPADs with high efficiency, low noise, and high gain for different application scenarios, such as visible

Citation: Xun, M.; Li, Y.; Feng, J.; He, C.; Liu, M.; Guo, Q. Effect of Proton Irradiation on Complementary Metal Oxide Semiconductor (CMOS) Single-Photon Avalanche Diodes. *Electronics* **2024**, *13*, 224. https://doi.org/10.3390/electronics13010224

Academic Editor: Domenico Caputo

Received: 25 November 2023
Revised: 29 December 2023
Accepted: 2 January 2024
Published: 4 January 2024

Copyright: © 2024 by the authors. Licensee MDPI, Basel, Switzerland. This article is an open access article distributed under the terms and conditions of the Creative Commons Attribution (CC BY) license (https://creativecommons.org/licenses/by/4.0/).

and infrared light. In 2019, NASA reported an avalanche photodiode (APD) focal plane array assembled with linear-mode photon-counting capability for space lidar applications. The APD array uses a high-density, vertically integrated photodiode frame structure, and a preamplifier in the ROIC is directly integrated under the APD array to reduce the transmission capacitance. A microlens array is used to improve the fill factor. Its spectral response ranges from 0.9- to 4.3-µm wavelengths, its photon detection efficiency is as high as 70%, and it has a dark count rate of <250 kHz at 110 K [3]. In 2022, silicon photomultipliers (SiPMs), which are SPAD arrays based on a standard 55 nm Bipolar–CMOS–DMOS (BCD) technology, were developed by the Ecole Polytechnique Federale de Lausanne (EPFL). SiPMs are integrated into a coaxial light detection and ranging (LiDAR) system with a time-correlated single-photon counting (TCSPC) module system. Each SPAD cell is passively quenched by a monolithically integrated 3.3 V-thick oxide transistor. The measured gain is 3.4×10^5 at a 5 V excess bias voltage. The single-photon timing resolution (SPTR) is 185 ps, and the multiple-photon timing resolution (MPTR) is 120 ps at a 3.3 V excess bias voltage. Under the condition of a 25 m distance, the accuracies of SPTR and MPTR are 2 cm and 2 mm [4]. In 2022, a best-performing CMOS SPAD with a peak photon detection probability (PDP) of 55% at 480 nm, spanning from the near ultraviolet (NUV) to near infrared (NIR) spectrum, and a normalized dark count rate (DCR) of 0.2 cps/μm^2 at an excess bias of 6 V was proposed. Its after-pulsing probability is about 0.1% at a dead time of ~3 ns, and its single-photon time resolution (SPTR) is 12.1 ps (FWHM) at a 6 V excess bias voltage with a diameter of 25 µm. SPADs operate over a wide range of temperatures, from −65 °C to 40 °C, reaching a normalized DCR of 1.6 mcps/μm^2 at a 6 V excess bias voltage and −65 °C [5]. Some big companies, such as STMicroelectronics, Sony, and HAMAMATSU, have also developed a series of SPADs for different applications.

In the field of radiation detection, SPAD arrays combined with different types of scintillators, which can absorb energy from radiation, are mainly used in high-sensitivity gamma-ray detectors and medical PET imaging [6–9]. Scintillator detectors are used for real-time radiation dose rate detection above the environmental background. PET imaging uses radioactive isotope tracing methods to display its location and concentration. By detecting the gamma photons generated by an isotope, the emission position of the photons can be reconstructed, and changes in metabolic processes and other physiological activities can be visualized. In addition, a single SPAD can be used to detect low-energy electrons and X-rays. A SPAD collects electrons generated by incident electrons and X-rays in the multiplication region instead of the photons emitted by scintillators. This makes detection faster and more accurate.

With the dramatic increase in interest in satellite-to-ground quantum communication and space environment detection, SPADs, with the advantages of high efficiency, low power consumption, easy integration, and anti-magnetic field performance, are more and more widely used in space and high-energy radiation detection [10–13]. But they are inevitably exposed to radiation environments, which can affect the performance of SPADs. Most satellite-to-ground quantum communication satellites are in near-earth orbit at an orbital altitude of 500 km, and the space radiation environment includes electrons and protons in the Van Allen radiation belt and high-energy protons in the South Atlantic Anomaly (SAA) region [14–17]. The space radiation environment during deep space exploration is dominated by high-energy galactic cosmic rays, including most of the particles in the periodic table from Z = 1 to Z = 92, with energies ranging from 1 MeV/n to 1 TeV/n. SiPMs were used in space-borne scintillation detectors for many space missions. For example, they have been used for the gain control system on board the Hard X-ray Modulation Telescope (HXMT), a Chinese X-ray space observatory launched in June 2017 [18]. The experiment GMOD assembled an SiPMs array with a CeBr3 scintillator and an Application-Specific Integrated Circuit (ASIC) to detect cosmic gamma-ray phenomena such as Gamma-Ray Bursts (GRBs) in space carried by the Educational Irish Research Satellite 1 (EIRSAT-1). This is a 2U cube satellite deployed from the International Space Station, and it remained in orbit at an altitude of 405 km and a tilt of 51.6 degrees for a year, which is a safe space environment to avoid serious damage to SiPMs. SiPMs are also used in the Large Hadron Collider CMS, LHCb, and the proposed International Linear Collider (ILC) at the European Organization for Nuclear Research (CERN), which reaches 10^{14} p/cm^2 [19,20].

In radiation environments, protons, electrons, γ rays, and heavy ions can cause certain parameters, such as the breakdown voltage, leakage current, DCR, gain, and photon detection efficiency, to deteriorate at different levels through the displacement damage dose (DDD) effect and the Total Ionizing Dose (TID) effect [21,22]. This is due to the point defects in silicon and the interface defects at the Si-SiO$_2$ interface near STI. These defects include the vacancy (V_{Si}), the substitutional phosphorus (P_{Si}), the interstitial oxygen (Oi), the double vacancy ($V_{Si}V_{Si}$), the A-center ($V_{Si}O_i$), and the E-center ($P_{Si}V_{Si}$). They are electrically active and act as efficient generation–recombination centers which cause leakage currents and DCR increases [23–27].

To study the SPAD radiation effect of protons, a SPAD of 180 nm standard CMOS technology with a P-I-N structure and radiation tolerance design is used in this experiment. The sensitivity of ionization radiation damage and displacement radiation damage for SPADs is investigated using γ rays and protons beams. Two types of SPADs, with and without Shallow Trench Isolation (STI), are also designed and compared to study the influence of the Si-SiO$_2$ interface defects near the STI after radiation. The dark current, breakdown voltage, DCR, and photon detection probability (PDP) of the SPADs before and after irradiation are measured, and the radiation damage mechanism of the CMOS SPAD is analyzed.

2. Experimental Design

The CMOS SPAD, as shown in Figure 1, is based on the P-I-N structure, with a P-well epitaxial layer and n-type buried channel, and designed by the 180 nm CMOS process. Figure 2 is the cross-section of the SPAD [28,29]. In this design, the n-type buried channel ensures isolation from the substrate, while the deep n-well structure provides contact from N+ to the n-type buried channel. The lateral diffusion and light doping of the P epitaxial layer can avoid premature breakdown at the edge of the junction depletion region. The p+/DNW junction, enabling wider depletion, along with novel guard ring designs, facilitate device operation at up to 10 V of excess bias. The DCR is mainly caused by the tunneling noise at an excess bias of 10 V, but in this design, a P-I-N structure with standard CMOS technology is used to reduce the tunneling noise, resulting in better noise performance. The CMOS SPAD had a photon detection probability (PDP) greater than 40% from 440 to 620 nm, and the dark count rate (DCR) was 12.85 cps/μm^2. In addition, due to the use of n-type buried channels, the peak electric field of the detector is concentrated between the n-type buried channels and the P epitaxial layer.

Figure 1. SPAD for 180 nm CMOS technology.

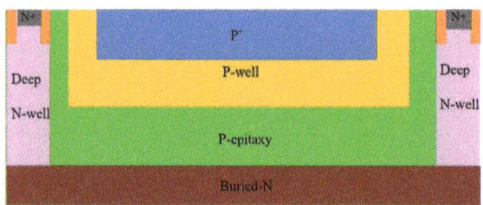

Figure 2. Cross-section of the P-I-N-structure SPAD.

Proton irradiation experiments at two different energies of 30 MeV and 52 MeV were conducted on the Cyclotron Proton Accelerator in the air, and a laser pointing system was used to align it with the beamline center. The proton beam region is 5 cm × 5 cm, and the uncertainty of the beam intensity had a variation of ±5%. The proton line energy transport (LET), Nonionizing Energy Loss (NIEL), Total Ionizing Dose (TID), and displacement damage dose (DDD) are shown in Table 1. The LET data come from the NIST stopping power and range tables for the protons program PSTAR, and the NIEL data come from Ref. [30]. The DDD of the SPAD in LEO orbit with an altitude of 400 km and an inclination of 51.6° was 19.6 TeV/g with a 2 mm shielding thickness of aluminum [31]. As a contrast, we chose a proton fluence of 5×10^{10} p/cm^2 at energies of 30 MeV and 52 MeV. All the SPAD pins were shorted and connected to ground during the irradiation, and I–V characteristics and PDP measurements were performed before and after irradiation. The parameter testing system for a SPAD includes a Keysight semiconductor parameter analyzer and a DCR and PDP measurement system. The DCR and PDP measurement system consists of a light source, a filter, a spectrograph, an integrating sphere, a sample chamber, and a light source calibration and computer control system, as shown in Figure 3. The halogen lamps can provide a stable light source with a wavelength range of 350 nm–1100 nm. Monochromatic light with specific frequencies can be generated after light passes through the filters and spectrometers. The integrating sphere can reduce small errors caused by an uneven distribution of incident light sources on the detector or beam offset during measurement, thus improving the accuracy of a measurement. The output trigger pulse count of the device is read by an oscilloscope, and then the data are statistically analyzed to obtain the DCR and PDP. A passive quenching circuit with a 50 kΩ resistor was used to measure the DCR, and the DCR is defined as the average counts of pulses per second (cps) in a 1 min measurement in darkness.

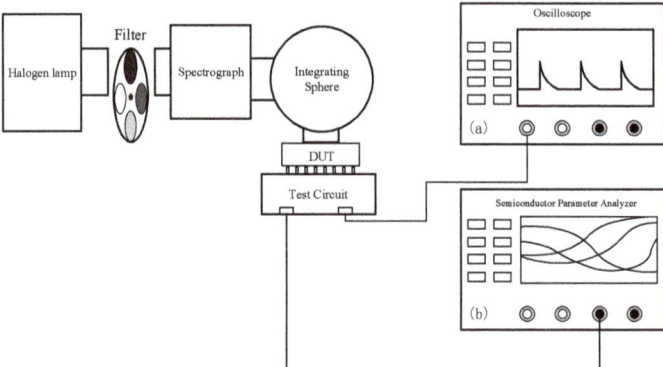

Figure 3. The CMOS SPAD parameter testing system with (**a**) oscilloscope for DCR and PDP measurement and (**b**) semiconductor parameter analyzer for I–V characteristics measurement.

Table 1. LET, NIEL, TID, and DDD of 30 MeV and 52 MeV protons.

Proton Energy (MeV)	LET (MeV/(g/cm^2))	NIEL (MeV/(g/cm^2))	Fluence (p/cm^2)	TID (krad)	DDD (MeV/g)
30	1.47×10	5.63×10^{-3}	5.00×10^{10}	11.8	2.82×10^8
52	9.58×10	3.37×10^{-3}	5.00×10^{10}	7.66	1.69×10^8

3. Results

3.1. I–V Characteristics

Figure 4 shows the I–V characteristics of the CMOS SPAD. The leakage current of the SPAD after 30 MeV proton irradiation did not increase significantly before reaching the avalanche breakdown voltage. When a bias voltage of 26 V was applied, the reverse current increased from 5.47 mA to 5.71 mA. Table 2 shows a comparison of the SPAD breakdown voltages. The breakdown voltage increased by only 20 mV after 30 MeV proton irradiation, while it remained unchanged after 52 MeV proton irradiation. This indicates that proton displacement damage can lead to a slight increase in the breakdown voltage, but not significantly. This is similar to the results of SPAD γ experiments based on the same P-I-N structure in Ref. [28], indicating that the leakage current and breakdown voltage (V_B) are almost insensitive to ionization damage and displacement damage.

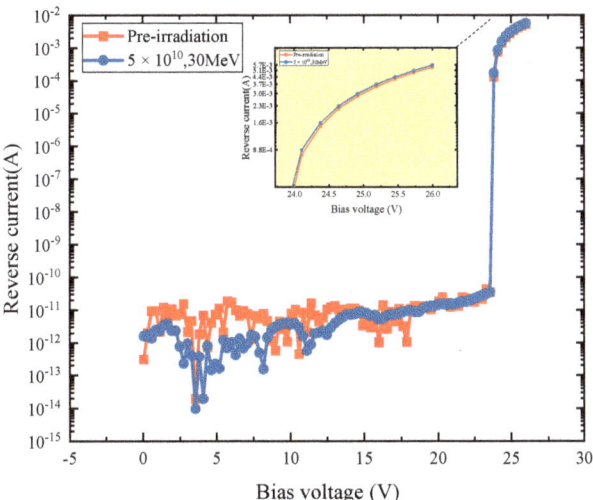

Figure 4. I–V characteristics for CMOS SPAD.

Table 2. Comparison of the SPAD breakdown voltage.

Proton Energy (MeV)	V_B (Fresh) (V)	V_B (5×10^{10} p/cm^2) (V)
30	24.23	24.25
52	24.18	24.18

3.2. DCR

Figure 5 shows the DCR data under different excess voltages before and after irradiation with a passive quenching circuit at 23 °C. It can be seen that before irradiation, the DCR increases with the increase in excess voltage from 74cps@2V to 256cps @10V. But after proton irradiation, the increase in the DCR is very significant. The DCR increases from 74cps@2V before irradiation to 520cps@2V after 52 MeV proton irradiation. This is consistent with the trend of DCR change with excess voltage before irradiation. However, due to the increase in displacement damage defects in the junction depletion region caused by proton irradiation, the trend of DCR increase with higher excess voltage is significantly enhanced. The DCR increased form 580cps@2V to 11102cps@10V after 30 MeV proton irradiation. Under the same proton fluence of 5×10^{10} p/cm^2, the change caused by 30 MeV proton irradiation is greater than that caused by 52 MeV proton irradiation. This is due to the NIEL of low-energy protons being higher than that of high-energy protons, resulting in a DDD of 2.82×10^8 MeV/g for the 30 MeV protons, which is higher than the DDD of 1.69×10^8 MeV/g for the 52 MeV protons. So, the displacement damage defects generated by the 30 MeV protons in the junction depletion region resulted in a larger DCR at the same excess voltage.

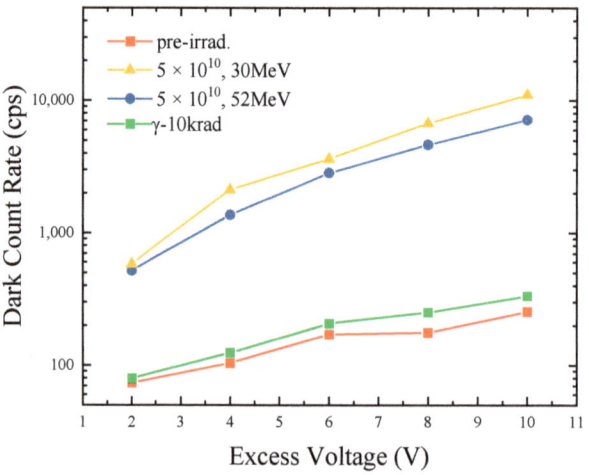

Figure 5. DCR of different excess voltages after proton and γ irradiation.

The TID and DDD caused by proton irradiation resulted in Si-SiO$_2$ interface defects and junction depletion region defects. In order to confirm the main reason for the increase in the DCR, γ irradiation was conducted with a TID of 10 krad(Si), while the TID of the 52 MeV proton was only 7.66 krad(Si). The DCR after γ irradiation increased from 256 cps to 336 cps at an excess voltage of 10 V. However, after proton irradiation, the DCR increased to 7160 cps, which is approximately 20 times greater than that of γ irradiation, indicating that the increase in the DCR is mainly caused by the displacement damage of proton irradiation, and the TID effect is not obvious.

3.3. PDP

The PDP is defined as the ratio of the SPAD-detected photons to the incident photons and reflects the generation of photo-generated carriers. It reflects the photosensitivity of the SPAD. The PDP depends on two main parameters: the absorption probability and the triggering efficiency. The absorption probability is the probability of photons being absorbed in the depletion region, and it depends on the reflectivity, the depth of the junction, and the thickness of the depletion region, while the triggering efficiency is the probability of photo-generated electron–hole pairs triggering a self-sustaining avalanche process, which depends on the electric field [1]. Figure 6 shows a comparison of the PDP curve at an excess voltage of 6 V before and after 30 MeV proton irradiation. It can be seen that there is no significant change in the PDP in the wavelength range of 400 nm–800 nm. The maximum PDP of the SPAD is 37.9% and 38.6% at a wavelength of 500 nm. This indicates that the depletion region defects caused by a proton displacement irradiation damage dose of 2.82×10^8 MeV/g have no effect on the absorption of incident photons and the generation of photo-generated carriers, so the PDP does not change. Ref. [28] reports that the PDP also has no changes after γ irradiation, which also means that the PDP is not an important radiation-sensitive parameter to consider in radiation environments [32].

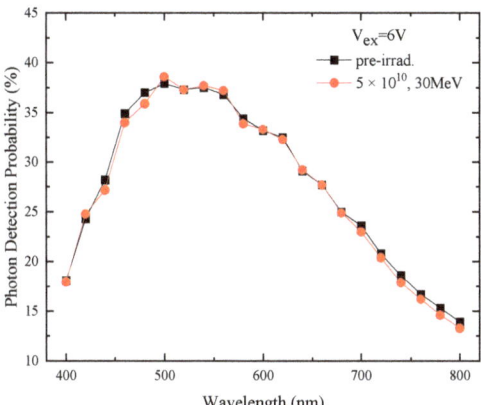

Figure 6. PDP value of 30 MeV protons.

4. Discussion

From the previous results, it can be concluded that the SPAD leakage current, breakdown voltage, and PDP are not sensitive to proton displacement damage, but the DCR is very sensitive. The DCR reflects the inherent noise inside a single-photon detector. The sources of noise in silicon devices include thermal noise, tunneling-assisted noise, and trap-assisted noise. Among them, thermal noise and tunneling-assisted noise are related to the operating temperature, doping concentration, and excess voltage of the device. Trap-assisted noise is related to defects introduced during the CMOS manufacturing process, and trap defects introduced during irradiation also produce trap-assisted noise [33–37]. The thermal generation and band-to-band tunneling effects of free carriers within the depletion region collectively contribute to the DCR, which is largely dependent on temperature. When the temperature increases by 10 °C near the room temperature of 23 °C, the DCR usually increases by more than double [38,39]. At this temperature, thermal generation is the main noise source. The thermal generation of free carriers is closely related to the presence of impurities and crystal defects, which introduce local energy levels near the middle of the band gap. According to the Shockley–Read–Hall theory, electron–hole pairs are generated sequentially through generation–recombination (G-R) centers. The proton irradiation of silicon-based devices can also form deep-energy-level defects near the center of the energy band in the depletion region, which can trap or emit electrons and become the trap center. These irradiated deep-energy-level defects contribute to the increase in the dark count rate. Besides the thermal effect, another major contributor to the DCR are Poole–Frenkel effects and rap-assisted tunneling, but the DCR does not significantly increase with temperature but instead increases with excess bias, which usually happens in high-doping junctions. The effects of displacement damage on semiconductor materials and devices can be understood in terms of the energy levels introduced in the bandgap. Those radiation-induced levels result in the following effects: the recombination lifetime and diffusion length are reduced; the generation lifetime decreases; majority-carrier and minority-carrier trapping increase; the majority-carrier concentration changes; the thermal generation of electron–hole pairs is enhanced in the presence of a sufficiently high electric field; tunneling at junctions is enabled; and radiation-induced defects reduce the carrier mobility and can exhibit metastable configurations [40,41].

To analyze the displacement damage of protons in the SPAD, we used The Stopping and Range of Ions in Matter Software (SRIM) to simulate the proton transportation process, which is a program written by J.F. Ziegler, M.D. Ziegler, and J.P. Biersack to simulate the interaction process of ion beams with solids, and the Monte Carlo method was used to calculate details such as vacancies, energy deposition, and particle positions during the collision process [42].

The simulation results of the proton trajectory, stopping power, and energy deposition per unit distance in the photo collector region are shown in Figure 7. At the SPAD surface, the stopping power of silicon for the 30 MeV protons is higher than that for the 52 MeV protons, indicating that the damage caused by the 30 MeV protons on silicon is more severe under the fluence of 5×10^{10} p/cm^2. As the incident depth increases, the deposited energy of the 30 MeV protons in silicon increases significantly, which is manifested as an enhancement of proton scattering in the same number of particle trajectories, resulting in a larger projected area of the incident direction. The calculated number of vacancies in silicon is shown in Table 3. Comparing the difference between the vacancy numbers and the ΔDCR under different excess voltages, it is found that the number of vacancies is 24 for each 30 MeV proton and 13.7 for each 52 MeV proton, and the ratio is 1.75. However, the ratio of DCR increase at the five excess voltages varies from 0.93 to 1.59, which is a little lower than the ratio of vacancy. This may be because not all trap vacancies contribute to the generation of carriers, which will cause the DCR in the depletion region to increase; some vacancies are used for the carrier's recombination. The DCR increase is defined as

$$\Delta DCR = DCR \text{ after irradiation} - DCR \text{ before irradiation} \quad (1)$$

Table 3. The ratio of vacancies and ΔDCR at different excess voltages.

Proton Energy (MeV)	Total Vacancies (/ion)	ΔDCR@2V (cps)	ΔDCR@4V (cps)	ΔDCR@6V (cps)	ΔDCR@8V (cps)	ΔDCR@10V (cps)
30	24	506	2020	3468	6584	10,846
52	13.7	446	1268	2686	4488	6904
ratio	1.75	1.13	1.59	1.29	1.47	1.57

In order to analyze the influence of Si-SiO$_2$ interface defects on the STI structure, two types of SPAD units, with and without an STI structure, were designed on the same chip. The simulation results of the SPAD design and the electric field distribution with and without an STI structure are shown in Figure 8. Between the p-well and deep n-well, we designed an STI structure using silicon dioxide as an insulating layer. It can be seen that the electric field distribution near the STI structure changes significantly. After ionizing radiation, interface charges accumulate at the Si-SiO$_2$ interface near the STI structure.

After proton irradiation with the same fluence, the DCR was measured under different excess voltages. The results are shown in Figure 9. The comparison of the results with and without an STI structure before irradiation shows that the presence of the STI structure increases the DCR from 256 cps to 362 cps under a 10 V bias. The interface defects in the STI structure before irradiation increase the DCR by 41.4%. Ref. [28] has proven that the increase in the DCR caused by ionizing irradiation is mainly due to the induced Si-SiO$_2$ interface traps near the STI structure. The DCR of the SPAD with and without an STI structure significantly increased after irradiation. At 10 V, the DCR increased by 30 times. However, when the excess voltage was 0–8 V, the DCR of the SPAD without an STI structure was higher than that of the SPAD with an STI structure. When the excess voltage was greater than 8 V, the DCR of the SPAD without the STI structure was lower. A possible reason for this is that the Si-SiO$_2$ interface defect charges near the STI structure will be released from the interface and drift into the depletion region when the electric field exceeds a certain value, and then an avalanche process is formed, resulting in an increase in the DCR.

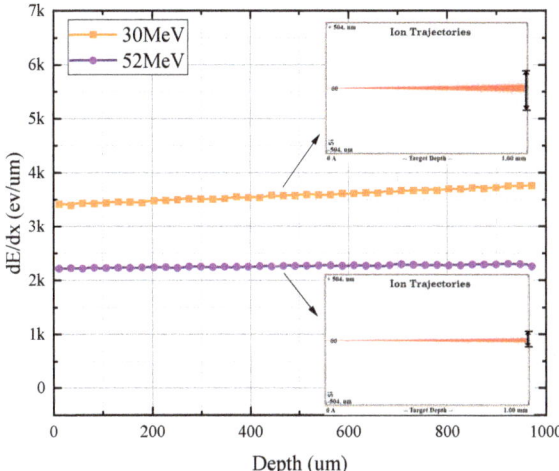

Figure 7. Stopping power and trajectory of protons in silicon.

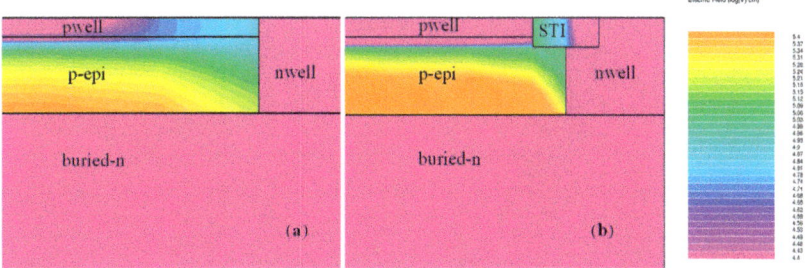

Figure 8. SPAD and electric field distribution without STI structure (**a**) and with STI structure (**b**).

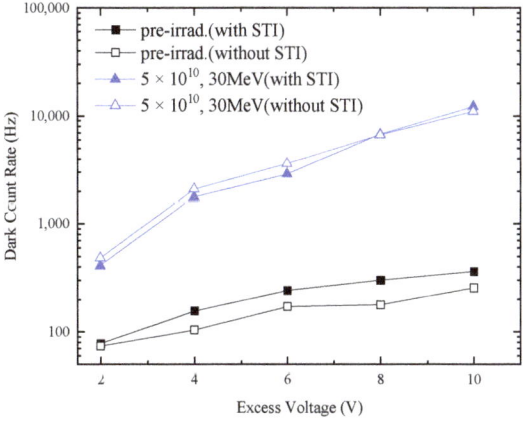

Figure 9. Comparison of DCR irradiation with and without STI structure.

5. Conclusions

Proton irradiations on the CMOS SPAD of 30 MeV and 52 MeV are studied in this article. The leakage current, breakdown voltage, and PDP before and after proton irradiations were measured and compared with γ rays. SPAD units with and without an STI structure were designed and simulated, and we summarize our results as follows:

1. After the 30 MeV proton radiation, the reverse current increased from 5.47 mA to 5.71 mA at a bias voltage of 26 V. The breakdown voltage increased by only 20 mV after the 30 MeV proton irradiation, while it remained unchanged after the 52 MeV proton irradiation. The reported results of the SPAD γ experiments based on the same P-I-N structure in Ref. [17] indicate that the breakdown voltage (V_B) is almost insensitive to ionization damage and displacement damage.

2. Before irradiation, the DCR increased with an increase in the excess voltage from 74cps@2V to 256cps@10V, but the DCR increased rapidly to 520cps@2V and 7160cps@10V after the 52 MeV proton irradiation. For the 30 MeV proton irradiation, the DCR increased form 580cps@2V to 11102cps@10V. The displacement damage defects generated by the 30 MeV protons with a DDD of 2.82×10^8 MeV/g resulted in a larger DCR increase than a DDD of 1.69×10^8 MeV/g for the 52 MeV protons at the same excess voltage. The trend of the DCR increasing with a higher excess voltage is significantly enhanced due to the displacement damage defects in the junction depletion region. A comparison of γ irradiation with a TID of 10 krad (Si) and the 52 MeV protons with a TID of 7.66 krad (Si) shows that the increase in the DCR is mainly caused by the displacement damage of proton irradiation instead of the TID effect.

3. The SPAD units with and without an STI structure also show that the main reason for the DCR increase is the depletion region defects caused by proton displacement damage rather than the $Si-SiO_2$ interface trap generated by ionization.

4. The comparison of the leakage current, breakdown voltage, and PDP shows that the design of the SPAD based on the standard CMOS process exhibits good radiation hardening, but the process of the depletion region should be improved to reduce the DCR after irradiation.

Author Contributions: Conceptualization, M.X. and Y.L.; methodology, M.X. and Y.L.; software, M.X.; validation, M.X. and Y.L.; formal analysis, Y.L.; investigation, Y.L.; resources, M.L. and C.H.; data curation, J.F.; writing—original draft preparation, M.X.; writing—review and editing, M.X. and Y.L.; visualization, M.X.; supervision, Y.L.; project administration, Q.G.; funding acquisition, Y.L. and Q.G. All authors have read and agreed to the published version of the manuscript.

Funding: This research was funded by the West Light Talent Training Plan of the Chinese Academy of Sciences under grant No. 2021-XBQNXZ-020, the Tianshan Innovation Team Program of Xinjiang Uygur Autonomous Region No. 2022D14003, the Fund of Robot Technology Used for Special Environment Key Laboratory of Sichuan Province No. 21kftk03, and the "Light of West China" Program of the Chinese Academy of Sciences under grant No. 2020-XBQNXZ-004.

Data Availability Statement: Data are contained within the article.

Conflicts of Interest: The authors declare no conflicts of interest.

References

1. Cusini, I.; Berretta, D.; Conca, E.; Incoronato, A.; Madonini, F.; Maurina, A.A.; Nonne, C.; Riccardo, S.; Villa, F. Historical Perspectives, State of art and Research Trends of Single Photon Avalanche Diodes and Their Applications (Part 1: Single Pixels). *Front. Phys.* **2022**, *10*, 607. [CrossRef]
2. Cusini, I.; Berretta, D.; Conca, E.; Incoronato, A.; Madonini, F.; Maurina, A.A.; Nonne, C.; Riccardo, S.; Villa, F. Historical Perspectives, State of Art and Research Trends of SPAD Arrays and Their Applications (Part II: SPAD Arrays). *Front. Phys.* **2022**, *10*, 906671. [CrossRef]
3. Sun, X.; Abshire, J.B.; Krainak, M.A.; Lu, W.; Beck, J.D.; Sullivan, W.W.; Mitra, P.; Rawlings, D.M.; Fields, R.A.; Hinkley, D.A.; et al. HgCdTe avalanche photodiode array detectors with single photon sensitivity and integrated detector cooler assemblies for space lidar applications. *Opt. Eng.* **2019**, *58*, 067103. [CrossRef]
4. Zhao, J.X.; Milanese, T.; Gramuglia, F.; Keshavarzian, P.; Tan, S.S.; Tng, M.; Lim, L.; Dhulla, V.; Quek, E.; Lee, M.J.; et al. On Analog Silicon Photomultipliers in Standard 55-nm BCD Technology for LiDAR Applications. *IEEE J. Sel. Top. Quantum Electron.* **2022**, *28*, 10. [CrossRef]
5. Gramuglia, F.; Wu, M.L.; Bruschini, C.; Lee, M.J.; Charbon, E. A Low-Noise CMOS SPAD Pixel with 12.1 Ps SPTR and 3 Ns Dead Time. *IEEE J. Sel. Top. Quantum Electron.* **2022**, *28*, 9. [CrossRef]

6. Therrien, A.C.; Bérubé, B.L.; Thibaudeau, C.; Charlebois, S.; Lecomte, R.; Fontaine, R.; Pratte, J.F. Modeling of Single Photon Avalanche Diode Array Detectors for PET Applications. In Proceedings of the IEEE Nuclear Science Symposium/Medical Imaging Conference (NSS/MIC)/18th International Workshop on Room-Temperature Semiconductor X-Ray and Gamma-Ray Detectors, Valencia, Spain, 23–29 October 2011.
7. Tétrault, M.A.; Lamy, É.; Boisvert, A.; Fontaine, R.; Pratte, J.F. Low Dead Time Digital SPAD Readout Architecture for Realtime Small Animal PET. In Proceedings of the 60th IEEE Nuclear Science Symposium (NSS)/Medical Imaging Conference (MIC)/20th International Workshop on Room-Temperature Semiconductor X-ray and Gamma-ray Detectors, Seoul, Republic of Korea, 27 October–2 November 2013.
8. Braga, L.H.C.; Perenzoni, M.; Stoppa, D. Effects of DCR, PDE and Saturation on the Energy Resolution of Digital SiPMs for PET. In Proceedings of the 60th IEEE Nuclear Science Symposium (NSS)/Medical Imaging Conference (MIC)/20th International Workshop on Room-Temperature Semiconductor X-ray and Gamma-ray Detectors, Seoul, Republic of Korea, 27 October–2 November 2013.
9. Braga, L.H.C.; Pancheri, L.; Gasparini, L.; Perenzoni, M.; Walker, R.; Henderson, R.K.; Stoppa, D. A CMOS mini-SiPM detector with in-pixel data compression for PET applications. In Proceedings of the IEEE Nuclear Science Symposium/Medical Imaging Conference (NSS/MIC)/18th International Workshop on Room-Temperature Semiconductor X-Ray and Gamma-Ray Detectors, Valencia, Spain, 23–29 October 2011.
10. Llosá, G.; Belcari, N.; Bisogni, M.G.; Collazuol, G.; Del Guerra, A.; Marcatili, S.; Moehrs, S.; Piemonte, C. Silicon photomultipliers and SiPM matrices as photodetectors in nuclear medicine. In Proceedings of the IEEE Nuclear Science Symposium/Medical Imaging Conference, Honolulu, HI, USA, 26 October–3 November 2007.
11. Gundacker, S.; Heering, A. The silicon photomultiplier: Fundamentals and applications of a modern solid-state photon detector. *Phys. Med. Biol.* **2020**, *65*, 17TR01. [CrossRef] [PubMed]
12. Eisaman, M.D.; Fan, J.; Migdall, A.; Polyakov, S.V. Invited Review Article: Single-photon sources and detectors. *Rev. Sci. Instrum.* **2011**, *82*, 071101. [CrossRef] [PubMed]
13. Hadfield, R.H. Single-photon detectors for optical quantum information applications. *Nat. Photonics* **2009**, *3*, 696–705. [CrossRef]
14. Domingos, J.; Jault, D.; Pais, M.A.; Mandea, M. The South Atlantic Anomaly throughout the solar cycle. *Earth Planet. Sci. Lett.* **2017**, *473*, 154–163. [CrossRef]
15. Badavi, F.F. Validation of the new trapped environment AE9/AP9/SPM at low Earth orbit. *Adv. Space Res.* **2014**, *54*, 917–928. [CrossRef]
16. Bühler, P.; Zehnder, A.; Kruglanski, M.; Daly, E.; Adams, L. The high-energy proton fluxes in the SAA observed with REM aboard the MIR orbital station. *Radiat. Meas.* **2002**, *35*, 489–497. [CrossRef] [PubMed]
17. Koshiishi, H. Space radiation environment in low earth orbit during influences from solar and geomagnetic events in December 2006. *Adv. Space Res.* **2014**, *53*, 233–236. [CrossRef]
18. Li, Z.W.; Liu, C.Z.; Xu, Y.P.; Yan, B.; Li, Y.G.; Lu, X.F.; Li, X.F.; Zhang, S.; Chang, Z.; Li, J.C.; et al. A novel analog power supply for gain control of the Multi-Pixel Photon Counter (MPPC). *Nucl. Instrum. Meth. A* **2017**, *850*, 35–41. [CrossRef]
19. Chatrchyan, S.; Hmayakyan, G.; Khachatryan, V.; Sirunyan, A.M.; Adam, W.; Bauer, T.; Bergauer, T.; Bergauer, H.; Dragicevic, M.; Erö, J.; et al. The CMS experiment at the CERN LHC. *J. Instrum.* **2008**, *3*, 361.
20. Musienko, Y.; Heeling, A.; Ruchti, R.; Wayne, M.; Karneyeu, A.; Postoev, V. Radiation damage studies of silicon photomultipliers for the CMS HCAL phase I upgrade. *Nucl. Instrum. Meth. A* **2015**, *787*, 319–322. [CrossRef]
21. Musienko, Y.; Heering, A.; Ruchti, R.; Wayne, M.; Andreev, Y.; Karneyeu, A.; Postoev, V. Radiation damage in silicon photomultipliers exposed to neutron radiation. *J. Instrum.* **2017**, *12*, C07030. [CrossRef]
22. Jouni, A.; Sicre, M.; Malherbe, V.; Mamdy, B.; Thery, T.; Belloir, J.-M.; Soussan, D.; De Paoli, S.; Lorquet, V.; Lalucaa, V.; et al. Proton-Induced Displacement Damages in 2-D and Stacked CMOS SPADs: Study of Dark Count Rate Degradation. *IEEE Trans. Nucl. Sci.* **2023**, *70*, 515–522. [CrossRef]
23. Pelenitsyn, V.; Korotaev, P. First-principles study of radiation defects in silicon. *Comput. Mater. Sci.* **2022**, *207*, 111273. [CrossRef]
24. Wimbauer, T.; Ito, K.; Mochizuki, Y.; Horikawa, M.; Kitano, T.; Brandt, M.S.; Stutzmann, M. Defects in planar Si pn junctions studied with electrically detected magnetic resonance. *Appl. Phys. Lett.* **2000**, *76*, 2280–2282. [CrossRef]
25. Bychkova, O.; Garutti, E.; Popova, E.; Stifutkin, A.; Martens, S.; Parygin, P.; Kaminsky, A.; Schwandt, J. Radiation damage uniformity in a SiPM. *Nucl. Instrum. Methods Phys. Res. Sect. A Accel. Spectrometers Detect. Assoc. Equip.* **2022**, *1039*, 167042. [CrossRef]
26. Watts, S.J. Overview of radiation damage in silicon detectors-Models and defect engineering. *Nucl. Instrum. Meth. A* **1997**, *386*, 149–155. [CrossRef]
27. Ratti, L.; Brogi, P.; Collazuol, G.; Betta, G.F.D.; Ficorella, A.; Marrocchesi, P.S.; Morsani, F.; Pancheri, L.; Torilla, G.; Vacchi, C. DCR Performance in Neutron-Irradiated CMOS SPADs From 150-to 180-nm Technologies. *IEEE Trans. Nucl. Sci.* **2020**, *67*, 1293–1301. [CrossRef]
28. Li, Y.; Veerappan, C.; Lee, M.-J.; Wen, L.; Guo, Q.; Charbon, E. A Radiation-Tolerant, High Performance SPAD for SiPMs Implemented in CMOS Technology. In Proceedings of the IEEE Nuclear Science Symposium/Medical Imaging Conference/Room-Temperature Semiconductor Detector Workshop (NSS/MIC/RTSD), Strasbourg, France, 29 October–6 November 2016.
29. Veerappan, C.; Charbon, E. A Low Dark Count p-i-n Diode Based SPAD in CMOS Technology. *IEEE Trans. Electron. Dev.* **2016**, *63*, 65–71. [CrossRef]

30. Akkerman, A.; Barak, J.; Chadwick, M.B.; Levinson, J.; Murat, M.; Lifshitz, Y. Updated NIEL calculations for estimating the damage induced by particles and γ-rays in Si and GaAs. *Radiat. Phys. Chem.* **2001**, *62*, 301–310. [CrossRef]
31. Campajola, M.; Di Capua, F.; Fiore, D.; Nappi, C.; Sarnelli, E.; Gasparini, L. Long-Term Degradation Study of CMOS SPADs in Space Radiation Environment. In Proceedings of the 18th European Conference on Radiation and Its Effects on Components and Systems (RADECS), COBHAM, Goteburg, Sweden, 16–21 September 2018; IEEE: New York, NY, USA, 2018.
32. Liu, Q.L.; Zhang, H.Y.; Hao, L.X.; Hu, A.Q.; Wu, G.; Guo, X. Total dose test with γ-ray for silicon single photon avalanche diodes. *Chin. Phys. B* **2020**, *29*, 088501. [CrossRef]
33. Lee, J.; Bosman, G. 1/f γ drain current noise model in ultrathin oxide MOSFETs. *Fluct. Noise Lett.* **2004**, *4*, L297–L307. [CrossRef]
34. Kim, J.S.; Park, C.H.; Min, H.S.; Park, Y.J. Theory of 1/f noise currents in n+ p diodes, n+ p photodiodes, and Schottky diodes. In Proceedings of the 7th van der Ziel Symposium on Quantum 1/f Noise and Other Low Frequency Fluctuations in Electronic Devices, St Louis, MO, USA, 7–8 August 1998; Amer Inst Physics: Melville, SK, Canada, 1999.
35. Capan, I.; Janicki, V.; Jacimovic, R.; Pivac, B. C-V and DLTS studies of radiation induced Si-SiO$_2$ interface defects. *Nucl. Instrum. Methods Phys. Res. Sect. B-Beam Interact. Mater. At.* **2012**, *282*, 59–62. [CrossRef]
36. Bertuccio, G.; Pullia, A. A method for the determination of the noise parameters in preamplifying systems for semiconductor radiation detectors. *Rev. Sci. Instrum.* **1993**, *64*, 3294–3298. [CrossRef]
37. Fleetwood, D.M.; Winokur, P.S.; Reber, R.A.; Meisenheimer, T.L.; Schwank, J.R.; Shaneyfelt, M.R.; Riewe, L.C. Effects of Oxide Traps, Interface Traps, and Border Traps on Metal-Oxide-Semiconductor Devices. *J. Appl. Phys.* **1993**, *73*, 5058–5074. [CrossRef]
38. Richardson, J.A.; Grant, L.A.; Henderson, R.K. Low Dark Count Single-Photon Avalanche Diode Structure Compatible with Standard Nanometer Scale CMOS Technology. *IEEE Photonics Technol. Lett.* **2009**, *21*, 1020–1022. [CrossRef]
39. Moscatelli, F.; Marisaldi, M.; Maccagnani, P.; Labanti, C.; Fuschino, F.; Prest, M.; Berra, A.; Bolognini, D.; Ghioni, M.; Rech, I.; et al. Radiation tests of single photon avalanche diode for space applications. *Nucl. Instrum. Meth. A* **2013**, *711*, 65–72. [CrossRef]
40. Srour, J.R.; Palko, J.W. Displacement Damage Effects in Irradiated Semiconductor Devices. *Ieee T Nucl. Sci.* **2013**, *60*, 1740–1766. [CrossRef]
41. Wu, M.-L.; Ripiccini, E.; Kizilkan, E.; Gramuglia, F.; Keshavarzian, P.; Fenoglio, C.A.; Morimoto, K.; Charbon, E. Radiation Hardness Study of Single-Photon Avalanche Diode for Space and High Energy Physics Applications. *Sensors* **2022**, *22*, 2919. [CrossRef] [PubMed]
42. Ziegler, J.F.; Ziegler, M.D.; Biersack, J.P. SRIM-The stopping and range of ions in matter (2010). *Nucl. Instrum. Methods Phys. Res. Sect. B-Beam Interact. Mater. At.* **2010**, *268*, 1818–1823. [CrossRef]

Disclaimer/Publisher's Note: The statements, opinions and data contained in all publications are solely those of the individual author(s) and contributor(s) and not of MDPI and/or the editor(s). MDPI and/or the editor(s) disclaim responsibility for any injury to people or property resulting from any ideas, methods, instructions or products referred to in the content.

Article

Evaluation of Single Event Upset on a Relay Protection Device

Hualiang Zhou [1,2,*], Hao Yu [3], Zhiyang Zou [1,2], Zhantao Su [1,2], Qianyun Zhao [1,2], Weitao Yang [3,*] and Chaohui He [3]

1. NARI Group Corporation (State Grid Electric Power Research Institute), Nanjing 211106, China
2. NARI Technology Co., Ltd., Nanjing 211106, China
3. School of Nuclear Science and Technology, Xi'an Jiaotong University, Xi'an 710049, China
* Correspondence: zhouhualiang@sgepri.sgcc.com.cn (H.Z.); wtyang@stu.xjtu.edu.cn (W.Y.)

Abstract: Traditionally, studies have primarily focused on single event effects in aerospace electronics. However, current research has confirmed that atmospheric neutrons can also induce single event effects in China's advanced technology relay protection devices. Spallation neutron irradiation tests on a Loongson 2K1000 system-on-chip based relay protection device have revealed soft errors, including abnormal sampling, refusal of operation and interlock in the relay protection device. Given the absence of standardized evaluation methods for single event effects on relay protection devices, the following research emphasizes the use of Monte Carlo simulation and software fault injection. Various types of single event upsets, such as single bit upsets, dual bit upsets, and even eight bit upsets, were observed in Monte Carlo simulations where atmospheric neutrons hit the chip from different directions (top and bottom). The simulation results indicated that the single event effect sensitivity of the relay protection device was similar whether the neutron hit from the top or the bottom. Through software fault injection, the study also identified soft errors caused by neutron induced single event upsets on the Loongson 2K1000 system, including failure to execute, system halt, time out, and error result. And the soft error number of system halts and error results exceeded that of time outs and failures to execute in all three tested programs. This research represents a preliminary assessment of single event effects on relay protection devices and is expected to provide valuable insights for evaluating the reliability of advanced technology relay protection devices.

Keywords: relay protection device; Monte Carlo; fault injection; single event effect; soft error

1. Introduction

Relay protection devices are crucial components in power systems, serving the important function of swiftly disconnecting faults and maintaining the stability of the grid [1–3]. The reliability of these devices has a direct impact on the overall stability of the grid [4–6]. Relay protection devices commonly embrace emerging applications and advanced semiconductor technologies [7,8]. However, these advancements also introduce new challenges, such as the susceptibility to single event effects (SEE) induced by atmospheric neutrons.

In the field of aerospace electronic systems, significant attention has traditionally been given to SEE due to the presence of energetic particles. These particles can deposit energy and cause single event upsets (SEU) and other effects [9,10]. However, there is relatively little focus on the impact of SEE on advanced technology relay protection devices. When evaluating factors influencing the reliability of advanced technology relay protection devices, the emphasis has typically been on voltage, temperature, electromagnetic interference, and others rather than SEE [11]. In [12], the overvoltage and undervoltage effect on relay protection devices was discussed. In [13], the impact of a static var compensator on a distance protection relay was evaluated. In [14], a test device simulating live verification relay protection as designed. In [15], an intelligent relay protection system was developed, and the system can automatically select a relay protection set point basing initial data on weather conditions, time of year, soil resistance, current, voltage, etc. In [16], the failure causes of relay protection switching power supply were explored. In [17], the author provided a reliable quantitative basis for relay protection systems' operating maintenance by the aid of a semi-supervised Mahalanobis distance machine learning algorithm. And in [18], authors subdivided the influence factors of incorrect actions on relay protection devices of the State Grid Corporation of China from 2006 to 2017; they considered the causes of incorrect actions mainly from defects in relay protection devices, secondary circuits or communication systems. In [19,20], the outstanding engineers, K. Zimmerman and D. Haas from Schweitzer Engineering Laboratories, appealed to the manufactures and end users to continuously monitor and work toward improving overall system design to mitigate single event effects. All these facts indicate that the field of relay protection device currently lacks consideration of single event effects, as it primarily concentrates on conventional factors. Especially with the increase use of advanced technology semiconductor devices in the field of relay protection, the continued neglect of single event effects on relay protection devices may lead to unpredictable consequences. Therefore, it is crucial to urgently conduct research on single event effects of advanced technology relay protection devices in the present and near future.

Relay protection devices typically operate in terrestrial environments where they are exposed to atmospheric neutrons. These neutrons possess a broad energy spectrum, ranging from meV to GeV [21]. When these neutrons interact with atomic nuclei in semiconductors, they can induce SEE. For example, high-energy neutrons may react with silicon and produce secondary high-energy heavy ions, while thermal neutrons can interact with boron contamination and generate energetic secondary particles. These energetic secondary ions/particles can deposit energy in the semiconductor and result in SEE [22]. It can be speculated that as more advanced semiconductor devices are utilized in relay protection devices, the risk of SEE also increases. Therefore, it becomes crucial to pay more attention to this issue. Notably, there have been recorded incidents of SEE in Chinese relay protection devices in 2018 and 2020 [23,24]. These incidents highlight the importance of assessing the impact of SEE on relay protection devices in China. As the largest supplier of complete electric power equipment in China and an active participant in the global power industry, the NARI Group Corporation (NARI) has an obligation and responsibility to acquire knowledge of atmospheric neutron SEE on advanced technology relay protection devices in China [25]. As a result, our current research is dedicated to addressing the influence of SEE on Chinese relay protection devices.

The spallation neutron source is an excellent candidate for conducting atmospheric neutron induced SEE evaluation [26]. With the operation of the China Spallation Neutron Source (CSNS), it has become feasible to study atmospheric neutron SEE in China [27,28]. Due to factors such as uncertainty in irradiation tests, irradiation hours, and cost, the current study primarily focuses on using spallation neutron irradiation to confirm whether SEE can affect the target relay protection device, specifically the Loongson 2K1000 system-on-chip based development kit. Once this confirmation is established, greater emphasis and effort are placed on software fault injection. Compared to irradiation testing, software fault injection allows for more detailed insights that may be challenging to extract solely through irradiation [29]. Additionally, the fault injection technique relies on the results of Monte Carlo simulations, which utilize models constructed from the tested chip. Through these efforts, detailed soft errors induced by atmospheric neutron SEE on the advanced technology relay protection devices can be examined and evaluated.

The structure of the paper is as follows: Section 2 provides an introduction to relay protection architecture. Section 3 introduces SEE assessment framework on relay protection device. Section 4 briefs the spallation neutron source irradiation, and Section 5 presents the Monte Carlo simulation. Then, Section 6 details the fault injection based on Monte Carlo outcomes, and Section 7 analyzes the results. Finally, we draw conclusions based on our findings in Section 8.

2. Relay Protection Architecture

A relay protection device generally consists of various modules that perform different functions and are interconnected through buses or interfaces. Some common modules found in a relay protection device include the input module, protection module, management module, power supply module, etc.

These modules primarily consist of three types of CPUs (Central Processing Units): protection CPU, startup CPU, and management CPU. The protection and startup CPUs are responsible for signal sampling, protection processing, and trip control. The management CPU handles recording, human–machine interface communication, and other related tasks. The architecture of the key CPUs can be observed in the left section of Figure 1.

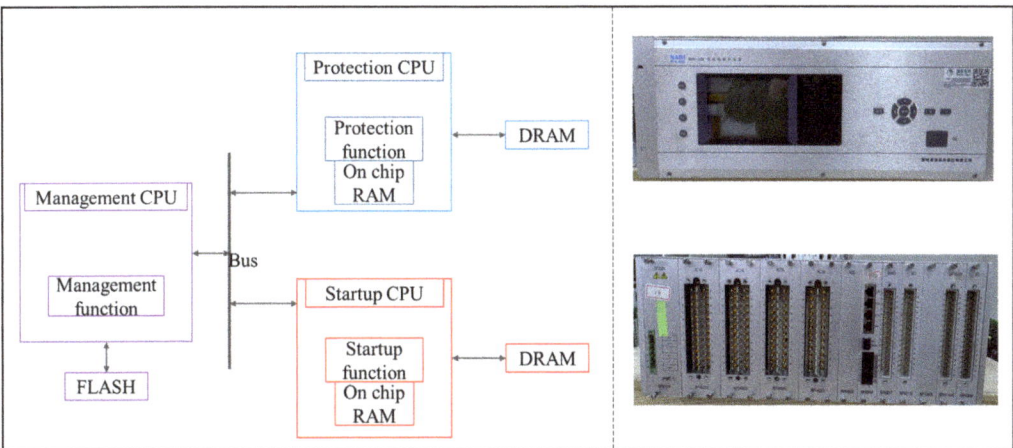

Figure 1. Architecture of the key CPUs in relay protection devices and the equipment photo of the Loongson 2K1000 system-on-chip based relay protection equipment; the left part is the architecture of the key CPUs, and the right part shows the photo of the front side and the back side of the terminal connector in the top and bottom.

The Loongson 2K1000 system-on-chip development kit plays a crucial role in the relay protection system of the Chinese power grid, particularly within the integrated dual CPUs. The kit features the dual-GS264 processor, which operates at a maximum frequency of 1 GHz. Each core of the processor is equipped with independent two level instruction and data caches and on-chip random access memory (RAM). Furthermore, the processor incorporates a diverse array of high-speed interfaces [30]. In the Loongson 2K1000 system-on-chip development kit, one processor serves as the management CPU, while the other is multiplexed to act as the startup CPU during the launch stage. Once launched, it assumes the role of the protection CPU. The right section of Figure 1 shows the front and back sides of the relay protection equipment.

In the context of relay protection architecture, the on-chip RAM (random access memory), DRAM (dynamic random access memory), and Flash serve as essential data storage media. However, these storage media are susceptible to SEE [31,32]. In addition, the registers in the CPUs may also suffer from SEE. The occurrence of SEE in these memories can lead to unexpected outcomes in the power grid, potentially resulting in incalculable losses. Thus, the fault injection and the Monte Carlo simulations were mainly performed on the memory block.

3. SEE Assessment Framework on Relay Protection Devices

As mentioned above, there can be a lack of outcomes in relay protection SEE assessment. To address this issue, considering the operations of CSNS, we proposed a research framework that combined spallation neutron source irradiation testing, Monte Carlo simulations, and software fault injection to assess SEE influence on relay protection devices. Figure 2 shows the framework of the current study in which the irradiation test checked whether the atmospheric neutron could induce SEE on relay protection. Then, the Monte Carlo simulation provided details about SEE, such as the distribution of multi bit upsets. At the same time, the Monte Carlo simulation also provides the upset information during software fault injection.

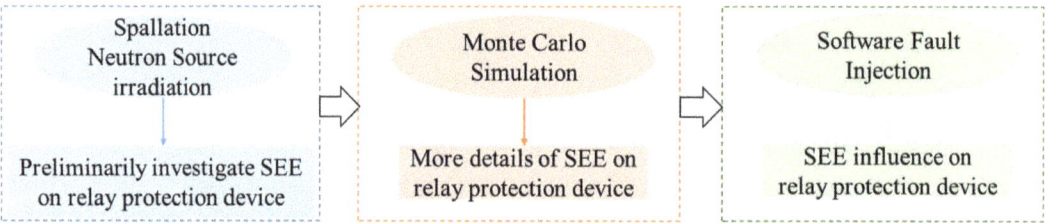

Figure 2. Framework of SEE assessment on relay protection devices.

4. Spallation Neutron Source Irradiation

In the absence of an established standard for spallation neutron source SEE evaluation in relay protection devices and limited research in this area, it was crucial to investigate whether any SEE can be detected during the spallation neutron irradiation process.

The primary objective of the irradiation was to examine the occurrence of SEE in relay protection devices when the device under test (DUT) was exposed to atmospheric neutron irradiation at the CSNS end. In the irradiation test, the DUT was placed at the distance of 17.5 cm from the terminal in the irradiation room. The neutron spectrum was derived from the actual atmospheric fluence with a significant magnification factor. At CSNS, the synchrotron accumulated and accelerated the proton beam to 1.6 GeV. Then, the beam was extracted in a single turn and was delivered to the metal target through the ring-to-target beam transport. The ultra-high-energy protons impinged on the metal target and produced spallation neutrons applied in irradiation tests [33]. During irradiation, the equivalent high-energy neutron fluence was about 3×10^7 neutrons/(cm^2·s) with an intended continuous exposure time of 10 min. If a soft error was detected during this process, a new round irradiation test was initiated. Ultimately, soft errors, including abnormal sampling, relay protection refusal to operate, and relay protection device interlock, were detected. They are defined as follows:

- ➤ Abnormal sampling: the sample value is out of range as expected during irradiation;
- ➤ Protection refusal to operate: it fails to perform its intended protective function even when it receives a fault signal;
- ➤ Relay protection device interlock: the device is intentionally prevented from tripping or operating in response to a fault signal.

These findings demonstrate that atmospheric neutrons cause SEE in relay protection devices and indeed result in unexpected outcomes. And it emphasizes the urgent need to conduct more detailed research about SEE assessment on relay protection device. This also highlights that our current research is valuable and has practical significance.

5. Monte Carlo Simulation

5.1. Simulation Construction

The Geant4 simulation was performed on the target [34,35]. Even though the incoming direction and angle minimally impact the interaction between the high-energy neutron and atomic nuclei, the chip's structure varies between the top and bottom scenes. Consequently, two types of simulations were performed: one involved neutron particles striking from the passivation layer of the chip (referred to as 'From top' or 'First case'), and the other entailed neutron particles incoming from the silicon substrate (referred to as 'From bottom' or 'Second case'). Figure 3 illustrates the schematic of these two simulation scenarios. Except for the incoming location, all other parameters remained consistent for both simulations. Table 1 provides the architectural details of the constructed target in Geant4. It is noteworthy that the B layer above the silicon substrate, following the sequence in Table 1, served as an equivalent layer for boron contamination within the chip, as it could be introduced during the semiconductor contact and doping processes.

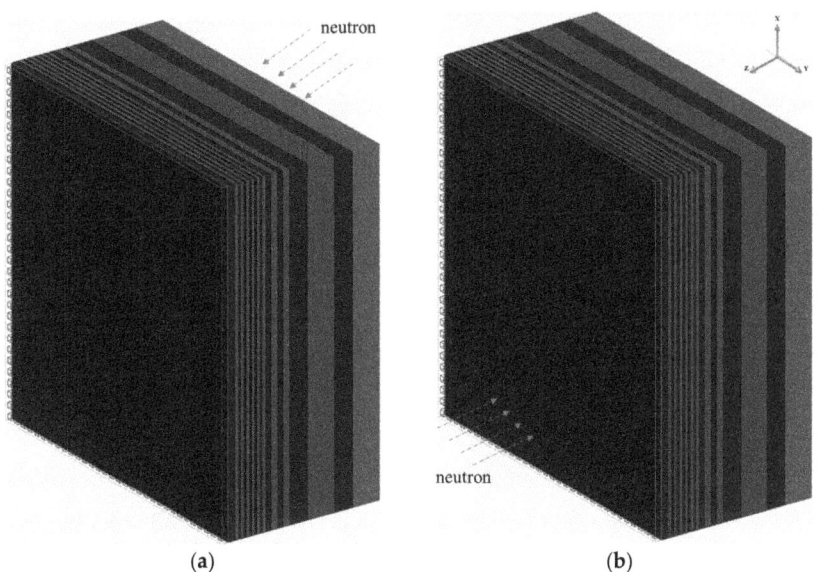

Figure 3. Neutron impinging from top and bottom schematic diagrams, (**a**) from top and (**b**) from bottom.

Table 1. Architectural details of the constructed target in Geant4.

Layer	Material	Thickness/nm
1	Al	880
2	SiO2	600
3	Cu	880
4	SiO2	600
5	Cu	215
6	SiO2	180
7	Cu	215
8	SiO2	180
9	Cu	100
10	SiO2	80
11	Cu	100
12	SiO2	80
13	Cu	100
14	SiO2	80
15	Cu	100
16	SiO2	80
17	Cu	100
18	SiO2	80

Table 1. *Cont.*

Layer	Material	Thickness/nm
19	Cu	105
20	SiO2	230
21	B	2
22	Si	1200

In the simulation, a 32 × 32 array of sensitive volumes was positioned, with each sensitive volume measuring 160 nm × 160 nm × 160 nm in size. The critical charge was 3820 eV. A total of 10^8 impinging neutrons were generated from a planar source with a size of 10,080 nm × 10,080 nm. The neutron spectrum was derived from the terminal of the China Spallation Neutron Source, as depicted in Figure 4. It can be observed that the neutron spectrum at the CSNS terminal was similar to the spectrum at ground level in Beijing, but with an amplification factor of 10^9 from its actual fluence.

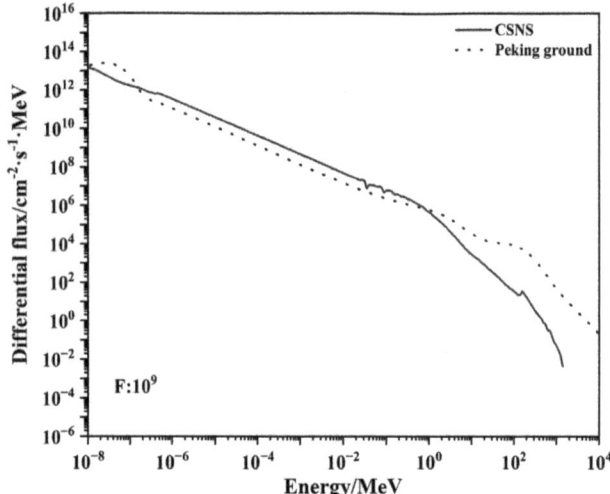

Figure 4. The differential flux of the neutron beam at CSNS applied in Monte Carlo simulation.

5.2. Simulation Results

In the first case, a total of five SEUs were detected, while in the second case, six SEUs were detected. The details of these SEUs are listed in Tables 2 and 3, respectively. It was observed that when the neutron struck from the top direction, a maximum of five bits were affected and experienced flipping. Conversely, when the neutron struck from the bottom direction, the number of affected bits increased to eight.

Table 2. SEE of simulation when atmospheric neutron struck from top.

SEU	Count
1	130
2	26
3	13
4	2
5	3

Table 3. SEE of simulation when atmospheric neutron struck from bottom.

SEU	Count
1	118
2	33
3	9
4	2
5	2
8	1

According to flipping cell coordinates, the specific distributions of these multi bit upset events could also be extracted. Figure 5 represents the flipping bit distribution schematic diagrams of a part of multi bit upset events, including three bits in (a), four bits in (b), five bits in (c), and eight bits in (d).

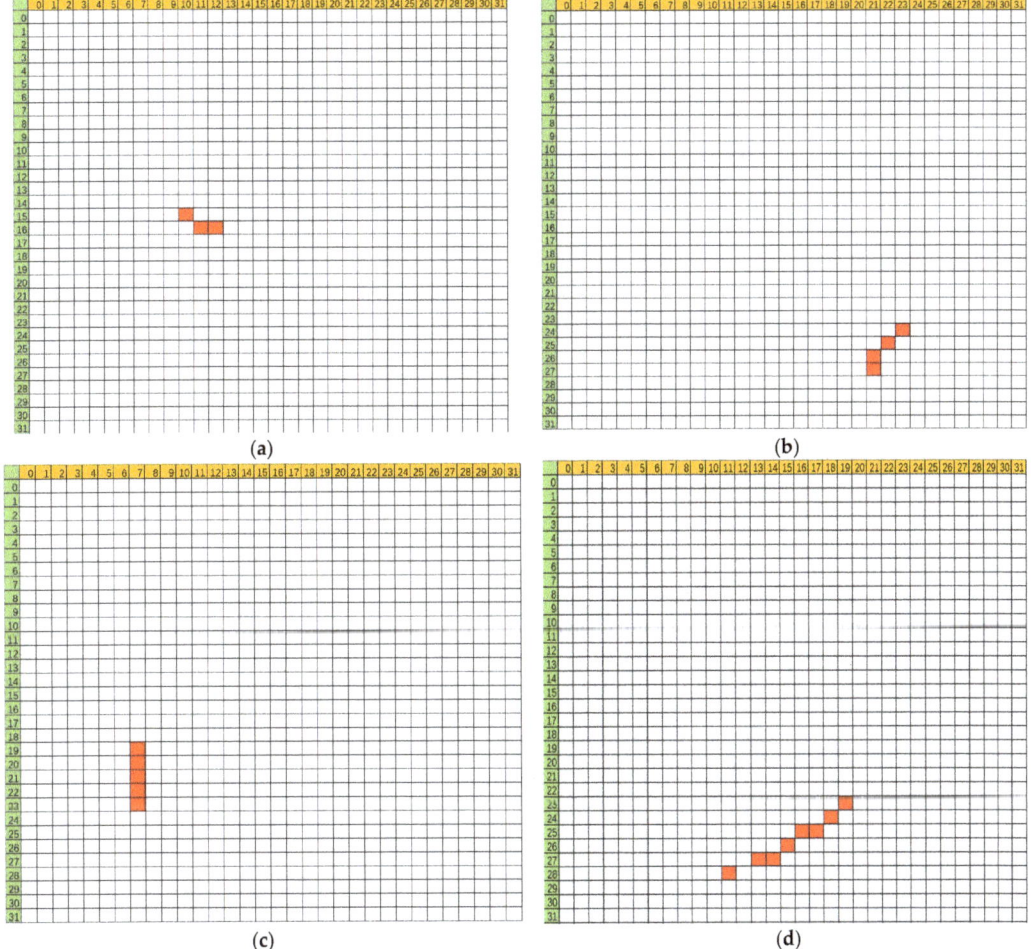

Figure 5. Part of multi bit upset distribution, (**a**) three bit upset, (**b**) four bit upset, (**c**) five bit upset, (**d**) eight bit upset. The red stands for upset information.

Meanwhile, the percentages of different types of upset were obtained. Figure 6 depicts the proportion of them in two cases, where (a) is for the first case while (b) is for the second case.

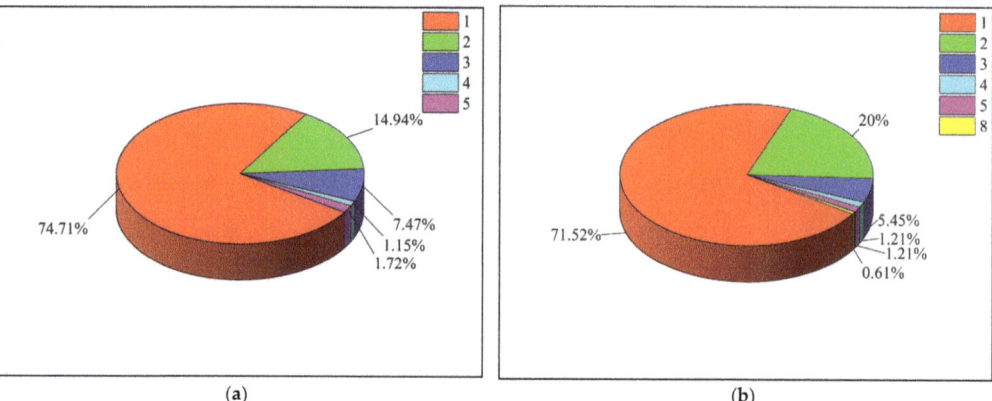

Figure 6. Percentage of different upset in two cases, (**a**) first case and (**b**) second case.

6. Fault Injection Based on Monte Carlo Outcomes

Based on the results of the Monte Carlo simulation, more specific fault injection could be performed. During the fault injection process, the upset information aligned with the distribution and percentage detected in the Monte Carlo simulation.

6.1. Fault Injection Design

To enhance the efficiency of the fault injection software, a management terminal was developed in Python language. Figure 7 presents the graphical user interface of this terminal. Within this interface, users could input the target injection location, upset categories, and proportions for the fault injection, based on results of the Monte Carlo simulation. This information was then utilized to effectively carry out the fault injection.

Three general test programs were developed by us to evaluate the performance of the relay protection Loongson 2K1000 system-on-chip development kit. These programs are as follows:

Fibonacci sequence: two sets of sequences are tested, the first set consisting of 10 terms and the second consisting of 15 terms;

Matrix Multiply: performing matrix multiplication on the two matrices [3][2] and [2][3];

Management operation: verifying if the entered username and password match the set username and password, and output the result.

During the execution of each test program in Linux, faults were injected into the corresponding code segment. The code segment's address ranged from 0x120,000,000 to 0x120,004,000, totaling 16,385 bytes. The details of the fault injection process were as follows, and Figure 8 depicts a diagram illustrating the fault injection.

Firstly, the test program was compiled into an executable file in the Linux environment and downloaded.

Secondly, the random injection time points were generated. Since there were 16,385 bytes in the code segment, a total of 16,385 time points were created in an operation duty for one tested program. At this point, the code segment was injected byte by byte.

Thirdly, a type of upset was extracted from the obtained flipping categories which were derived from the Monte Carlo simulation. If the upset information was extracted from the first case simulation, it corresponded to a neutron striking from the top. In contrast, if it was extracted from the second case, it represented a neutron hitting from the bottom.

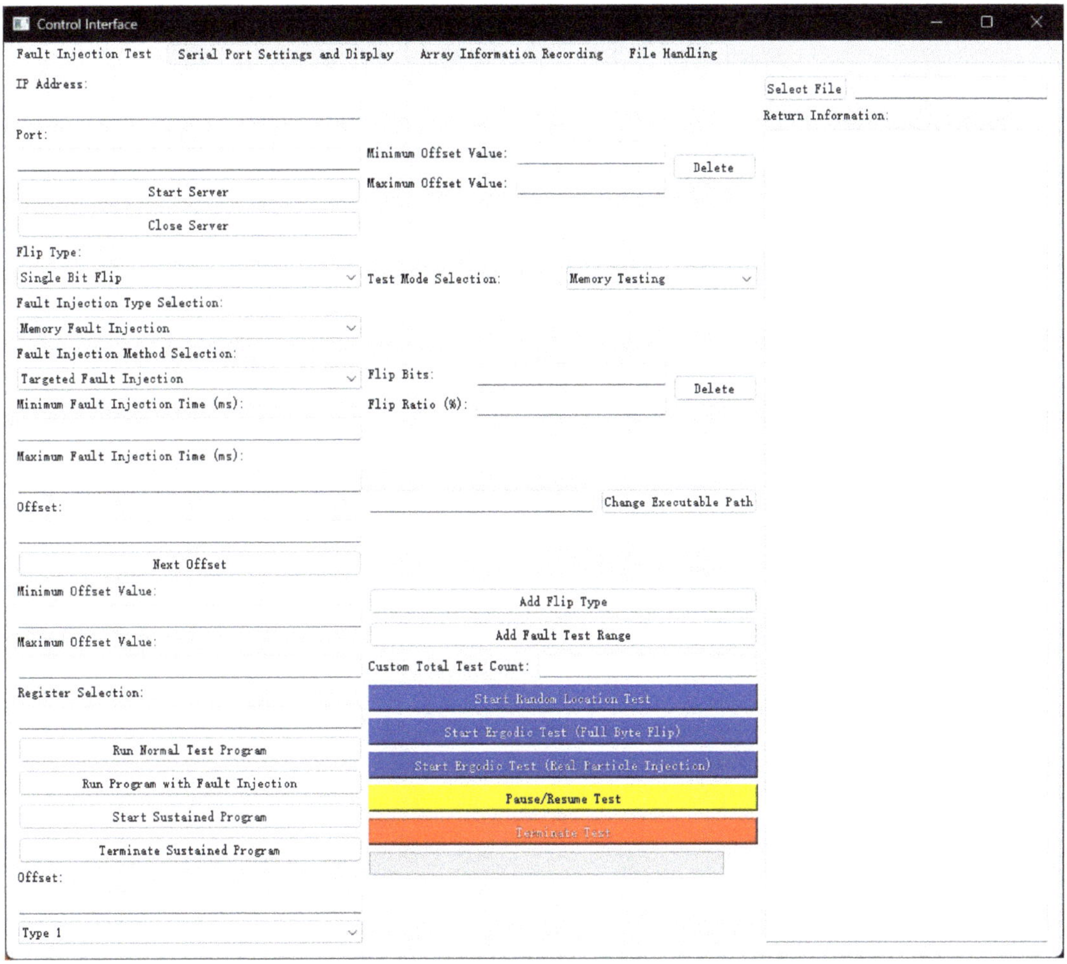

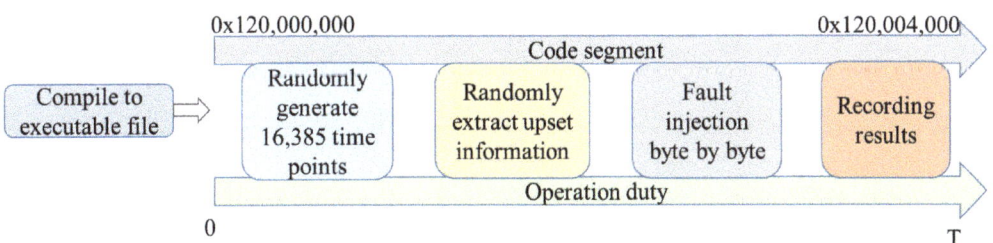

Figure 7. Window of the indigenous-designed fault injection terminal.

Figure 8. Fault injection diagram.

Fourth, to perform a fault injection, a series of system calls were utilized. First and foremost, the fault injection program initiated a new process as a child process using the fork() system call. The child process's ID was then retrieved. The child process took charge of interrupting, modifying, and monitoring the program under test. Then, when a specified injection time point was reached, the PTRACE() (an abbreviation of "process trace") system call was employed to manipulate the fault injection in the child process. As a core modification method, PTRACE() allowed for bit-flip modifications in the memory corresponding to the test program's code or data.

Lastly, it recorded the final results from the fault injection.

6.2. Fault Injection Results

Throughout the fault injection process, a total of five types of results were obtained. They included failure to execute (FE), system halt (SH), time out (TO), error result (ER) and normal. Among them, the first four soft errors were abnormal for the relay protection device. The results are defined as follows:

Failure to execute (FE): program's exit code experiences an abnormality and cannot start to execute;

System halt (SH): program execution is halted;

Time out (TO): program execution is out of the expected duration;

Error result (ER): the execution results are different from the expected results;

Normal: the injected faults have no visible influence on the tested program's execution.

Concerning neutron striking from the top and bottom directions, the fault injection was performed for both cases. And the results for the two cases were recorded. For the Fibonacci sequence, 284 and 303 soft errors are detected based on the neutron from top and bottom striking simulation results, respectively. Figure 9 depicts the fault injection results of the Fibonacci sequence, showing that the SH and ER soft errors were much higher than FE and TO. Additionally, 546 and 552 soft errors were observed in the Matrix Multiply fault injection from the neutron top and bottom hitting simulation results. Figure 10 shows the results of the fault injection on Matrix Multiply, indicating no FE soft errors in this test. Similarly, the SH and ER soft errors were more significant. When it came to the Management operation fault injection, 292 and 277 soft errors were obtained relying on the results from the first and second case simulation, respectively. Figure 11 displays the outcomes of the Management operation fault injection, showing results similar to those of the Fibonacci sequence.

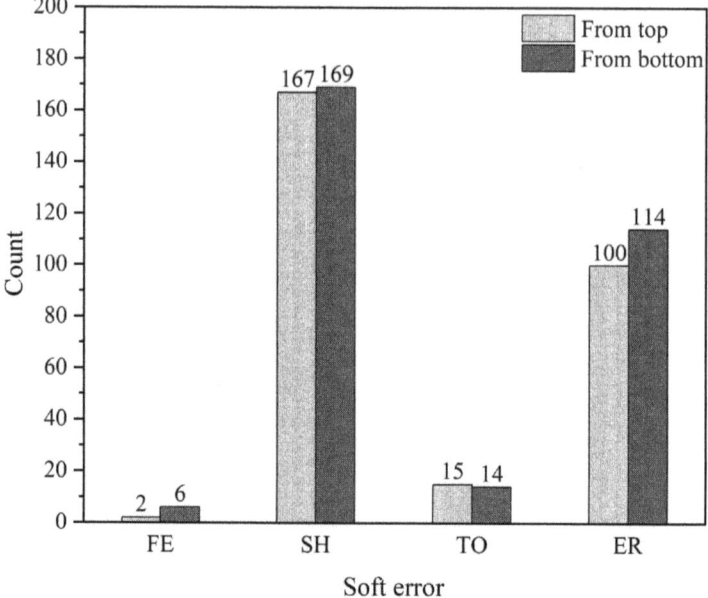

Figure 9. Soft error in fault injection results of the Fibonacci sequence.

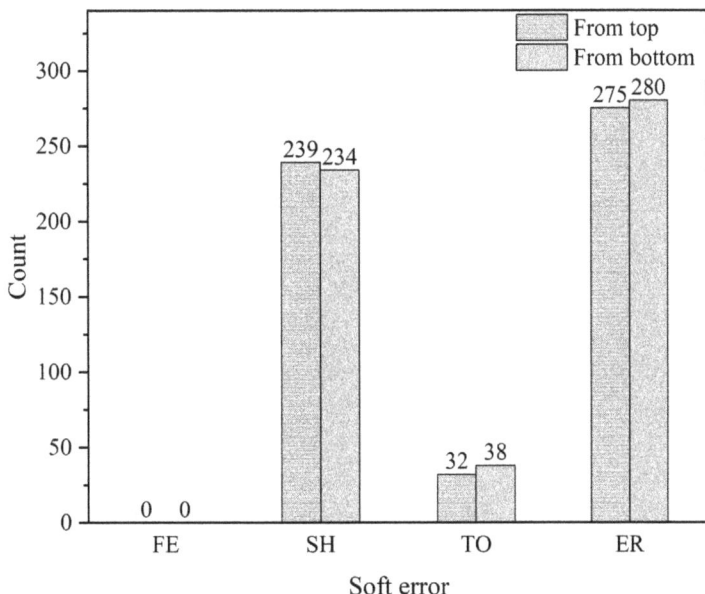

Figure 10. Soft error in fault injection results of the Matrix Multiply.

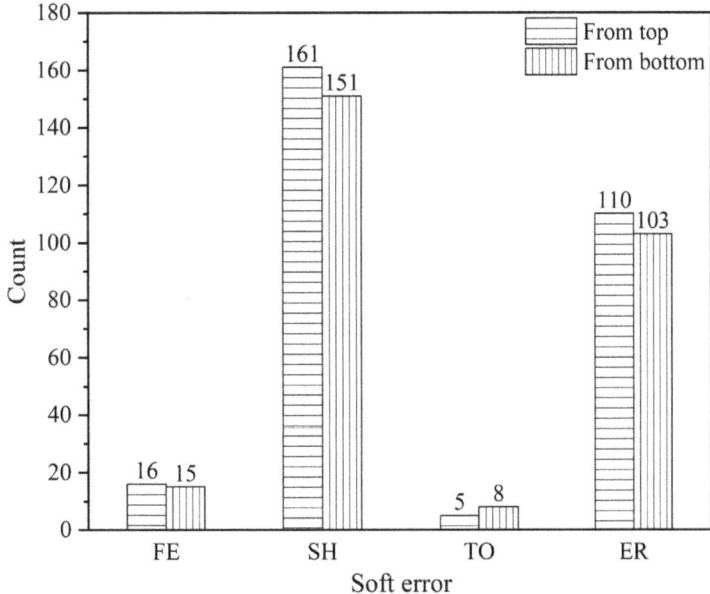

Figure 11. Soft error in fault injection results of the Management operation.

7. Results Analysis

The atmospheric neutron spectrum encompasses a range of neutron energies from meV to GeV, including thermal (n_{th}) and high-energy neutrons. It is important to note that born contamination is still considered to exist within advanced semiconductors, even after eliminating the boro-phospho-silicate glass packages [21,22]. This suggests that one possible cause of SEE induced by atmospheric neutrons comes from the following reactions: $^{10}B+n_{th} \rightarrow {^7Li}(1.01\ MeV) + \alpha(1.78\ MeV)$ and $^{10}B+nth \rightarrow {^7Li}(0.84\ MeV) +$

α(1.47 MeV) + γ(0.48 MeV), in which the probability of the later reaction is more than 90%. Additionally, the linear energy transfers of the ^{7}Li(0.84 MeV) and α(1.47 MeV) are 2.10 and 1.15 MeV·cm^2·mg^{-1}, respectively, which are sufficient to induce SEE within relay protection devices [21]. Another source of single event effects (SEE) arises from high-energy neutrons interacting with silicon nuclei. When an energetic neutron collides with silicon nuclei, various reactions can occur, such as Si(n, α), Si(n, p), Si(n, d), Si(n, n-α). The generated secondary energetic particles in these processes can induce SEE within relay protection devices [36].

From the obtained results in Tables 2 and 3, it can be observed that neutrons striking from the top or bottom of the chip can result in multiple types of SEU. Further, the cross section for the first case was 2.42 × 10^{-15} cm^2·bit^{-1} while for the second case was 2.35 × 10^{-15} cm^2·bit^{-1}. These are close to the results obtained in spallation neutron source irradiation on similar technology memory. In the irradiation test, the SEE cross section was 1.50 × 10^{-15} cm^2·bit^{-1} [37]. It demonstrates that the Monte Carlo simulation and the fault injection results are credible. Meanwhile, the soft error rate for them at Beijing ground (9.5 n/(cm^2·h)) were 22.99 and 22.33 FIT/Mbit when neutrons struck from top and bottom, respectively. These indicate that the SEE sensitivity of relay protection devices are almost close whether a neutron hits from the top or bottom.

Furthermore, Figures 9–11 show that soft errors occur in proximity to each other when neutrons strike from the top or bottom in each test program. This suggests that the discrepancy in SEE influence between neutrons hitting from the top and bottom is minimal. The results indicate the need for solutions to mitigate SEE on relay protection devices in the current and near future, such as the error correcting code in memory, the redundancy in data and code, the rollback examination in software, lockstep in dual cores, or others.

Although these findings regarding single event effects and soft errors derived from atmospheric neutron studies, it is reasonable to speculate that they can be applied to the evaluation of SEE induced by energetic protons in space. This suggests that when energetic protons impact the chip from both the top and bottom in aerospace applications, the resulting SEE sensitivity and occurrence of soft errors may be relatively similar under these two conditions.

From Figure 6, it can be observed that single bit upsets account for about 70% of flipping. For these single bit upsets, they can be addressed by techniques, such as error checking code. More seriously, this means almost 30% of upsets are difficult to mitigate. This also indicates much more SEE research on relay protection devices is required.

In the fault injection results of different test programs, a common phenomenon is that the number of "SH" and "ER" exceeds that of "TO" and "FE". Software fault injection simulates soft errors caused by bit flips during program execution, resulting in data or instruction errors. These errors can lead to outcomes such as the loss of data integrity, alteration of instruction flow, and triggering of exception signals. Among these, abnormal exit status codes (FE) often occur when soft errors cause the program to jump to incorrect code paths or error handling routines, while timeout (TO) is caused by the program entering into an unintended, prolonged wait state or getting caught in a loop condition. These two outcomes typically occur when soft errors do not compromise data integrity, trigger exception signals, or lead to broader errors. Therefore, the majority of soft errors are more likely to cause program execution halting (SH) or error results (ER).

The main objective of this research is to preliminarily assess the impact of SEE induced soft errors on relay protection devices in China using spallation neutron source, which has been successfully achieved. Although energy levels above 1 MeV or 10 MeV are usually used to evaluate SEE in atmospheric neutron irradiation, the contribution of thermal neutrons to SEE in advanced technology relay protection devices remains unclear. In the future, by leveraging thermal neutron absorptions or other techniques, we can further evaluate the relative contributions of different energy levels of neutrons from atmospheric neutron to SEE and soft errors in relay protection devices.

8. Conclusions

Atmospheric neutrons are confirmed to induce soft errors in relay protection devices using China spallation neutron source irradiation. For the core processor of the relay protection Loongson 2K1000 system-on-chip, the Monte Carlo simulation was performed and single event effects were obtained when neutrons struck from the top and bottom of the chip. Simulation results demonstrated that the single event effect vulnerability was close for neutrons hitting from the top and bottom. The fault injection was performed on three general test programs relying on the single event upset information from Monte Carlo simulations. Soft errors, including failure to execute, system halt, time out, and error result were obtained and the occurrence of system halts and error results were higher than failures to execute and time outs. The fault injection results mean that the effects are almost the same for the relay protection device when a neutron hits from the top or bottom.

Author Contributions: Conceptualization, H.Z., W.Y. and C.H.; methodology, H.Z. and Z.Z.; software, H.Y.; validation, Z.S.; formal analysis, Q.Z.; resources, H.Z.; writing—original draft preparation, W.Y. and H.Z. All authors have read and agreed to the published version of the manuscript.

Funding: Project supported by the Science and Technology Project of NARI Technology Co., Ltd. (Grant No. SGNRGF00XAJS2301697) and the National Natural Science Foundation of China (Grant Nos. 11835006, 11690040, and 11690043).

Data Availability Statement: The data used to support the findings of this study are available from the corresponding author upon request.

Acknowledgments: Thanks for the support from the NARI Group Corporation (State Grid Electric Power Research Institute), and NARI Technology Co., Ltd.

Conflicts of Interest: The authors H. Zhou, Z. Zou, Z. Su, and Q. Zhao were employed by the company NARI Group Corporation (State Grid Electric Power Research Institute), and NARI Technology Co., Ltd. The remaining authors declare that the research was conducted in the absence of any commercial or financial relationships that could be construed as a potential conflict of interest.

References

1. Tian, M.; Zhang, L.; Guo, P.; Zhang, H.; Chen, Q.; Li, Y.; Xue, A. Data dependence analysis for defects data of relay protection devices based on apriori algorithm. *IEEE Access* **2020**, *8*, 120647–120653. [CrossRef]
2. Zhang, B.; Hao, Z.; Bo, Z. New development in relay protection for smart grid. *Prot. Control. Mod. Power Syst.* **2016**, *1*, 14. [CrossRef]
3. Wang, J.F.; Yang, P.W.; Chen, X.L. State evaluation of relay protection system for state maintenance. *IOP Conf. Ser. Earth Environ. Sci.* **2019**, *354*, 012120. [CrossRef]
4. Esfahani, M.M.; Mohammed, O. An intelligent protection scheme to deal with extreme fault currents in smart power systems. *Electr. Power Energy Syst.* **2020**, *115*, 105434. [CrossRef]
5. Kiliçkiran, H.C.; Şengör, İ.; Akdemir, H.; Kekezoğlu, B.; Erdinç, O.; Paterakis, N.G. Power system protection with digital overcurrent relays: A review of nonstandard characteristics. *Electr. Power Syst. Res.* **2018**, *164*, 89–102. [CrossRef]
6. Zamani, M.A.; Sidhu, T.S.; Yazdani, A. A protection strategy and microprocessor-based relay for low-voltage microgrids. *IEEE Trans. Pow. Deliv.* **2011**, *26*, 1873–1883. [CrossRef]
7. Wei, D.; Lu, Y.; Jafari, M.; Skare, P.M.; Rohde, K. Protecting Smart Grid Automation Systems Against Cyberattacks. *IEEE Trans. Smart Grid* **2011**, *2*, 782–795. [CrossRef]
8. Liang, W.; Ouyang, F.; Wang, S.; Zhu, W.; Li, G. Smart Operation and Maintenance Platform of Protection Relay Based on Mobile Sensing and Big Data. *IOP Conf. Ser. Earth Environ. Sci.* **2021**, *632*, 042021. [CrossRef]
9. Luza, L.M.; Wrobel, F.; Entrena, L.; Dilillo, L. Impact of Atmospheric and Space Radiation on Sensitive Electronic Devices. In Proceedings of the IEEE European Test Symposium, Barcelona, Spain, 23–27 May 2022.
10. Reed, R.A.; Kinnison, J.; Pickel, J.C.; Buchner, S.; Marshall, P.W.; Kniffin, S.; LaBel, K.A. Single-event effects ground testing and on-orbit rate prediction methods: The past, present, and future. *IEEE Trans. Nucl. Sci.* **2003**, *50*, 622–634. [CrossRef]
11. Liu, H.; Zhang, Y.; Li, W.; Zhang, X.; Wang, H. Research on the influence and test of core components on relay protection device. *IOP Conf. Ser. Earth Environ. Sci.* **2021**, *631*, 012089. [CrossRef]
12. Li, Y.; Zhi, L.H.; Ping, L.D.; Bo, Z.H.; Zhang, L.; Hua, T.X. Analysis of interference factors of relay protection devices in high altitude areas and research on solutions. *Rev. Adhes. Adhes.* **2023**, *11*, 125–139.
13. Ngo, M.; Nguyen, H.; Dinh, T. A study of SVC's impact simulation and analysis for distance protection relay on transmission lines. *Int. J. Electr. Comput. Eng.* **2017**, *7*, 1686–1695.

14. Ji, B.; Yang, X.; Liu, C.; Xu, H.; Ji, Y.; Tao, S. Develop a test device to check the charged relay protection. In Proceedings of the 2021 IEEE International Conference on Power, Intelligent Computing and Systems, Shenyang, China, 29–31 July 2021.
15. Dikarev, P.V.; Shilin, A.A.; Ahmedova, O.O. Circuit breaker control of intelligent relay protection system. In Proceedings of the 2021 International Conference on Industrial Engineering, Applications and Manufacturing, Sochi, Russia, 17–21 May 2021.
16. Zhang, H.; Liu, Z.; Fan, Z.; Song, H.; Niu, Z.; Xiong, Z.; Deng, M.; Shuai, J.; Li, W.; Li, X. Failure causes and solutions of relay protection switching power supply. *J. Phys. Conf. Ser.* **2022**, *2196*, 012039. [CrossRef]
17. Ying, L.; Jia, Y.; Li, W. Research on state evaluation and risk assessment for relay protection system based on machine learning algorithm. *IET Gener. Transm. Distrib.* **2020**, *14*, 3619–3629. [CrossRef]
18. Chen, Q.; Zhang, L.; Guo, P.; Zhang, H.; Tian, M.; Li, Y.; Xue, A. Modeling and Analysis of Incorrect Actions of Relay Protection Systems Based on Fault Trees. *IEEE Access* **2020**, *8*, 114571–114579. [CrossRef]
19. Haas, D.; Zimmerman, K. Single Event Upsets in SEL Relays. March 2018. Available online: https://selinc.com (accessed on 5 May 2023).
20. Zimmerman, K.; Haas, D. Impacts of single event upsets on protective relays. In Proceedings of the 72nd Annual Conference for Protective Relay Engineers, College Station, TX, USA, 25–28 March 2019.
21. Yang, W.; Li, Y.; Li, Y.; Hu, Z.; Cai, J.; He, C.; Wang, B.; Wu, L. Neutron Irradiation Testing and Monte Carlo Simulation of a Xilinx Zynq-7000 System on Chip. *Electronics* **2023**, *12*, 2057. [CrossRef]
22. Weulersse, C.; Houssany, S.; Guibbaud, N.; Segura-Ruiz, J.; Beaucour, J.; Miller, F.; Mazurek, M. Contribution of Thermal Neutrons to Soft Error Rate. *IEEE Trans. Nucl. Sci.* **2018**, *65*, 1851–1857. [CrossRef]
23. Li, Y.; Zhou, H.; Zheng, Y. Error-tolerant Design and Application of Relay Protection Device against Unexpected Memory Bit Change. *Autom. Electr. Power Syst.* **2021**, *45*, 155–162. (In Chinese)
24. Hao, Z.; Lei, S.; Tao, P.; Songze, L. Analysis and countermeasures of single event upset soft errors in a relay protection device. *Power Syst. Prot. Control.* **2021**, *49*, 144–149. (In Chinese)
25. NARI Group Corporation Brief Introduction. Available online: https://www.china-power-contractor.cn/NARI-Group-Corporation-brief-introduction.html (accessed on 12 July 2023).
26. Andreani, C.; Senesi, R.; Paccagnella, A.; Bagatin, M.; Gerardin, S.; Cazzaniga, C.; Frost, C.D.; Picozza, P.; Gorini, G.; Mancini, R.; et al. Fast neutron irradiation tests of flash memories used in space environment at the ISIS spallation neutron source. *AIP Adv.* **2018**, *8*, 025013. [CrossRef]
27. Chen, H.; Chen, Y.; Wang, F.; Liang, T.; Jia, X.; Ji, Q.; Hu, C.; He, W.; Yin, W.; He, K.; et al. Target station status of China Spallation Neutron Source. *Neutron News* **2018**, *29*, 2–6. [CrossRef]
28. Tang, J.; Liu, R.; Zhang, G.; Ruan, X.; Wu, X.; An, Q.; Bai, J.; Bao, J.; Bao, Y.; Cao, P.; et al. Initial years' neutron-induced cross-section measurements at the CSNS Back-n white neutron source. *Chin. Phys. C* **2021**, *45*, 062001. [CrossRef]
29. Yang, W.; Li, Y.; He, C. Fault injection and failure analysis on Xilinx 16 nm FinFET Ultrascale+ MPSoC. *Nucl. Eng. Technol.* **2022**, *54*, 2031–2036. [CrossRef]
30. Loongson. *Loongson 2K1000 Processor User Manual v1.4*; Loongson: Beijing, China, 2021.
31. De Sio, C.; Azimi, S.; Sterpone, L.; Codinachs, D.M. Analysis of Proton-induced Single Event Effect in the On-Chip Memory of Embedded Processor. In Proceedings of the 2022 IEEE International Symposium on Defect and Fault Tolerance in VLSI and Nanotechnology Systems (DFT), Austin, TX, USA, 19–21 October 2022.
32. Koga, R.; George, J.; Bielat, S. Single Event Effects Sensitivity of DDR3 SDRAMs to Protons and Heavy Ions. In Proceedings of the 2012 IEEE Radiation Effects Data Workshop, Miami, FL, USA, 16–20 July 2012.
33. Wei, J.; Fang, S.; Cao, J.; Chi, Y.; Deng, C.; Dong, H.; Dong, L.; Fu, S.; Kang, W.; Li, J.; et al. China spallation neutron source: Accelerator design iterations and R&D status. *J. Korean Phys. Soc.* **2007**, *50*, 1377–1384.
34. Agostinelli, S.; Allison, J.; Amako, K.A.; Apostolakis, J.; Araujo, H.; Arce, P.; Asai, M.; Axen, D.; Banerjee, S.; Barrand, G.J.N.I.; et al. GEANT4-a simulation toolkit. *Nucl. Instrum. Methods Phys. Res. A* **2003**, *506*, 250–303. [CrossRef]
35. Yang, W.; Li, Y.; Zhang, W.; Guo, Y.; Zhao, H.; Wei, J.; Li, Y.; He, C.; Chen, K.; Guo, G.; et al. Electron inducing soft errors in 28 nm system-on-Chip. *Radiat. Eff. Defects Solids* **2020**, *175*, 745–754. [CrossRef]
36. Casolaro, P.; Campajola, L.; De Luca, D. Neutrons for studies of radiation hardness. *Il Nuovo Cimento C* **2020**, *43*, 57.
37. Yang, W.; Li, Y.; Li, Y.; Hu, Z.; Xie, F.; He, C.; Wang, S.; Zhou, B.; He, H.; Khan, W.; et al. Atmospheric neutron single event effect test on Xilinx 28 nm system on chip at CSNS-BL09. *Microelec. Reliab.* **2019**, *99*, 119–124. [CrossRef]

Disclaimer/Publisher's Note: The statements, opinions and data contained in all publications are solely those of the individual author(s) and contributor(s) and not of MDPI and/or the editor(s). MDPI and/or the editor(s) disclaim responsibility for any injury to people or property resulting from any ideas, methods, instructions or products referred to in the content.

Article

Machine Learning-Based Soft-Error-Rate Evaluation for Large-Scale Integrated Circuits

Ruiqiang Song [1,2,*], Jinjin Shao [1], Yaqing Chi [1,2,*], Bin Liang [1,2], Jianjun Chen [1,2] and Zhenyu Wu [1,2]

[1] College of Computer, National University of Defense Technology, Changsha 410073, China; shaojinjin308@163.com (J.S.); liangbin@nudt.edu.cn (B.L.); cjj192000@163.com (J.C.); wuzhenyu@nudt.edu.cn (Z.W.)
[2] Key Laboratory of Advanced Microprocessor Chips and Systems, Changsha 410073, China
* Correspondence: songrq07@nudt.edu.cn (R.S.); yqchi@nudt.edu.cn (Y.C.)

Abstract: Transient pulses generated by high-energy particles can cause soft errors in circuits, resulting in spacecraft malfunctions and posing serious threats to the normal operation of spacecraft. For integrated circuits used in space applications, it is necessary to first evaluate soft errors caused by transient pulses. Conventional soft-error-rate evaluation tools are designed to simulate the generation of transient pulses using many accurate models, while the propagation of transient pulses is primarily simulated by circuit-level simulation tools. Due to the limitations of simulation tools, conventional evaluation approaches are limited to the circuit scale. The simulation runtime is unbearable for large-scale integrated circuits. This paper presents an approach for evaluating the soft error rate using machine learning. A back propagation neural network is implemented in the proposed approach. It helps to determine the probability of transient pulse propagation. Compared with the conventional soft-error-rate evaluation results, the proposed approach demonstrates a strong correlation in both trend and magnitude. The average difference between the results obtained using the proposed evaluation method and the experimental results is 23.5%, which is 7.5% higher than that between the results obtained using the conventional evaluation method and the experimental results. Compared to the conventional evaluation method, the proposed approach improves the runtime by an order of magnitude. The proposed approach also benefits the locating of highly sensitive circuit nodes in large-scale integrated circuits. Circuit design and radiation hardening are both useful applications.

Keywords: machine learning; single event transient; soft error rate; transient pulse propagation

1. Introduction

When a high-energy particle passes through an integrated circuit in the space radiation environment, it loses energy along its path [1,2]. The lost energy is transferred to the semiconductor material, ionizing electrons of silicon atoms [3]. These ionized electron-hole pairs are subject to both drift and diffusion. They move throughout the entire semiconductor material and are collected by transistors [4–6]. The collected electron-hole pairs produce unexpected transient pulses in circuit nodes [7]. These transient pulses propagate along the circuit path and cause soft errors [8–10]. A soft error is a significant threat to integrated circuits. It alters the logic function and can potentially lead to catastrophic consequences for an entire chip, system, or even a spacecraft.

To mitigate soft errors in integrated circuits for space applications, it is crucial to evaluate the soft error rate (SER) during the circuit design phase. In previous works, several circuit-level evaluation approaches have been proposed to investigate the SER of integrated circuits [11–21]. These works have proposed many accurate models to generate transient pulses in circuit nodes. Then, they utilize simulation tools, such as the Simulation Program with Integrated Circuit Emphasis (SPICE) and Technology Computer-Aided Device (TCAD), to simulate transient pulse propagation and capture. Based on the simulated results, conventional evaluation approaches determine soft errors and calculate the SER of integrated circuits. Due to the limitations of simulation tools, conventional evaluation approaches are limited to the circuit scale. The simulation runtime is unbearable for large-scale integrated circuits [22].

This paper presents a novel approach for evaluating SER in order to reduce the simulation runtime. A back propagation neural network (BPNN) is implemented to determine the probability of transient pulse propagation. The SER can be determined based on the probability of propagation. The proposed approach does not require determining the probability value of transient pulse propagation to flip-flops through actual circuit-level simulation. Instead, it takes the probability of pulse propagation for each instance in the data path as the input value. This input is then fed into a machine learning model, and the propagation probability value is obtained through the calculation of the machine model. A chip with three test circuits was designed using commercial CMOS technology to investigate the accuracy of the proposed approach. The proposed approach achieves a good consistency in both trend and order of magnitude.

2. SER Evaluation Overview

In previous works, several approaches for evaluating soft error rate (SER) have been proposed. These approaches are used to evaluate key circuits, including combination circuits, flip-flops, and SRAM. In general, the existing soft error evaluation approaches are mainly divided into three categories: SPICE-level evaluation approaches, TCAD-level evaluation approaches, and Monte Carlo-based evaluation approaches.

The SPICE-level evaluation approach is widely used. Based on the SPICE device model and the netlist of the evaluated circuit, a separate current source is introduced directly at the sensitive node of the circuit to simulate the transient current caused by incident particles [23–25]. Then, it simulates the corresponding circuit response to obtain soft errors. Correas et al. simulated the evaluation of a 90 nm SRAM circuit using the SPICE circuit-level soft error evaluation tool. The evaluation results obtained are in good agreement with the experimental results [12]. Shambhulingaiah et al. utilized the same tool to simulate sequential instances, such as flip-flops, and identified the sensitive nodes of the flip-flop [13]. Wang and Du et al. utilized the SPICE circuit-level simulation tool to simulate the propagation process of single-event transients in large-scale combinational circuits. Their objective was to assess the impact of single-event transient pulses on soft errors in these circuits [14,15]. Li et al. simulated and analyzed the reliability of integrated circuits using the SPICE tool and proposed a corresponding evaluation process [16].

The TCAD-level evaluation approach differs different from the SPICE-level evaluation approach. It first constructs the TCAD model based on the layout structure and manufacturing process parameters of the circuit instance. Then, the TCAD model simulates the ionization of electron-hole pairs in the incident particles using a specific numerical distribution, such as exponential or Gaussian distribution. The transport process of electron-hole pairs in the TCAD model is calculated using the carrier drift diffusion and other models embedded in the TCAD simulation tool. It simulates the charge collection of the instance and the instantaneous response of the circuit node, and determines whether the circuit instance produces soft errors. Yoni et al. constructed a comprehensive 3D TCAD model using the layout structure of D flip-flops. They subsequently simulated the circuit's response when the D flip-flop cell was exposed to terrestrial neutrons [17]. Xu et al. utilized TCAD simulation tools to investigate the mechanism of soft errors in standard instances. They also employed a combination of TCAD simulation tools and SPICE circuit-level simulation tools [18].

Recently, the Monte Carlo-based evaluation approach has become an important evaluation approach. It utilizes Monte Carlo tools, such as Geant4 and SRIM, to simulate and calculate the interaction between the incident particles and the semiconductor material. Then, it converts the charge accumulated by the incident particles in the material into charge and transient current collected by the device through charge transport and charge collection mechanisms. Finally, it simulates the transient response of the circuit using additional simulation tools and determines whether a soft error occurs. Many Monte Carlo-based evaluation approaches have been proposed to evaluate the SER of circuits, such as MRED [11], MUSCA SEP3 [22], PHITS-HyENEXSS [19], and IRT [20].

3. The SER Evaluation Using Machine Learning Models

3.1. The Transient Pulse Propagation Probability

A conventional register-to-register circuit path in integrated circuits is shown in Figure 1. It is used to explain the evaluation of transient pulse propagation using machine learning models. When a high-energy particle strikes this circuit, some logic instances (such as C0) collect the ionized electron-hole pairs and produce a transient pulse at circuit nodes. Then, the transient pulse propagates to flip-flops along circuit paths. When the transient pulse arrives at the input pin of instance C1, it propagates directly, and the probability of transient pulse propagation P_{C1} is equal to 1. However, when the transient pulse arrives at the input pin of instances C2 and C3, it may not propagate due to logic masking. For instance, C2 is an OR-gate instance. The transient pulse can only propagate when the value of the other input pin is 0. Similarly, C3 is an AND-gate instance. The transient pulse is able to propagate only when the value of the other input pin is 1. Therefore, the transient pulse propagation probabilities P_{C2} and P_{C3} depend on the instance type and input pin values. The values are determined using the following equations:

$$P_{C2} = 1 - P_{otherpin,1} \tag{1}$$

$$P_{C3} = P_{otherpin,1} \tag{2}$$

where $P_{otherpin,1}$ is the probability when the value of the other pin is 1. If the input vectors are random, $P_{otherpin,1}$ is equal to 0.5. The transient pulse that can propagate to flip-flops is determined by the propagation probabilities along the circuit path. For instance, the transient pulse that can propagate to flip-flop 1 (FF1) is determined by P_{C1}, P_{C2}, and P_{C3}. If a relationship between P_{FF1} and P_{C1}–P_{C3} can be determined, the transient pulse propagation can be easily evaluated.

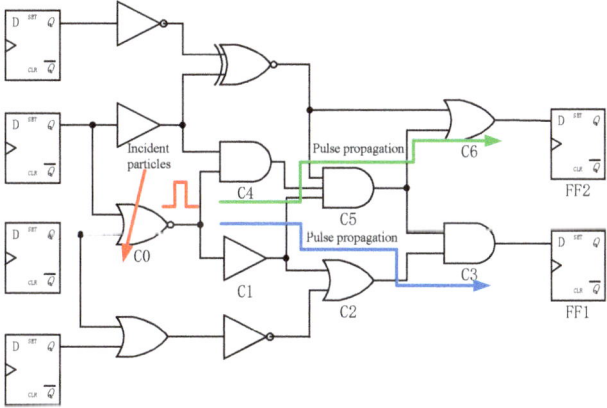

Figure 1. A conventional registers-to-registers circuit path.

Unfortunately, determining the relationship between flip-flops and logic cells along the circuit path is challenging due to the complexity of circuit structures. A simple fitting equation may not be suitable for all circuit structures. Recently, machine learning has been widely used in integrated circuit design [18,26]. Some machine learning models are used to analyze circuit structures in order to identify inherent connections between circuit cells. In this paper, a machine learning model (BPNN) is used to determine the relationship between flip-flops and logic instances along the circuit path. The BPNN is first described in [27], and its basic structure consists of neurons, interconnection layers, and connection weight values. In this paper, the BPNN model consists of an input interconnection layer, a hidden interconnection layer, and an output interconnection layer. The basic structure is shown in Figure 2. The input layer consists of 20 neurons, which is determined by the maximum number of stages in the data paths. The number of neurons in a hidden layer and the number of hidden layers are adjustable parameters. They can impact the prediction accuracy of the BPNN model. The transient pulse propagation probability calculated by the BPNN model is significantly different from the results of SPICE-level simulations when the number of neurons in one hidden layer is lower (5 to 10). The prediction accuracy is less than 0.6. With an increase in the number of neurons in the hidden layer, the model's prediction accuracy is significantly improved. When the number of neurons exceeds 15, the calculation accuracy of the BPNN model can approach 0.9. However, the prediction accuracy does not improve any further when the number of neurons exceeds 20. Instead, the training time and prediction time increase significantly. In particular, when the number of neurons in the hidden layer reaches 23, the model's prediction accuracy is reduced by 2% to 5%. There was overfitting during the training of the BPNN model. Therefore, the hidden layer consists of 15 neurons. This ensures that the calculations are highly accurate and also allows for a more efficient training and prediction time. Furthermore, the model's prediction accuracy does not significantly improve as the number of hidden layers increases. However, including additional connection weight values required significant adjustments, which led to a substantial increase in training time. Therefore, there is only a single hidden layer used to construct the BPNN model. The output layer consists of 10 neurons, which is determined by the range of propagation probability. The input vectors of the BPNN are the probabilities of transient pulse propagation along the circuit path. The output value of the BPNN is the probability that a transient pulse can propagate to a flip-flop. The basic neuron is activated by the sigmoid function:

$$f(x_i) = \frac{1}{1+e^{-x}} \qquad (3)$$

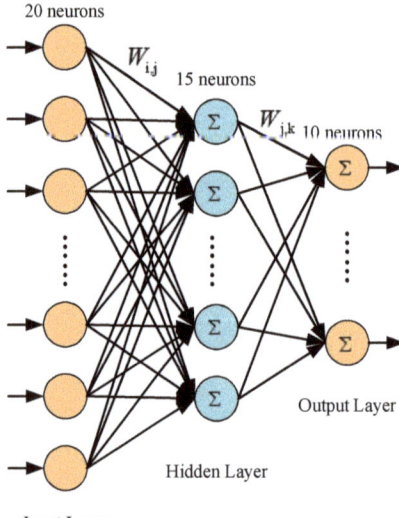

Figure 2. The BPNN structure used in the paper.

Based on Equation (3), the neurons in the input layer can be calculated using the following equation:

$$f(P_{Ci}) = \frac{1}{1 + e^{-P_{Ci}}} \tag{4}$$

The input values of neurons in the hidden layer are determined by the output values of neurons in the input layer. The neurons in the hidden layer can be calculated by the following equation:

$$N_{j,hidden} = \sum_{i=0}^{M-1} w_{ij} f(P_{Ci}) \tag{5}$$

where w_{ij} represents the connection weight value between the input layer and the hidden layer. $N_{j,hidden}$ represents the output value of the hidden layer neuron. Similarly, the neurons in the output layer can be calculated using Equation (5), and the transient pulse propagation probability to flip-flops can be determined:

$$P_{FF_k} = \sum_{j=0}^{N-1} w_{jk} N_{j,hidden} \tag{6}$$

The most important aspect of BPNN is training the connection weight value w_{ij} and w_{jk}. These connection weight values significantly affect the accuracy of evaluating transient pulse propagation. In this paper, several benchmark circuits from the ISCAS 85 suite are selected. The ISCAS '85 benchmark circuits are ten combinational circuits provided to authors at the 1985 International Symposium on Circuits and Systems. They have subsequently been used by many researchers as a basis for comparing results in test generation. The selected circuits are used to create a training set. The training set is used to calibrate connection weight values. A benchmark circuit structure is shown in Figure 1. It is used to illustrate how to generate the training set. Firstly, one circuit instance is randomly selected, such as C0. Based on the C0, circuit instances C1, C2, and C3 are extracted because they are part of the data path. The probabilities of transient pulse propagation for P_{C1}, P_{C2}, and P_{C3} are determined. Secondly, a SPICE-level simulation tool is used to simulate the propagation of transient pulses in the benchmark circuit. The input value of the circuit changes randomly with each clock cycle, allowing the value in each circuit instance to be altered. A dual exponential current source is then injected into the C0. The equation of the dual exponential current source can be shown in our previous work [28]. The SPICE-level simulation tool is used to determine whether the transient pulses can propagate to flip-flop 1. This process is repeated multiple times (such as 1000 times) to count the number of pulses that successfully propagate to flip-flop 1. The propagation probability $P_{FF_1,simulation}$ is obtained by dividing the count data by the total number of injected transient pulses. Finally, P_{C1}, P_{C2}, P_{C3}, and $P_{FF_1,simulation}$ constitute one datum in the training set. Then, another circuit instance is selected, such as C4, C5 or C6. The above steps are repeated to obtain more data in the training set. During BPNN training, P_{C1}, P_{C2}, and P_{C3} are used as input data. BPNN calculates the transient pulse propagation probability $P_{FF_1,prediction}$ through Equations (4)–(6). This data will be different from $P_{FF_1,simulation}$. Based on the prediction results $P_{FF_1,prediction}$ and the simulation results $P_{FF_1,simulation}$, connection weight values are calibrated using the following equations:

$$E(w) = \frac{1}{2} \sum_{k=0}^{N-1} \left(P_{FF_k,prediction} - P_{FF_k,simulation} \right)^2 \tag{7}$$

$$w_{ij,new} = w_{ij,old} - \eta_1 \frac{\partial E(w)}{\partial w_{ij,old}} \tag{8}$$

$$w_{jk,new} = w_{jk,old} - \eta_1 \frac{\partial E(w)}{\partial w_{jk,old}} \tag{9}$$

In this paper, parallel simulation is used to accelerate the generation of training data. The cost to obtain the training data is no more than 9 h for each test circuit. Approximately 12,000 training data were generated through the simulations mentioned above. In total, 60% of the data was used to train the BPNN model, and the remainder was used for model validation. The prediction accuracy of the model is calculated using the Equation (10). P represents the precision value of the prediction. TP represents the number of positive predictions that are correct, while FP represents the number of positive prediction errors. The calculation results are detailed in Table 1.

$$P = \frac{TP}{TP + FP} \tag{10}$$

Table 1. The results of the training.

Transient Pulse Propagation Probability Range	Prediction Precision Value
0.1	0.914
0.2	0.922
0.3	0.904
0.4	0.893
0.5	0.877
0.6	0.911
0.7	0.864
0.8	0.844
0.9	0.857
1.0	0.821

3.2. The Transient Pulse Capture Evaluation

Another important aspect of SER evaluation is the capture of transient pulses. When a transient pulse arrives at the input pin of flip-flops, such as FF1 in Figure 1, it needs to meet a certain signal–clock relationship to be captured. If the transient pulse is not captured by flip-flops, it will not alter the stored value of the flip-flops and will not result in a soft error. Figure 3 shows the relationship between the transient pulse and the clock waveform. The capture of transient pulses depends on both the width of the pulse and the period of the clock. The probability of capturing transient pulses in flip-flops can be calculated using the following equation:

$$P_{capture, FF_k} = \frac{T_{width}}{T_{period}} P_{FF_k, prediction} \tag{11}$$

where $P_{FF_k, prediction}$ is calculated by the BPNN. T_{period} represents the clock period, while T_{width} denotes the transient pulse width. Since the incident time of high-energy particles is random, it is also random whether the transient pulse and the clock period satisfy the signal-clock relationship. Therefor, a random function in the range of 0 to 1 is used to determine whether the transient pulse is captured by flip-flops. For each flip-flop affected by the transient pulse, the random function generates a value. If the random value is lower than the transient pulse capture probability $P_{capture, FF_k}$, the transient pulse can be captured by flip-flops. On the contrary, the transient pulse is not captured when the random value exceeds the probability of capturing the transient pulse.

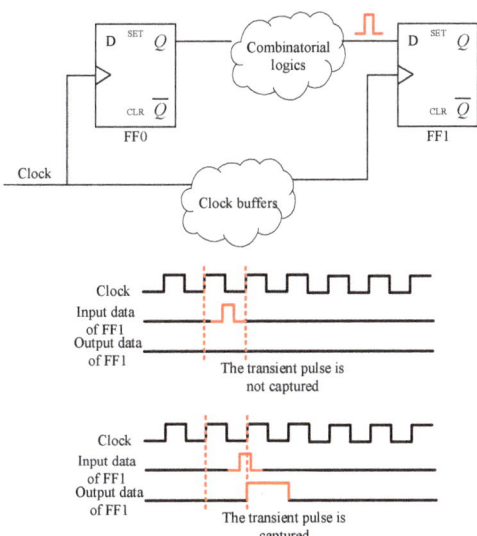

Figure 3. The signal–clock relationship to capture a transient pulse.

4. The SER Evaluation Approach

Based on the above SER evaluation principles, a machine learning-based SER evaluation approach is proposed. The basic flow of the proposed approach is shown in Figure 4.

The gate-level netlist of the circuit serves as an input file for the proposed evaluation approach. Before starting SER evaluation, the number of transient pulses that need to be injected is determined based on both the incident particle flux and the area of the circuit (the layout area or the sum of all instance areas). For instance, if the flux of the incident particles is 1×10^7 ions/cm^2 and the circuit's layout area is 1 mm^2, it indicates that 1×10^5 ions will strike the circuit. Therefore, when performing SER evaluation, the proposed approach also needs to evaluate the soft errors that occur in the circuit after 1×10^5 transient pulse injections. In addition, since the location of the incident particle is random, its impact on the circuit instance is also random. Therefore, for each transient pulse injection, the proposed approach first randomly selects a circuit instance based on the circuit netlist. This means that the circuit instance is affected by the incident particles, resulting in the production of a transient pulse. Secondly, for each data path, all connected logic instance types from the selected instance to flip-flops are extracted. Each logic instance type is then converted into a probability of transient pulse propagation. It is important to note that since the selected logical instance may affect multiple flip-flops, it is often possible to generate multiple input data for the machine learning model in the second step. Thirdly, the data are inputted into the calibrated BPNN model, and the propagation probability of the transient pulse to a flip-flop is calculated using Equations (4)–(6). Fourthly, Equation (11) is used to determine whether the transient pulse will be captured by the flip-flop based on the calculated propagation probability. If the result calculated in Equation (11) is less than the random value generated by the random function, the transient pulse is considered to be captured by the flip-flop. The number of soft errors has increased by one. Fifthly, once all affected flip-flops have been traversed, the number of soft errors in the circuit caused by a transient pulse can be determined. After calculating all transient pulse injections, the total number of soft errors can be determined under a specific incident flux condition. The soft error rate of the circuit can be calculated.

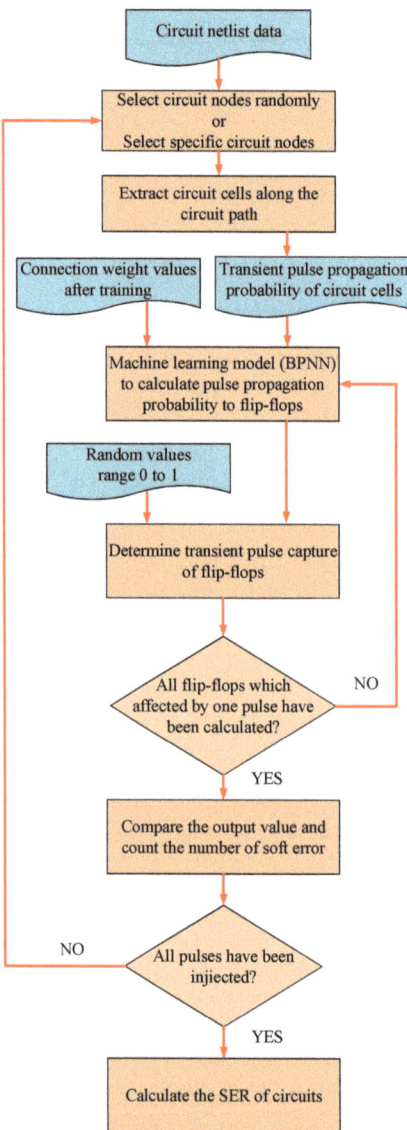

Figure 4. The basic flow of the proposed approach.

A comparison between the proposed approach and other conventional evaluation approaches is shown in Table 2. The proposed approach only uses the BPNN model to obtain SER evaluation results. It does not realistically simulate the transient pulse propagation and capture using circuit-level simulation tools. The proposed approach can reduce the run time of SER evaluation and is not limited by the size of the circuit. It is worth noting that the proposed approach not only obtains the SER of the evaluated circuit. It is also useful for locating highly sensitive circuit nodes in the evaluated circuit. For instance, specific circuit nodes are selected in the proposed approach. Then, the SER flow runs to obtain the SER of specific circuit nodes. Compared to the calculated results, the nodes of the high-sensitive circuit are located.

Table 2. Comparison of the proposed approach with other state-of-the-art approaches.

SER evaluation approach	The transient pulse generation evaluation	The transient pulse propagation evaluation
SPICE-level simulation approach	The dual exponential current source, etc.	Circuit-level simulation tools
TCAD-level simulation approach	Ionization charge distribution model Carrier transport equation, etc.	Circuit-level simulation tools
Monte Carlo-based simulation approach	Nested sensitive volumes model Drift diffusion equation, etc.	Circuit-level simulation tools
The proposed approach	Pulse width data that vary with LET	Machine learning model

5. The SER Evaluation Approach Validation

5.1. Test Chip Design and Experimental Setup

A SER test structure was designed using commercial CMOS technology to investigate the accuracy of the proposed approach. The schematic of the SER test structure is shown in Figure 5. It consists of one random vector generator, two test circuits, and one SER detection circuit. The random vector generator (Linear Feedback Shift Register, LFSR) is used to create the input vectors. The test circuit consists of combinational logic instances and flip-flops. Note that two test circuits have the same topology and layout structures. However, they are spaced out widely to ensure that an incident particle only impacts one test circuit, as shown in Figure 6. The input pins of two test circuits are connected to the random vector generator. It ensures the test circuits have the same input vectors. The SER detection circuit consists of several XOR-gate instances and OR-gate instances. The XOR-gate instances compares the output values of two test circuits. If a test circuit is irradiated by incident particles, the output values are changed. However, the other test circuit is not impacted by incident particles; it can produce correct output values. The XOR-gate instance produces a 1 due to the different output values. The soft error induced by the incident particle is propagated to the SER counter circuit and satisfied.

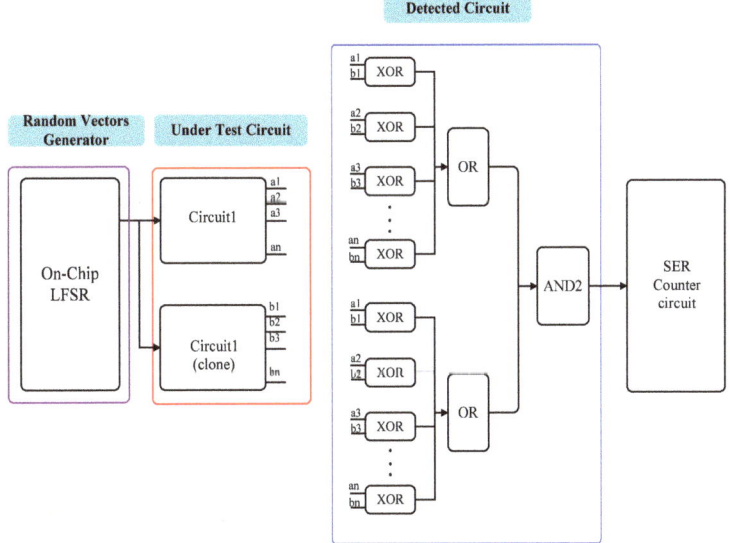

Figure 5. The schematic of the SER test structure.

A test chip with three SER test structures was fabricated using commercial 65 nm CMOS technology. The detailed test chip layout is shown in Figure 6. The test chip was irradiated with heavy ions. Four heavy ions with different parameters were chosen, as shown in Table 3. The heavy-ion dose rate was 1×10^4 ions/cm^2/s, and the flux was 1×10^7 ions/cm^2. Before the radiation experiment, the cover plate from the test chip was removed. The front side of the chip was positioned within the range of the ion beam's influence. During radiation experiments, the ion beams randomly strike their respective targets. Some ions strike the test chips and generate transient pulses. The test system consisted of a test chip and other necessary chips, such as field-programmable gate arrays (FPGAs) and serial communication chips [29,30]. FPGAs connected all signal ports (input, output, and clock) of the test chip to provide input and clock signals. They were also used to capture output signals when the test chip was irradiated. After conducting heavy-ion experiments, the error counts were exported to the computer using the serial communication interface.

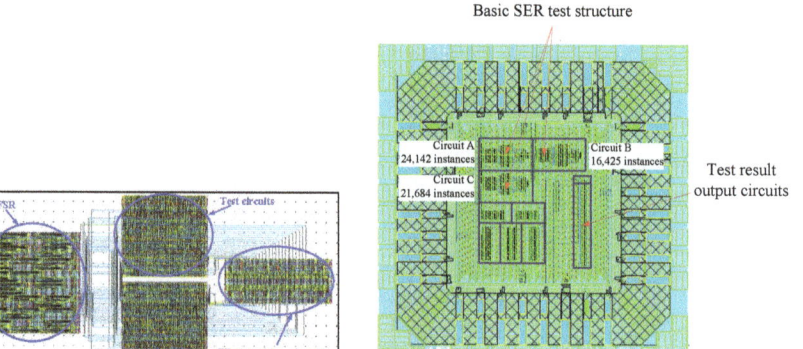

Figure 6. The basic SER test structure layout and the detailed test chip layout.

Table 3. Heavy ions used in the experiment.

Ion	Energy at the Silicon Surface (MeV)	Effective LET (MeV·cm^2/mg)	Range (um)
Cl	165	15.2	51.8
Ti	185	21.2	37.9
Ge	205	37.6	35.5
Kr	835.5	99.8	41.2

5.2. SER Evaluation Setup

The proposed evaluation approach and the SPICE-level evaluation approach were used to investigate the SER of test circuits. The gate-level netlist of three test circuits serves as an input file for the evaluation approaches. The number of transient pulse injections per test circuit is determined by dividing the particle flux by the layout area of the test circuit. The position and moment of transient pulse injection are random. This is due to the fact the location and momentum of the incident particles are random during radiation experiments. The width of transient pulses injected into the test circuit is set to 100 ps, 200 ps, 350 ps, and 500 ps, respectively. It represents the various pulse widths generated by different particles. These data were determined through our previous transient pulse measurements [31–33]. For each test circuit, the number of soft errors can be obtained when the evaluation tool completes the transient pulse width injection simulation. According to the number of soft errors, the soft error cross-section can be calculated using the following equation:

$$SER = \frac{N_{error}}{f_{ions} \times N_{instances}} \qquad (12)$$

where N_{error} represents the number of soft errors that is calculated by the evaluation approaches. f_{ions} represents the flux of ions. $N_{instances}$ represents the total instance number of test circuits.

5.3. SER Evaluation Results Comparison

The evaluation results of the proposed approach and the conventional SER evaluation approach are compared first. The connection weight values were saved after training the BPNN model. They are imported again into the BPNN model during the evaluation of test circuits. The BPNN model is used to calculate the pulse propagation probabilities in circuits A, B, and C. The calculated results are compared with the transient pulse propagation probabilities simulated by the circuit-level simulation tool. The prediction accuracy is calculated using Equation (10) and the results are shown in Table 4. The average prediction accuracy of the three test circuits is approximately 0.8. Although the three circuits have different circuit structures, the trained BPNN model can still accurately calculate the probability of transient pulse propagation.

Table 4. The prediction accuracy of three test circuits.

Transient Pulse Propagation Probability Range	Circuit A Prediction Precision Value	Circuit B Prediction Precision Value	Circuit C Prediction Precision Value
0.1	0.832	0.807	0.821
0.2	0.829	0.811	0.825
0.3	0.813	0.824	0.817
0.4	0.824	0.803	0.813
0.5	0.809	0.795	0.808
0.6	0.805	0.792	0.811
0.7	0.789	0.801	0.804
0.8	0.775	0.789	0.796
0.9	0.764	0.773	0.792
1	0.765	0.768	0.788

Figure 7 shows the comparison between the evaluation results and experimental results. The evaluation results obtained by the proposed approach show good consistency in both trend and order of magnitude. The difference between the results obtained using the proposed approach and the experimental results is calculated. The average value is 23.5%, which is 7.5% higher than that between the conventional evaluation approach and the experimental results. When the LET value is 15.2 MeV·cm^2/mg, there is a greater discrepancy between the simulation results and the experimental results. With the increase in LET value, the simulation results are in good agreement with the experimental results. Some reasons may cause this difference. The first reason is that the proposed approach does not consider the transient pulse reshaping and reconvergence. Transient pulse reshaping or reconvergence results in a change in the width of the transient pulse, which in turn change the data value of T_{width} in Equation (11). Circuit-level simulations are used to investigate the difference in circuit A at low LET values. When particles strike most instances, it only generates a transient pulse that propagates to the input of the flip-flop. However, when a particle strikes a specific instance with a large fanout, although it only produces one transient pulse, more than one transient pulse is propagated to the input of the flip-flop due to pulse reconvergence, as shown in Figure 8. For the proposed approach, it is still evaluated based on the pulses that propagate independently on different data paths. As a result, the evaluation results may not accurately reflect soft errors and may differ from the experimental results. When the LET value increases, the transient pulse width generated by the ions also increases. It reduces the change in the width and number of transient pulses caused by the pulse reconvergence. The evaluation results obtained using the proposed evaluation method are closer to the experimental results.

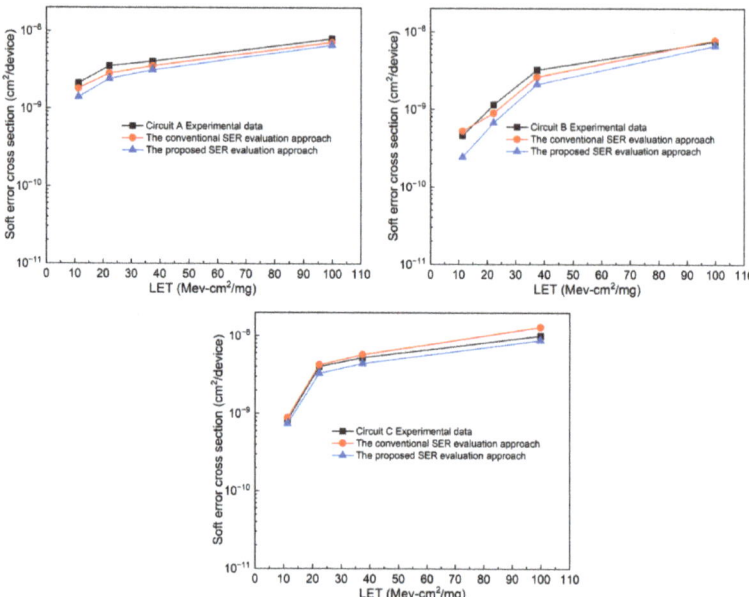

Figure 7. The simulated SER results with different evaluation approaches.

The other reason for this difference is the evaluation accuracy of the transient pulse propagation. The transient pulse propagation is evaluated using the BPNN. The training accuracy of connection weight values is the key factor that affect the evaluation accuracy. In our previous works, we observed a significant decrease in the prediction precision value when the probability of transient pulse propagation exceeded 0.7. It indicates that the propagation probability obtained by the BPNN are significantly different from the conventional transient pulse simulation results [34]. Because fewer combinational logic cells with large circuit stages have a high probability of transient pulse propagation, the training set does not include enough data, and the connection weight values are not effectively trained for this situation. Increasing the train set data can improve the training accuracy of connection weight values. It may be an effective way to solve this difference.

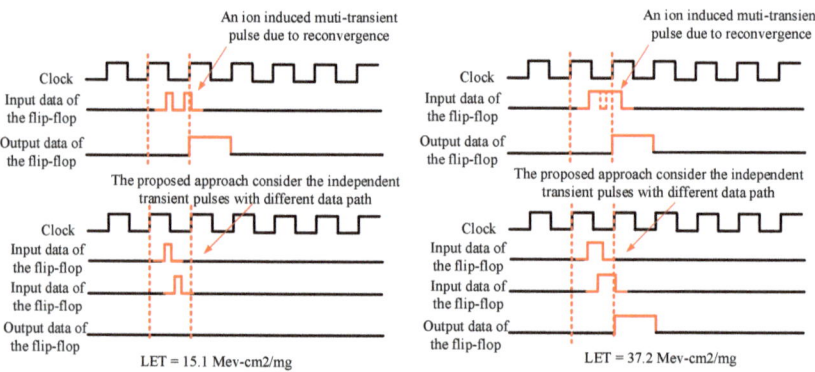

Figure 8. The width and number variation of transient pulses due to pulse reconvergence.

Figure 9 shows the average simulation runtime of the proposed approach and the conventional SER evaluation approach. For the conventional evaluation approach, the circuit-level simulation tool is used to simulate the propagation of transient pulses. This simulation helps determine if the flip-flops can capture the transient pulses. Although the size of the test circuit is only 10,000 instances, it results in a significant increase in time cost for a single circuit-level simulation. The significant time cost greatly reduces the evaluation performance of the conventional evaluation approach. For the proposed approach, a machine learning model (Equations (4)–(6)) is utilized to calculate the probability of transient pulse propagation. Subsequently, the transient pulse is captured using the transient pulse probability equation (Equation (11)). The proposed method can determine the soft error of the circuit solely through equation calculations. Therefore, the proposed method can significantly reduce the time required and enhance the performance of soft error evaluation.

In addition, Circuit C has only 5000 more instances than Circuit B. However, the simulation time for the conventional evaluation methods is nearly doubled. When the circuit size increases further, the evaluation time of traditional evaluation methods becomes unacceptable, limiting the size of the circuit that can be supported by this approach. For the proposed approach, as the circuit size increases, the evaluation time only shows a slight improvement. The proposed approach can support larger scale circuits, which improves the performance of the soft error evaluation method.

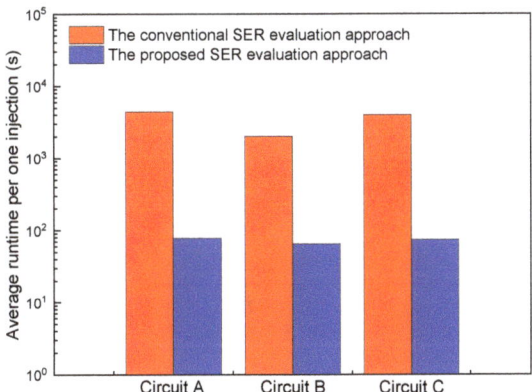

Figure 9. The average simulation runtime of two SER evaluation approaches.

6. Conclusions

This paper has presented an approach to evaluate the SER of integrated circuits. A machine learning model (BPNN) is implemented in the proposed approach. It helps to evaluate the transient pulse propagation and capture. Some commercial integrated circuits are designed and fabricated to validate the capability of the proposed approach. Compared to experimental data and the conventional SER evaluation results, the proposed approach also demonstrates a strong correlation in terms of trend and magnitude with the improvement in simulation runtime. The proposed evaluation tool has been used to evaluate the SER of circuits with more than 10,000 gates, which demonstrates that the proposed model can be applied to evaluate logic circuits with more than 10,000 gates. It is useful for circuit design and radiation hardening.

Author Contributions: Conceptualization, test chip design, writing—original draft preparation, R.S.; writing—review and editing, J.S.; heavy-ion experiment, Y.C. and J.C.; experimental results analysis and validation, B.L. and Z.W. All authors have read and agreed to the published version of the manuscript.

Funding: This research was funded by the National Natural Science Foundation of China (Grant No. 62174180) and the National University of Defense Technology research project (Grant No. ZK20-11).

Institutional Review Board Statement: Not applicable.

Data Availability Statement: The datasets used and/or analyzed during the current study are available from the corresponding author on reasonable request.

Acknowledgments: We wish to acknowledge the HI-13 team and the HIRFL team for heavy-ion experiment support.

Conflicts of Interest: The authors declare no conflict of interest.

Abbreviations

The following abbreviations are used in this manuscript:

BPNN   back propagation neural network
SER   soft error rate
SPICE Simulation Program with Integrated Circuit Emphasis
LFSR   Linear Feedback Shift Register
TCAD   Technology Computer-Aided Device

References

1. Zhang, Z.; Arehart, A.; Cinkilic, E.; Chen, J.; Zhang, E.; Fleetwood, D.; Schrimpf, R.; McSkimming, B.; Speck, J.; Ringel, S.A. Quasimonotonicity, regularity and duality for nonlinear systems of partial differential equations. *Appl. Phys. Lett.* **2013**, *103*, 042102. [CrossRef]
2. Fleetwood, D.; Shaneyfelt, M.; Schwank, J. Estimating oxide-trap, interface-trap, and border-trap charge densities in metal-oxide-semiconductor transistors. *Appl. Phys. Lett.* **1994**, *64*, 1965. [CrossRef]
3. Jun, B.; White, Y.; Schrimpf, R.; Fleetwood, M.; Brunier, F.; Bresson, N.; Cristoloveanu, S. Characterization of multiple Si/SiO_2 interfaces in silicon-on-insulator materials via second-harmonic generation. *Appl. Phys. Lett.* **2004**, *85*, 3095. [CrossRef]
4. Sun, Q.; Guo, Y.; Liang, B.; Tao, M.; Chi, Y.; Huang, P.; Wu, Z.; Luo, D.; Chen, J. Higher NMOS Single Event Transient Susceptibility Compared to PMOS in Sub-20 nm Bulk FinFET. *IEEE Electron Device Lett.* **2023**, *44*, 1712–1715. [CrossRef]
5. Pieper, N.J.; Xiong, Y.; Feeley, A.; Pasternak, J.; Dodds, N.; Ball, D.R.; Bhuva, B.L. Study of Multicell Upsets in SRAM at a 5-nm Bulk FinFET Node. *IEEE Trans. Nucl. Sci.* **2023**, *70*, 401–409. [CrossRef]
6. He, X.; Yue, D.; Huang, P.; Zhao, Z. Experimental Investigation of Charge Sharing Induced SET Depending on Transistors in Abutted Rows in 65 nm Bulk CMOS Technology. *IEEE Access* **2022**, *10*, 57362–57368. [CrossRef]
7. Ferlet-Cavrois, V.; Massengill, L.W.; Gouker, P. Single Event Transients in Digital CMOS—A Review. *IEEE Trans. Nucl. Sci.* **2013**, *60*, 1767–1790. [CrossRef]
8. Uemura, T.; Lee, S.; Min, D.; Moon, I.; Lee, S.; Pae, S. SEIFF: Soft Error Immune Flip-Flop for Mitigating Single Event Upset and Single Event Transient in 10 nm FinFET. In Proceedings of the 2019 IEEE International Reliability Physics Symposium (IRPS) 2019, Monterey, CA, USA, 31 March–4 April 2019; pp. 1–6. [CrossRef]
9. Mochizuki, A.; Onizawa, N.; Tamakoshi, A.; Hanyu, T. Multiple-event-transient soft-error gate-level simulator for harsh radiation environments In Proceedings of the TENCON 2015—2015 IEEE Region 10 Conference 2015, Macao, China, 1–4 November 2015; pp. 1–6. [CrossRef]
10. Azimi, S.; De Sio, C.; Portaluri, A.; Rizzieri, D.; Vacca, E.; Sterpone, L.; Merodio Codinachs, D. Exploring the Impact of Soft Errors on the Reliability of Real-Time Embedded Operating Systems. *Electronics* **2023**, *12*, 169. [CrossRef]
11. Black, J.D.; Dame, J.A.; Black, D.A.; Dodd, P.E.; Shaneyfelt, M.R.; Teifel, J.; Salas, J.G.; Steinbach, R.; Davis, M.; Reed, R.A.; et al. Using MRED to Screen Multiple-Node Charge-Collection Mitigated SOI Layouts. *IEEE Trans. Nucl. Sci.* **2019**, *66*, 233–239. [CrossRef]
12. Correas, V.; Saigne, F.; Sagnes, B.; Wrobel, F.; Boch, J.; Gasiot, G.; Roche, P. Prediction of Multiple Cell Upset Induced by Heavy Ions in a 90 nm Bulk SRAM. *IEEE Trans. Nucl. Sci.* **2009**, *56*, 2050–2055. [CrossRef]
13. Shambhulingaiah, S.; Lieb, C.; Clark, L.T. Circuit Simulation Based Validation of Flip-Flop Robustness to Multiple Node Charge Collection. *IEEE Trans. Nucl. Sci.* **2015**, *62*, 1577–1588. [CrossRef]
14. Wang, F.; Xie, Y.; Rajaraman, R.; Vaidyanathan, B. Soft Error Rate Analysis for Combinational Logic Using An Accurate Electrical Masking Model. In Proceedings of the 20th International Conference on VLSI Design held jointly with 6th International Conference on Embedded Systems (VLSID'07), Bangalore, India, 6–10 January 2007; pp. 165–170. [CrossRef]
15. Chen, S.; Du, Y.; Liu, B.; Qin, J. Calculating the Soft Error Vulnerabilities of Combinational Circuits by Re-Considering the Sensitive Area. *IEEE Trans. Nucl. Sci.* **2014**, *61*, 646–653. [CrossRef]
16. Li, X.; Qin, J.; Huang, B.; Zhang, X.; Bernstein, J. A new SPICE reliability simulation method for deep submicrometer CMOS VLSI circuits. *IEEE Trans. Device Mater. Reliab.* **2006**, *6*, 247–257. [CrossRef]
17. Xiong, Y.; Chiang, Y.; Pieper, N.J.; Ball, D.R.; Bhuva, B.L. Soft Error Rate Predictions for Terrestrial Neutrons at the 3-nm Bulk FinFET Technology. In Proceedings of the 2023 IEEE International Reliability Physics Symposium (IRPS), Monterey, CA, USA, 26–30 March 2023; pp. 1–6. [CrossRef]
18. Xu, C.; Liu, Y.; Liao, X.; Cheng, J.; Yang, Y. Machine Learning Regression-Based Single-Event Transient Modeling Method for Circuit-Level Simulation. *IEEE Trans. Electron Devices* **2021**, *68*, 5758–5764. [CrossRef]

19. Abe, S.I.; Watanabe, Y.; Shibano, N.; Sano, N.; Furuta, H.; Tsutsui, M.; Uemura, T.; Arakawa, T. Multi-Scale Monte Carlo Simulation of Soft Errors Using PHITS-HyENEXSS Code System. *IEEE Trans. Nucl. Sci.* **2012**, *59*, 965–970. [CrossRef]
20. Foley, K.; Seifert, N.; Velamala, J.B.; Bennett, W.G.; Gupta, S. IRT: A modeling system for single event upset analysis that captures charge sharing effects. In Proceedings of the 2014 IEEE International Reliability Physics Symposium, Waikoloa, HI, USA, 1–5 June 2014; pp. 5F.1.1–5F.1.9. [CrossRef]
21. Xiong, X.; Du, X.; Zheng, B.; Chen, Z.; Jiang, W.; He, S.; Zhu, Y. Soft Error Sensitivity Analysis Based on 40 nm SRAM-Based FPGA. *Electronics* **2022**, *11*, 3844. [CrossRef]
22. Artola, L.; Gaillardin, M.; Hubert, G.; Raine, M.; Paillet, P. Modeling Single Event Transients in Advanced Devices and ICs. *IEEE Trans. Nucl. Sci.* **2015**, *62*, 1528–1539. [CrossRef]
23. Aneesh, Y.M.; Sriram, S.R.; Pasupathy, K.R.; Bindu, B. An Analytical Model of Single-Event Transients in Double-Gate MOSFET for Circuit Simulation. *IEEE Trans. Electron Devices* **2019**, *66*, 3710–3717. [CrossRef]
24. Yi, B.; Lee, B.J.; Oh, J.H.; Kim, J.S.; Kim, J.H.; Yang, J.W. Physics-Based Compact Model of Parasitic Bipolar Transistor for Single-Event Transients in FinFETs. *IEEE Trans. Nucl. Sci.* **2018**, *65*, 866–870. [CrossRef]
25. Sheng, L.; He, W.; Zhang, Z.; He, L.; Cao, J.; Wu, Q. Investigation of double peak voltage in pulse quenching effect on the single-event transient. In Proceedings of the 2016 IEEE International Nanoelectronics Conference (INEC), Chengdu, China, 9–11 May 2016; pp. 1–2. [CrossRef]
26. Vibhu, V.; Mittal, S.; Kumar, V. Machine Learning-based model for Single Event Upset Current Prediction in 14 nm FinFETs. In Proceedings of the 2023 36th International Conference on VLSI Design and 2023 22nd International Conference on Embedded Systems (VLSID), Hyderabad, India, 8–12 January 2023; pp. 1–6. [CrossRef]
27. Rumelhart, D.; Hinton, G.; Williams, R. Learning representations by back-propagating errors. *Nature* **1986**, *323*, 533–536. [CrossRef]
28. Song, R.; Chen, S.; Du, Y.; Huang, P.; Chen, J.; Chi, Y. PABAM: A Physics-Based Analytical Model to Estimate Bipolar Amplification Effect Induced Collected Charge at Circuit Level. *IEEE Trans. Device Mater. Reliab.* **2015**, *15*, 595–603. [CrossRef]
29. Shao, J.; Song, R.; Chi, Y.; Liang, B.; Wu, Z. TAISAM: A Transistor Array-Based Test Method for Characterizing Heavy Ion-Induced Sensitive Areas in Semiconductor Materials. *Electronics* **2022**, *11*, 2043. [CrossRef]
30. Chi, Y.; Cai, C.; He, Z.; Wu, Z.; Fang, Y.; Chen, J.; Liang, B. SEU Tolerance Efficiency of Multiple Layout-Hardened 28 nm DICE D Flip-Flops. *Electronics* **2022**, *11*, 972. [CrossRef]
31. Yibai, H.; Shuming, C. Simulation Study of the Selectively Implanted Deep-N-Well for PMOS SET Mitigation. *IEEE Trans. Device Mater. Reliab.* **2014**, *14*, 99–103. [CrossRef]
32. He, Y.; Chen, S.; Chen, J.; Chi, Y.; Liang, B.; Liu, B.; Qin, J.; Du, Y.; Huang, P. Impact of Circuit Placement on Single Event Transients in 65 nm Bulk CMOS Technology. *IEEE Trans. Nucl. Sci.* **2012**, *59*, 2772–2777. [CrossRef]
33. Chen, J.; Chen, S.; He, Y.; Qin, J.; Liang, B.; Liu, B.; Huang, P. Novel Layout Technique for Single-Event Transient Mitigation Using Dummy Transistor. *IEEE Trans. Device Mater. Reliab.* **2013**, *13*, 177–184. [CrossRef]
34. Song, R.; Shi, J.; Shao, J.; Zhang, X. Machine Learning based SET Propagation Prediction for Large Scale Integrated Circuits. In Proceedings of the 2021 IEEE 14th International Conference on ASIC (ASICON), Kunming, China, 26–29 October 2021; pp. 1–4. [CrossRef]

Disclaimer/Publisher's Note: The statements, opinions and data contained in all publications are solely those of the individual author(s) and contributor(s) and not of MDPI and/or the editor(s). MDPI and/or the editor(s) disclaim responsibility for any injury to people or property resulting from any ideas, methods, instructions or products referred to in the content.

Article

Refined Analysis of Leakage Current in SiC Power Metal Oxide Semiconductor Field Effect Transistors after Heavy Ion Irradiation

Yutang Xiang [1,2,3], Xiaowen Liang [1,2], Jie Feng [1,2], Haonan Feng [1,2,3], Dan Zhang [1,2], Ying Wei [1,2,*], Xuefeng Yu [1,2,*] and Qi Guo [1,2]

1. Xinjiang Technical Institute of Physics and Chemistry, Chinese Academy of Sciences, Urumqi 830011, China; xiangyutang21@mails.ucas.ac.cn (Y.X.); liangxw@ms.xjb.ac.cn (X.L.); fengjie@ms.xjb.ac.cn (J.F.); fenghaonan19@mails.ucas.edu.cn (H.F.); zhangdan@ms.xjb.ac.cn (D.Z.); guoqi@ms.xjb.ac.cn (Q.G.)
2. Xinjiang Key Laboratory of Electronic Information Material and Device, Urumqi 830011, China
3. University of Chinese Academy of Sciences, Beijing 100049, China
* Correspondence: weiying@ms.xjb.ac.cn (Y.W.); yuxf@ms.xjb.ac.cn (X.Y.)

Abstract: A leakage current is the most critical parameter to characterize heavy ion radiation damage in SiC MOSFETs. An accurate and refined analysis of the source and generation process of a leakage current is the key to revealing the failure mechanism. Therefore, this article finely tests the online and post-irradiation leakage changes and leakage pathways of SiC MOSFETs caused by heavy ion irradiation, analyzes the damaged location of the device in reverse, and discusses the mechanism of leakage generation. The experimental results further confirm that an increase in the leakage current of a device during heavy ion irradiation is positively correlated with the applied voltage of the drain, but the leakage path is not direct from the drain to the source. The experimental analysis of the source of the leakage current of the device after irradiation indicates that there is also a leakage current path between the device gate and source. The research results suggest that the experimental sample is more prone to a single-event gate rupture effect under this heavy ion radiation condition. The gate breakdown mainly occurs in the gate oxide layer at the neck region. This research can provide a theoretical basis for the radiation resistance reinforcement of SiC power devices.

Keywords: SiC MOSFET; single-event effect; single-event gate rupture; leakage current; heavy ion irradiation

1. Introduction

With the rapid development of China's aerospace technology, the demand for high-performance, high-power devices is becoming increasingly urgent [1,2]. SiC MOSFETs have shown broad application prospects in the aerospace field due to their superior performance in high temperature resistance, low loss, fast switching speed, and high blocking voltage [3,4]. However, high-energy particles present in space can cause radiation damage to electronic devices, thereby affecting device performance and reliability [5–7]. For SiC power devices, although SiC materials have a comprehensive bandgap structure and strong radiation resistance, they still exhibit significant single-event radiation effects due to the process structure and operating characteristics of the device, leading to severe leakage and even burnout [6,8–10], seriously hindering their rapid application in the aerospace field. At present, the damage mechanisms of single-event burnout (SEB) and single-event gate rupture (SEGR) induced by heavy ions in space are difficulties and hot topics in the study of single-event effects in SiC MOSFETs [11–13].

A leakage current is the most critical parameter for characterizing heavy ion radiation damage in SiC MOSFETs [14–16]. Existing research has shown that before the occurrence of SEB or SEGR in SiC MOSFETs, phenomena such as gate leakage current I_G and drain leakage current I_D increases occur [15,17,18]. Among them, the mechanism of an SEB effect is the occurrence of an abnormal bulk current, which leads to a sharp increase in the lattice temperature of the device, resulting in a local thermal burnout of the device [19–23]; the mechanism of the SEGR effect is that a transient additional electric field is generated in the gate dielectric layer that exceeds the critical electric field, causing the gate oxide layer to be broken down [14,24–27]. But, the specific occurrence of SEB or SEGR is closely related to the location of the formation of an internal leakage current in the device. Therefore, in-depth research and refined analysis of the source and location of SiC power MOSFET leakage caused by heavy ion radiation are crucial to revealing the mechanisms of SEB and SEGR effects.

The increase in the leakage current of SiC MOSFETs caused by single-event effects is mainly due to the radiation-induced gate current I_G and the drain current I_D. I_G especially comes from a breakdown current and an ionization current [16,19]; a breakdown current is a leakage current caused by the breakdown of the gate oxide layer caused by the electric field generated by the incident particles between the gate and drain [16,28]; an ionization current is the current generated by the excitation or ionization of atoms or electrons in the gate oxide layer by incident particles. The I_D mainly comes from the transport of holes and electrons in the channel and the increase in the drain leakage current caused by ionization and reverse breakdown effects in the drain structure caused by incident particles [28–31]. Although the values of I_G and I_D are related to the energy, angle, and position of the incident particles [14,32] and the process and structure of the device, the main influencing factor is the device's leakage source operating voltage V_{DS}. In general, the larger the V_{DS}, the greater the radiation-induced leakage of the device. Generally, when the V_{DS} is greater than 50% of the rated voltage, it will cause the device to experience SEB or SEGR effects [33].

In summary, although a large number of studies have identified an increase in the device leakage current during heavy ion radiation and even the occurrence of SEB and SEGR effects, these are mainly due to the decrease in gate oxide insulation performance (breakdown) and drain structure damage caused by heavy ion radiation. But, the master–slave or quantitative relationship between these two damage mechanisms and radiation environment, device technology, and structural changes is still unclear, and more experimental verification is needed. In addition, the location of gate oxide breakdown and leakage structure damage varies, and the leakage paths of the device will not be the same. As shown in Figure 1, the leakage paths for gate oxide leakage I_G include gate leakage I_{GD} and I_{GS} (Figure 1, ① and ②), while the leakage paths for drain leakage I_D include I_{GD} and I_{DS} (Figure 1, ③ and ④), red-dashed ellipses on the schematic diagram show the possible locations of damage. There is also a lack of detailed research and analysis of these leakage paths. The analysis of these leakage pathways and the determination of damage locations are the theoretical basis for strengthening SiC power MOSFETs against single-event radiation damage.

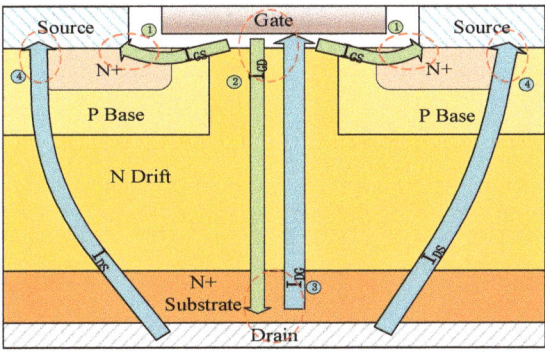

Figure 1. Schematic diagram of gate and drain current for SiC MOSFET.

Therefore, this experiment conducted single-event irradiation tests on SiC MOSFETs with different leakage source voltages and monitored the leakage current of the device during irradiation online. The I-V characteristics of the device before and after irradiation were compared, and the leakage current at each electrode of the device was finely tested. The path of a leakage current caused by heavy ions was analyzed and determined to determine the damage location of the device in reverse. The research results further deepen the understanding of the mechanism of a single-event radiation damage effect in SiC MOSFETs, providing a theoretical basis for the radiation hardening of SiC power devices.

2. Materials and Methods

The device used in the irradiation experiment is a typical planar gate structure, N-channel SiC power MOSFET packaged in TO-247L. The device parameters are V_{DS} = 1200 V, I_D = 40 A, and $R_{DS(on)}$ = 80 mΩ. A total of 10 test devices were used in the radiation test, including two for each of the three types of drain bias irradiation, floating irradiation and control devices. The device was unpacked before radiation, allowing heavy ions to irradiate the chip surface directly. Device packaging and internal chip morphology of the planar gate SiC MOSFET are shown in Figure 2. Electrical performance tests were conducted on the opened devices, and the results showed that opening did not significantly impact device performance.

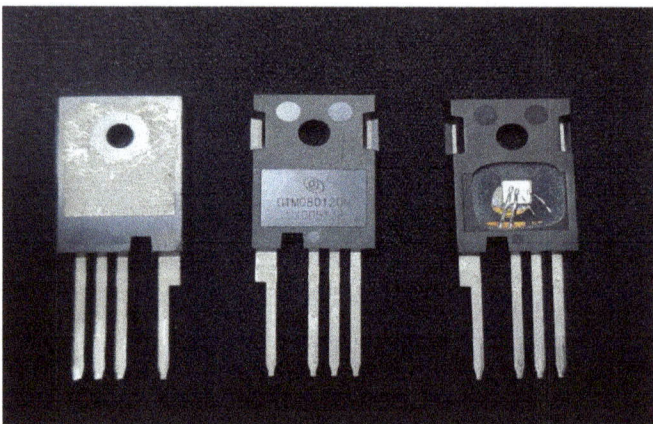

Figure 2. Device packaging and internal chip morphology of the planar gate SiC MOSFET.

The heavy ion test was completed at the Lanzhou Institute of Modern Physics, Chinese Academy of Sciences. The irradiated ion was ^{181}Ta ion, the total energy was 2369.8 MeV, the energy reaching the device surface was 1912.1 MeV, and the range was 111.3 μm. The LET value was 76.3 MeV/(mg/cm^2), and the beam spot area irradiated on the device surface in an atmospheric environment was about 4.4 cm^2, with the incident direction perpendicular to the device surface. The irradiation fluence rate was approximately 1.6×10^4 cm^2 s^{-1}, with a total injection rate of 1×10^6/cm^2.

During irradiation, in order to explore the changes in the leakage current of the device caused by heavy ions under different drain voltages, different voltages were applied to the drain electrodes of the device: V_{DS} = 60 V, 100 V, and 150 V, respectively, with zero bias on other electrodes. In addition, floating irradiation conditions, with no voltage applied to all electrodes, were also conducted. Under all irradiation bias conditions, a source meter was introduced between the source and drain of the device to monitor the leakage current I_{DSS} online. The schematic diagram of the SiC MOSFET heavy ion irradiation online testing system is shown in Figure 3.

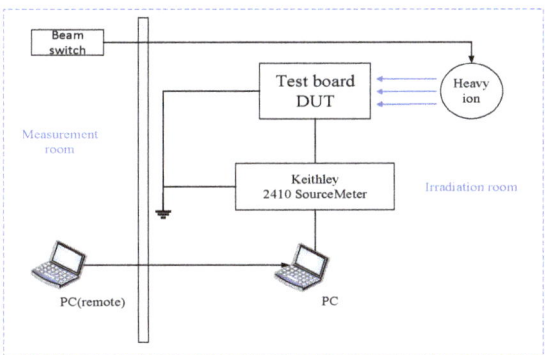

Figure 3. Schematic diagram of the online testing system for heavy ion irradiation of SiC MOSFET.

Before and after irradiation, the device's operating characteristics were tested using the B1500A semiconductor parameter analysis system, produced by Agilent Technologies, a company headquartered in California, United States. The test conditions were gate voltage V_{GS} = 20 V, source bias, and drain scanning from −10 V to 20 V. Current changes in the device under different operating conditions were observed. Also tested were the drain–source leakage current I_{DS}, gate–source leakage current I_{GS}, and the gate–drain leakage current I_{GD} data of the irradiated device; then, the specific damage location of the single-event effect in the device was analyzed.

3. Results

As shown in Figure 4, the device's drain–source leakage current I_{DSS}, detected during irradiation under different V_{DS} conditions, varies with irradiation time. The figure shows that when V_{DS} = 60 V, the current curve is almost a straight line, and the drain current jumps within an order of magnitude. When V_{DS} = 100 V and 150 V, the drain current during irradiation shows a linear increase trend with irradiation time, and as the drain bias increases, the slope of the drain current change increases, and the growth trend is significantly faster.

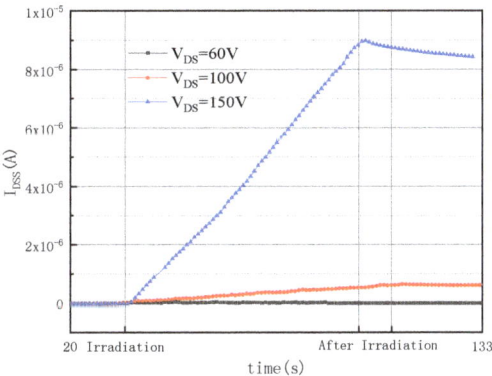

Figure 4. Time-dependent characteristics of drain leakage current for different V_{DS}.

After the irradiation stopped, the I_{DSS} of the device with a V_{DS} of 60 V returned to its initial value. In contrast, the I_{DSS} of the device remained unchanged under bias conditions of 100 V and 150 V, indicating that a single-event effect had occurred. As the drain voltage increased, the generated I_{DSS} also increased, showing a positive correlation. However, when the drain voltage was 150 V, the leakage current of the device did not reach the limit current, indicating that the device did not experience a typical single-event burnout phenomenon. Therefore, it can be considered that the I_{DSS} generated by the device under heavy ion irradiation at this time was caused by accumulated damage at the interface between the drain substrate and the drift region [34]. When Ta ions pass through SiC MOSFETs, they interact with the extranuclear electrons of the target atom and deposit energy. When the energy deposition exceeds a certain threshold, the material around the ion orbit will experience a melting phenomenon because the cooling rate of the material is fast, resulting in the melted material rapidly cooling into amorphous solid structures. These amorphous solids formed along the ion incidence path are localized defect clusters composed of high-concentration composite defects [34–36]. The high LET value of Ta ions can induce latent tracks in the active region of SiC MOSFETs [34,36,37]. These leakage channels, composed of defect clusters along the ion path, can cause the leakage current of the online monitoring point of the device to increase during continuous heavy ion radiation. So, it was determined that V_{DS} = 60 V did not cause a single-event effect and is the safe voltage of the device. It is preliminarily speculated that a small-current single-event burnout effect occurred between the drain and source electrodes in devices with biases equal to 100 V and 150 V [36]. Subsequent parameter tests were conducted to determine whether a single-event burnout occurred.

The electrical parameters of the device were tested before and after irradiation, and the transfer characteristics of the device under bias and floating conditions (I_{DS}-V_{GS}) are shown in Figure 5. Figure 5a shows that the device transfer characteristic curves and curves after irradiation under different drain–source bias conditions do not show significant drift. The device characteristic curve with V_{DS} of 60 V during irradiation coincides with the device characteristic curve under floating conditions, and there is no significant difference from the initial value; the transfer characteristics of two devices with V_{DS} of 100 V and 150 V during irradiation showed two significant current increases compared to devices with V_{DS} of 60 V, corresponding to gate voltages of −10 V to −2.5 V and 0 V to 5 V, respectively. I_{DS} increased by two to three orders of magnitude compared to devices with 60 V, and the larger the V_{DS}, the more significant the current increase. Figure 5b compares device transfer characteristic curves under different drain bias voltages in linear coordinates. It can be seen from the figure that the drain current of the device decreased after irradiation. Therefore, it is believed not only that the drain junction was damaged but there should also be current leakage at the gate.

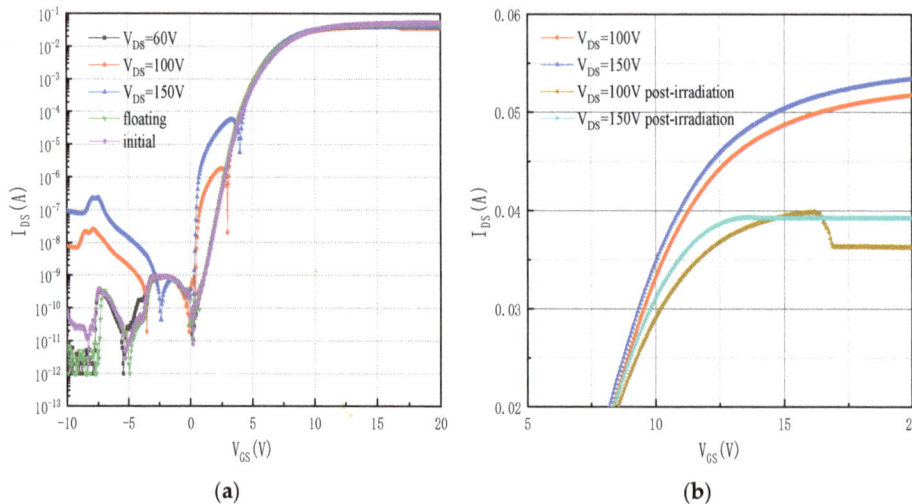

Figure 5. Transfer characteristic curve of devices with a flux of $1 \times 10^6/cm^2$: (**a**) logarithmic coordinate; (**b**) linear coordinate.

To finely analyze the path of the leakage current generated by heavy ion irradiation on the back of the device, which is the specific damage location, using a semiconductor parameter analysis system to test the electrode currents of each electrode of the device, the testing method is: floating either electrode of the device, applying a scanning voltage between the other two electrodes, and connecting a source meter in series at the tested electrode for current testing. Only one electrode is applied with voltage, and the other two electrodes are connected to ammeters and grounded. The current flowing through this electrode should be the sum of the currents at the other two electrodes. Therefore, it is believed that the existence of current leakage paths between the two electrodes of the device can be judged by the specific current division. Firstly, a scanning voltage was applied on the gate electrode and a test current between the gate–source and the gate–drain. Figure 6 shows the test results of the gate–source current I_{GS} and the gate–drain current I_{GD} of the device obtained using this method. The device irradiation conditions corresponding to each curve in Figures 4 and 5 are (A) $V_{DS} = 60$ V $1 \times 10^6/cm^2$, (B) $V_{DS} = 100$ V $1 \times 10^6/cm^2$, (C) $V_{DS} = 150$ V $1 \times 10^6/cm^2$, (D) device floating $1 \times 10^6/cm^2$, and (E) device initial value.

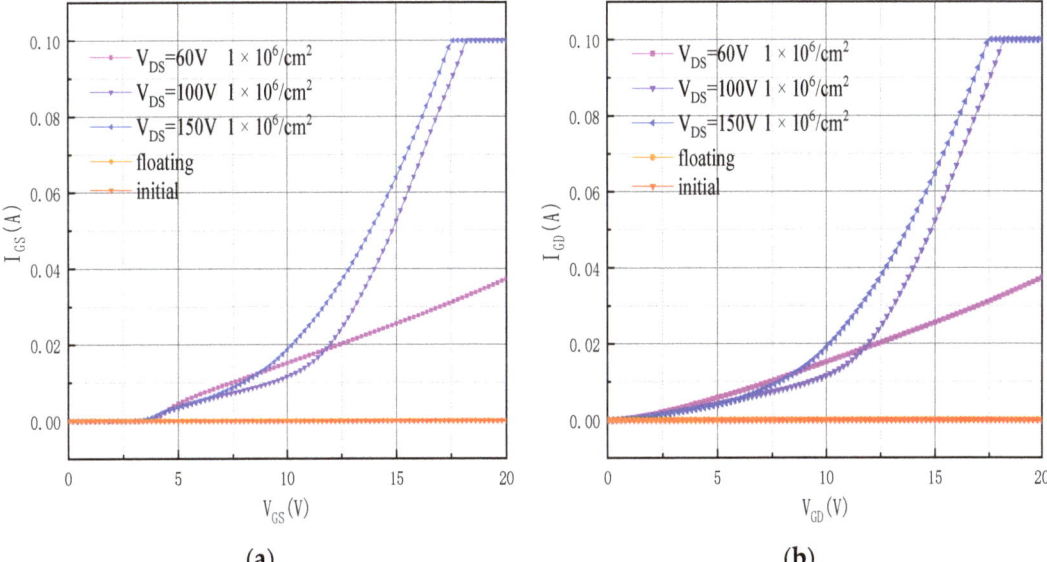

Figure 6. (**a**) Leakage current I_{GS} curve; (**b**) leakage current I_{GD} curve.

The trend of the I_{GS} leakage current at both electrodes of the gate–source in Figure 6a is almost consistent with that of the I_{GD} leakage current at both electrodes of the gate–drain in Figure 6b. After applying a scanning voltage to the gate, all devices with floating electrodes during irradiation did not experience an increase in current, which is consistent with non-irradiated devices; the current of devices with a drain bias voltage applied during irradiation shows an increasing trend. Among them, devices with irradiation biases V_{DS} of 150 V and 100 V reach current limits at a gate voltage of 17 V and 18 V, respectively. Devices with a bias V_{DS} of 60 V neither show any changes like floating devices nor increase the current limit slowly, unlike devices with a high drain bias.

Since the gate oxide layer is an insulating layer, it can be seen from the initial value test of the device before irradiation that the gate current hardly changes with an increase in gate voltage. However, an increased device leakage current after irradiation indicates shallow damage to the gate oxide layer. It is worth noting that the three I_{GS} curves in Figure 6a do not immediately increase after the scanning voltage is applied, indicating that there is no direct current leakage channel between the gate and source electrodes; that is, the damage location is not directly between the gate and source electrodes. All three curves gradually increase after the gate voltage reaches around 2.5 V because as the gate voltage gradually increases, which means V_{GS} is greater than the device threshold voltage; the channel region gradually depletes and approaches strong inversion and extends towards the neck region, forming an n-type layer along the surface coupled to the source electrode. When the gate oxide layer is damaged in the neck area, a leakage channel is formed and reaches the current limit. The I_{GD} curve in Figure 6b immediately increased after voltage was applied, indicating damage between the gate and drain electrodes of the device and a direct leakage channel for the current. Heavy ion irradiation caused damage and destruction of the gate oxide layer, resulting in a single-event gate rupture of the device, with the damage located between the gate and neck regions.

To determine whether a single-event burnout effect occurred between the source and drain of the device, a scanning voltage was applied to the drain electrode, a 0 V bias was applied to both the gate and source electrodes, an ammeter was connected in series on the gate and source electrodes were connected to the ground. The drain current I_D, gate current I_G, and source current I_S were measured. The relationship between them was analyzed, as shown in Figure 7.

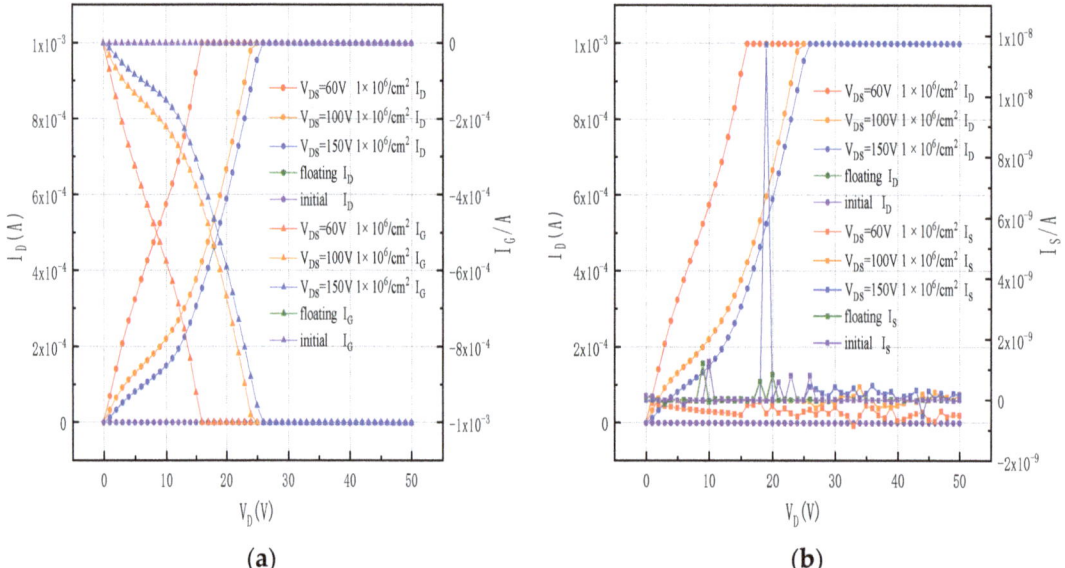

Figure 7. (a) Drain leakage current I_D and gate leakage current I_G curve; (b) frain leakage current I_D and source leakage current I_S curve.

Figure 7a shows that the gate current I_G and drain leakage current I_D of the device, subjected to bias voltage during irradiation, increased directly from 0 V, with almost the same magnitude and in opposite directions. At the same leakage voltage, the current limit is reached, indicating the existence of a current leakage channel between the two electrodes of the gate and drain. The leakage current curve of the floating irradiation device and the non-irradiation device is a straight line, with a value of almost 0. Figure 7b shows that the magnitude of the source current I_S is much smaller than that of the drain leakage current I_D, fluctuating in the 1×10^{-9} magnitude. From this, it can be inferred that there is no current leakage channel between the drain and sources. Because there is no damage location between the drain and source electrodes of the device, it can be determined that the test device did not experience single-event burnout.

Based on the above experimental results and analysis, it can be concluded that the leakage current I_{DSS} monitored online in heavy ion irradiation experiments is not caused by the increase in the drain–source current caused by single-event burnout but by the current path caused by a single-event gate rupture, forming a leakage current from the drain electrode to the gate electrode. It should be noted that the leakage current of the device, with an irradiation bias of 60 V, is inconsistent with the online monitoring performance and has reached saturation, which is believed to be caused by the activation of internal defects in the device due to multiple tests and power-ups. Moreover, the voltage at which the devices reach saturation current varies. Devices with an irradiation bias of 60 V first reach limit current, followed by devices with 100 V irradiation, and finally, devices with 150 V. It is believed that V_{DS} = 100 V, and 150 V devices experienced typical radiation-induced hard breakdown (RHB), but the device with V_{DS} = 60 V may have experienced radiation soft breakdown (RSB) [38–43]. The defect charges introduced by the heavy ion are closer to the conduction band and are more likely to be activated after multiple power tests [38]. So, the leakage current of the device reaches the limit saturation current first before V_{DS} = 100 V and V_{DS} = 150 V. The specific reason why the performance of powered irradiation devices is not positively correlated with the applied bias requires more experiments and simulations to verify.

In order to verify whether the analysis of the previous experimental data was reasonable and correct, that is, whether the device had experienced single-event gate rupture and whether the damage location was between the gate oxide layer and neck region, a cross-sectional analysis was conducted on the device. Firstly, there were no obvious burn marks on the surface of the chip, as shown in Figure 8a. Then, the specific damage location of the cover opening device was located. The damaged region was located by Optical Beam Induced Resistance Change (OBIRCH) analysis for the SiC MOSFET with single-event gate rupture, which was manifested as a "bright spot", as shown in Figure 8b. The "hole" was cross-sectioned using a focused ion beam (FIB) along the dotted line. FIB inspection on the gate oxide layer was performed to determine the specific damage location of the device, as shown in Figure 8c. Figure 8d shows the SEM image of the damaged region after striping the surface metal layer. A damaged area is observed at the poly gate. It shows that the gate oxide layer was broken due to the heavy ion irradiation. The damaged region covers one gate strip and connects with the neck region of the device. The conclusion of a damage location obtained through cross-sectional analysis is consistent with the previous data analysis through qualitative analysis of the leakage current, which can prove that the previous data analysis is reasonable; SiC MOSFET suffered single-event gate rupture. The appearance of the device did not show any signs of burning, and SEM did not find any burnt areas; the device did not experience single-event burnout.

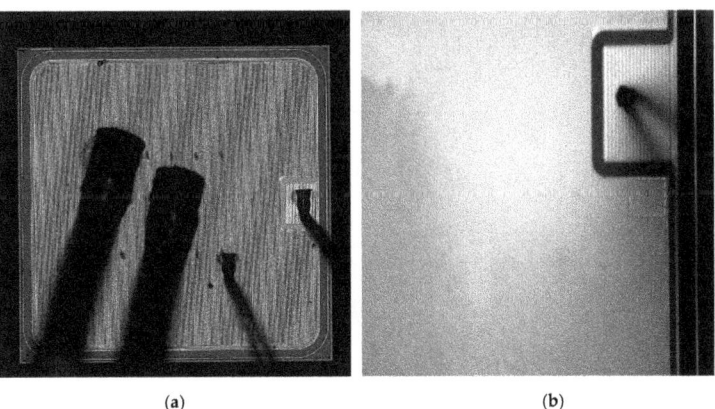

(a) (b)

Figure 8. *Cont.*

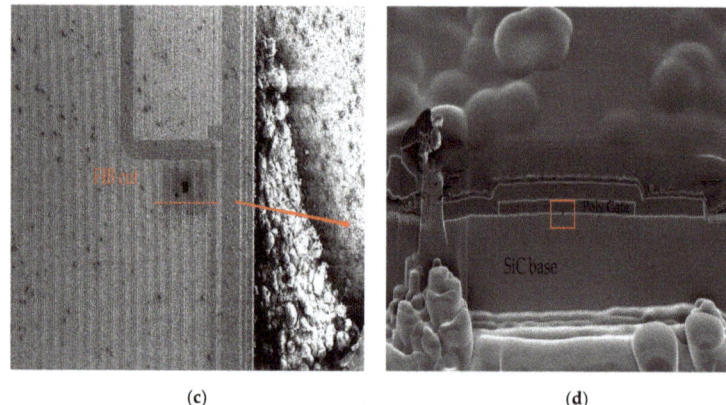

(c)   (d)

Figure 8. Failure analysis of the SiC MOSFET with SEGR. (**a**) Chip visual inspection, (**b**) OBIRCH analysis of SiC MOSFET. The "bright spot" area represents the damaged region. (**c**) Focused ion beam (FIB) cut area, (**d**) SEM diagram of the damaged region after striping metal layer.

4. Discussion

Based on the above experimental results, it can be indicated that this SiC MOSFET underwent a single-event gate rupture effect after being irradiated with ^{181}Ta ions, with the main leakage pathway being the gate oxide layer near the neck. When particles entered the SiC MOSFET device, the incident particles collided and ionized with SiC atoms in the SiC MOSFET device, generating a large number of electron–hole pairs [28,44,45]. Under the application of an electric field, electron–hole pairs drifted, and electrons gathered toward the drain region. Some ionized holes moved toward the source region. In contrast, the other holes mainly gathered at the junction of the gate oxide layer and the drift layer, generating a transient electric field below the gate dielectric layer, as shown in Figure 9 [28,44,45]. The generated transient electric field was superimposed with the original electric field. Once the superimposed electric field exceeded the intrinsic, breakdown electric field strength of the gate dielectric layer, the gate dielectric layer was broken down, causing it to lose its insulation effect and generate a conductive path in the gate dielectric layer [6,24,28,44–46]. The gate current rapidly increased, and finally, the gate was broken down, resulting in a single-event gate rupture effect. After multiple tests of the device, there was no recoverability of the damage, as the single-event gate rupture effect is a destructive effect.

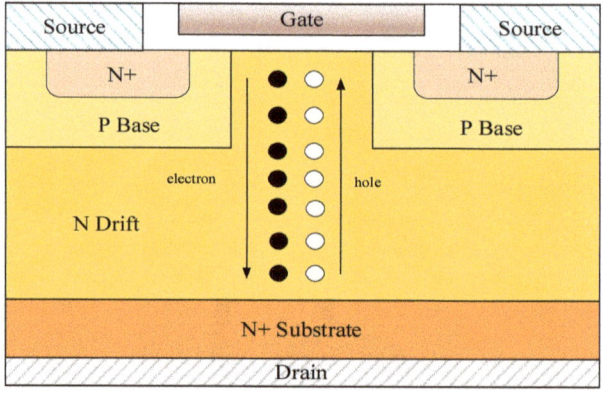

Figure 9. Schematic diagram of electron–hole movement after particle incident.

However, the reason why the device did not undergo single-event burnout and why the single-event gate rupture occurred is believed to be that when heavy ions irradiated the VDMOS device, the ions first came from the gate, and the induced current leakage path was from the drain to the gate, causing damage to the gate oxide layer of the device. When a higher bias voltage is selected, assuming that a voltage drop sufficient to open the channel partially is generated through the leakage of the gate oxide, the MOSFET is placed in a "partially conductive" condition, allowing a current to flow to the source. The leakage current is distributed between the drain–gate and drain–source, and the main leakage path of the current may change into drain to source, leading to single-event burnout. To verify whether the bias voltage is the main factor causing single-event burnout or single-event gate rupture of devices, a higher bias voltage will be selected for experiments under the same conditions, and some simulations will be performed for verification.

5. Conclusions

The experimental results indicate that the leakage current monitored online positively correlates with the voltage and will not return to its initial value after irradiation. The leakage current detection after irradiation proves that there is a leakage current path among drain–gate and gate–source of all biased devices, and there is a latent track in the 60 V biased device, with the damage location located between the gate and the neck region. No single-event burnout has occurred in this experiment. The subsequent device dissection analysis results also prove that SiC MOSFETs only caused a single-event gate rupture effect under this experimental condition. So, it can be explained that the increase in leakage current I_{DSS} monitored online during the heavy ion irradiation test process was not caused by single-event burnout but rather by a single-event gate rupture, with the leakage path of the current from the drain electrode to the gate electrode.

Author Contributions: Conceptualization, Y.X. and X.Y.; methodology, Y.X. and X.Y.; software, X.L.; validation, Y.X. and X.Y.; formal analysis, Y.X.; investigation, Y.X., D.Z., J.F. and H.F.; data curation, Y.X. and X.L.; writing, original draft preparation, Y.X.; writing, review and editing, Y.X., Y.W. and J.F.; visualization, Y.X.; supervision, X.Y. and Y.W.; project administration, X.Y. and Q.G.; funding acquisition, J.F. and Y.W. All authors have read and agreed to the published version of the manuscript.

Funding: This research was funded by the Young Scholars in Western China of the Chinese Academy of Sciences under grant No. 2021-XBQNXZ-021,the West Light Talent Training Plan of the Chinese Academy of Sciences under grant No. 2022-XBQNXZ-010, the Youth Science and Technology Talents Project of Xinjiang Uygur Autonomous Region No. 2022TSYCCX0094.

Data Availability Statement: Not applicable.

Acknowledgments: This work acknowledges the support of the Institute of modern physics, Chinese Academy of Sciences in the heavy ion experiment.

Conflicts of Interest: The authors declare no conflict of interest.

References

1. Zeng, Z.; Zhang, X.; Blaabjerg, F.; Miao, L. Impedance-oriented transient instability modeling of SiC MOSFET intruded by measurement probes. *IEEE Trans. Power Electron.* **2020**, *35*, 1866–1881. [CrossRef]
2. Sheng, K.; Guo, Q.; Zhang, J.; Qian, Z. Development and prospect of SiC power devices in power grid. *Proc. Chin. Soc. Electr. Eng.* **2012**, *32*, 1–7.
3. Hang, B.; Deng, X.; Zhang, Y. Recent Development and Future Perspective of Silicon Carbide Power Devices—Opportunity and Challenge. *J. China Acad. Electron. Inf. Technol.* **2009**, *4*, 111–118.
4. Shen, Z.; Xia, Y.; Yang, Y. Protection of Materials and Structures from Space Radiation Environments on Spacecraft. *Aerosp. Mater. Technol.* **2020**, *50*, 012089.
5. Zhiqiang, B. Research Progress on Reliability of 4H-SiC Power MOSFET. *Electron. Pack.* **2022**, *22*, 5–13. [CrossRef]
6. Liu, C.; Guo, G.; Li, Z. Recent research progress of single particle effect of SiC MOSFET. *Nucl. Tech.* **2022**, *45*, 3–16.
7. Tan, Z.; Wei, Z.; Sun, Y.; Wan, X.; Jin, H.; Wan, H.; Wang, J.; Liu, D.; Xu, J. Review of Radiation Effects in Power Semiconductor Devices. *Microelectronics* **2017**, *47*, 690–694. [CrossRef]

8. Chen, W.; Yang, H.L.; Guo, X.Q. The research status and challenge of space radiation physics and application. *Chin. Sci. Bull.* **2017**, *62*, 978–989. [CrossRef]
9. Ecoffet, R. Overview of in-orbit radiation induced spacecraft anomalies. *IEEE Trans. Nucl. Sci.* **2013**, *60*, 1791–1815. [CrossRef]
10. Ferlet-Cavrois, V.; Sturesson, F.; Zadeh, A.; Santin, G.; Truscott, P.; Poivey, C.; Schwank, J.R.; Peyre, D.; Binois, C.; Beutier, T.; et al. Charge collection in power MOSFETs for SEB characterization evidence of energy effects. *IEEE Trans. Nucl. Sci.* **2010**, *57*, 3515–3527. [CrossRef]
11. Huang, H.; Wang, N.; Wu, J.; Lu, T. Radiated disturbance characteristics of SiC MOSFET module. *J. Power Electron.* **2021**, *21*, 49–504. [CrossRef]
12. Lichtenwalner, D.J.; Akturk, A.; McGarrity, J.; Richmond, J.; Barbieri, T.; Hull, B.; Grider, D.; Allen, S.; Palmour, J.W. Reliability of SiC power devices against cosmic ray neutron single-event burnout. *Mater. Sci. Forum* **2018**, *924*, 559–562. [CrossRef]
13. Sato, I.; Tanaka, T.; Hori, M.; Yamada, R.; Toba, A.; Kubota, H. High power density inverter utilizing SiC MOSFET and interstitial via hole PCB for motor drive system. *Electr. Eng. Jpn.* **2021**, *214*, e23323. [CrossRef]
14. Ikpe, S.A.; Lauenstein, J.M.; Carr, G.A.; Hunter, D.; Ludwig, L.L.; Wood, W.; Iannello, C.J.; Del Castillo, L.Y.; Fitzpatrick, F.D.; Mojarradi, M.M.; et al. Long-term reliability of a hard-switched boost power processing unit utilizing SiC power MOSFETs. In Proceedings of the 2016 IEEE International Reliability Physics Symposium (IRPS), Pasadena, CA, USA, 17–21 April 2016; pp. ES-1–ES-18. [CrossRef]
15. Lauenstein, J.M. Getting SiC power devices off the ground: Design, testing, and overcoming radiation threats. In Proceedings of the 2018 Microelectronics Reliability and Qualification Workshop, El Segundo, CA, USA, 6–8 February 2018.
16. Martinella, C.; Stark, R.; Ziemann, T.; Alía, R.G.; Kadi, Y.; Grossner, U.; Javanainen, A. Current transport mechanism for heavy-ion degraded SiC MOSFETs. *IEEE Trans. Nucl. Sci.* **2019**, *66*, 1702–1709. [CrossRef]
17. Mizuta, E.; Kuboyama, S.; Abe, H.; Iwata, Y.; Tamura, T. Investigation of single-event damages on silicon carbide (SiC) power MOSFETs. *IEEE Trans. Nucl. Sci.* **2014**, *61*, 1924–1928. [CrossRef]
18. Martinella, C.; Ziemann, T.; Stark, R.; Tsibizov, A.; Voss, K.O.; Alia, R.G.; Kadi, Y.; Grossner, U.; Javanainen, A. Heavy-ion microbeam studies of single-event leakage current mechanism in SiC VD-MOSFETs. *IEEE Trans. Nucl. Sci.* **2020**, *67*, 1381–1389. [CrossRef]
19. Waskiewicz, A.E.; Groninger, J.W.; Strahan, V.H.; Long, D.M. Burnout of Power Mos Transistors with Heavy Ions of Californium-252. *IEEE Trans. Nucl. Sci.* **1986**, *33*, 1710–1713. [CrossRef]
20. Hohl, J.H.; Johnnson, G.H. Features of the Triggering Mechanism for Single Event Burnout of Power Mosfets. *IEEE Trans. Nucl. Sci.* **1989**, *36*, 2260–2266. [CrossRef]
21. Johnson, G.H.; Hohl, J.H.; Schrimpf, R.D.; Galloway, K.F. Simulating Single-Event Burnout of N-Channel Power MOSFET'S. *IEEE Trans. Electron Devices* **1993**, *40*, 1001–1008. [CrossRef]
22. Shoji, T.; Nishida, S.; Hamada, K.; Tadano, H. Analysis of Neutron-Induced Single-Event Burnout in SiC Power Mosfets. *Microelectron. Reliab.* **2015**, *55*, 1517–1521. [CrossRef]
23. Witulski, A.F.; Ball, D.R.; Galloway, K.F.; Javanainen, A.; Lauenstein, J.M.; Sternberg, A.L.; Schrimpf, R.D. Single-Event Burnout Mechanisms in Sic Power Mosfets. *IEEE Trans. Nucl. Sci.* **2018**, *65*, 1951–1955. [CrossRef]
24. Titus, J.L.; Su, Y.S.; Savage, M.W.; Mickevicius, R.V.; Wheatley, C.F. Simulation Study of Single-Event Gate Rupture Using Radiation-Hardened Stripe Cell Power Mosfet Structures. *IEEE Trans. Nucl. Sci.* **2003**, *50*, 2256–2264. [CrossRef]
25. Boruta, N.; Lum, G.; O'Donnell, H.; Robinette, L.; Shaneyfelt, M.; Schwank, J. A new physics-based model for understanding single-event gate rupture in linear devices. *IEEE Trans. Nucl. Sci.* **2001**, *48*, 1917–1924. [CrossRef]
26. Titus, J.L. An Updated Perspective of Single Event Gate Rupture and Single Event Burnout in Power MOSFETs. *IEEE Trans. Nucl. Sci.* **2013**, *60*, 1912–1928. [CrossRef]
27. Krishnamurthy, S.; Kannan, R.; Hussin, A.; Yahya, E.A. Investigation of Segr Effects on Power Vdmosfet for Various Heavy Ion Radiation. In Proceedings of the International Conference in Electrical, Pahang, Malaysia, 29 July 2019.
28. Fischer, T. Heavy-ion Induced Gate Rupture in Power MOSFETs. *IEEE Trans. Nucl. Sci.* **1987**, *34*, 1786–1791. [CrossRef]
29. Selva, L.E.; Swift, G.M.; Taylor, W.A.; Edmonds, L.D. On the role of energy deposition in triggering SEGR in power MOSFETs. *IEEE Trans. Nucl. Sci.* **1999**, *46*, 1403–1409. [CrossRef]
30. Wheatley, C.F.; Titus, J.L.; Burton, D.I. Single-Event Gate Rupture in Vertical Power MOSFETs: An Original Emperical Expression. *IEEE Trans. Nucl. Sci.* **1994**, *41*, 2152–2159. [CrossRef]
31. Titus, J.L.; Wheatley, C.F.; Burton, D.I.; Mouret, I.; Allenspach, M.; Brews, J.; Schrimpf, R.; Galloway, K.; Pease, R.L. Impact of Oxide Thickness on SEGR Failure in Vertical Power MOSFETs; Development of A Semi-Empirical Expression. *IEEE Trans. Nucl. Sci.* **1995**, *42*, 1928–1934. [CrossRef]
32. Titus, J.L.; Wheatley, C.F.; Allenspach, M.; Schrimpf, R.D.; Burton, D.I.; Brews, J.R.; Galloway, K.F.; Pease, R.L. Influence of Ion Beam Energy on SEGR Failure Thresholds of Vertical Power MOSFETs. *IEEE Trans. Nucl. Sci.* **1996**, *43*, 2938–2943. [CrossRef]
33. Lauenstein, J.M.; Casey, M.; Topper, A.; Wilcox, E.; Phan, A.; Ikpe, S.; LaBel, K. Silicon carbide power device performance under heavy-ion irradiation. In Proceedings of the 2015 IEEE Nuclear and Space Effects Conference, Boston, MA, USA, 16 July 2015.
34. Lei, Z.F.; Guo, H.X.; Tang, M.H.; Zeng, C.; Zhang, Z.G.; Chen, H.; En, Y.F.; Huang, Y.; Chen, Y.Q.; Peng, C. Degradation mechanisms of AlGaN/GaN HEMTs under 800 MeV Bi ions irradiation. *Microelectron. Reliab.* **2018**, *80*, 312–316. [CrossRef]
35. Ziegler, J.F.; Biersack, J.P. The stopping and range of ions in solids vol 1: The stopping and ranges of ions in matter. *Ion Implant. Sci. Technol.* **1985**, *10*, 51–108.

36. Ziwen, C. Study on the Irradiation Induced Damage of SiC Power VMOSFETs Devices. Ph.D. Thesis, Xiangtan University, Xiangtan, China, 2021. [CrossRef]
37. Bolotnikov, A.; Losee, P.; Permuy, A.; Dunne, G.; Kennerly, S.; Rowden, B.; Nasadoski, J.; Harfman-Todorovic, M.; Raju, R.; Tao, F.; et al. Overview of 1.2–2.2kV SiC MOSFETs targeted for industrial power conversion applications. In Proceedings of the 2015 IEEE Applied Power Electronics Conference and Exposition (APEC), Charlotte, NC, USA, 15–19 March 2015.
38. Schwank, J.R.; Shaneyfelt, M.R.; Fleetwood, D.M.; Felix, J.A.; Dodd, P.E.; Paillet, P.; Ferlet-Cavrois, V. Radiation Effects in MOS Oxides. *IEEE Trans. Nucl. Sci.* **2008**, *55*, 1833–1853. [CrossRef]
39. Sexton, F.W.; Fleetwood, D.M.; Shaneyfelt, M.R.; Dodd, P.E.; Hash, G.L.; Schanwald, L.P.; Loemker, R.A.; Krisch, K.S.; Green, M.L.; Weir, B.E.; et al. Precursor ion damage and angular dependence of single event gate rupture in thin oxides. *IEEE Trans. Nucl. Sci.* **1998**, *45*, 2509–2518. [CrossRef]
40. Massengill, L.W.; Choi, B.K.; Fleetwood, D.M.; Schrimpf, R.D.; Galloway, K.F.; Shaneyfelt, M.R.; Meisenheimer, T.L.; Dodd, P.E.; Schwank, J.R.; Lee, Y.M.; et al. Heavy-ion induced breakdown in ultra-thin gate oxides and high-k dielectrics. *IEEE Trans. Nucl. Sci.* **2001**, *48*, 1904–1912. [CrossRef]
41. Johnston, A.H.; Swift, G.M.; Miyahira, T.; Edmonds, L.D. Breakdown of gate oxides during irradiation with heavy ions. *IEEE Trans. Nucl. Sci.* **1998**, *45*, 2500–2508. [CrossRef]
42. Conley, J.F.; Suehle, J.S.; Johnston, A.H.; Wang, B.; Miyahara, T.; Vogel, E.M.; Bernstein, J.B. Heavy-ion induced soft breakdown of thin gate oxides. *IEEE Trans. Nucl. Sci.* **2001**, *48*, 1913–1916. [CrossRef]
43. Ceschia, M.; Paccagnella, A.; Sandrin, S.; Ghidini, G.; Wyss, J.; Lavale, M.; Flament, O. Low field leakage current and soft breakdown in ultra thin gate oxides after heavy ions, electrons, or X-ray irradiation. *IEEE Trans. Nucl. Sci.* **2000**, *47*, 566–573. [CrossRef]
44. Nichols, D.K.; Coss, J.R.; McCarty, K.P. Single-event gate rupture in commercial power MOSFET's. In Proceedings of the RADECS 93: Second European Conference on Radiation and Its Effects on Components and Systems, Saint Malo, France, 13–16 September 1993; p. 462467.
45. Busch, M.C.; Dooryhee, E.; Slaouri, A.; Toulemonde, M.; Mesli, A.; Siffert, P. Heavy ions induced electrical and structural defects in thermal SiO_2 films. In Proceedings of the RADECS '91, First European Conference Radiation and Its Effects on Components and Systems, La Grande-Motte, France, 9–12 September 1991; p. 484488.
46. Titus, J.L.; Wheatley, C.F. Experimental studies of single-event gate rupture and burnout in vertical power MOSFETs. *IEEE Trans. Nucl. Sci.* **1996**, *43*, 533–545. [CrossRef]

Disclaimer/Publisher's Note: The statements, opinions and data contained in all publications are solely those of the individual author(s) and contributor(s) and not of MDPI and/or the editor(s). MDPI and/or the editor(s) disclaim responsibility for any injury to people or property resulting from any ideas, methods, instructions or products referred to in the content.

Article

Recovery Effect of Hot-Carrier Stress on γ-ray-Irradiated 0.13 µm Partially Depleted SOI n-MOSFETs

Lan Lin [1,*], Zhongchao Cong [2,*] and Chunlei Jia [2]

1 School of Computer Science and Engineering, Southwest Minzu University, Chengdu 610041, China
2 China Academy of Launch Vehicle Technology, Beijing 100076, China; jiachun_lei@126.com
* Correspondence: linlan0921@163.com (L.L.); 17718451130@163.com (Z.C.)

Abstract: Many silicon-on-insulator (SOI) metal–oxide–semiconductor field-effect transistors (MOSFETs) are used in deep space detection systems because they have higher radiation resistance than bulk silicon devices. However, SOI devices have to face the double challenge of radiation and conventional reliability problems, such as hot carrier stress, at the same time. Thus, we wondered whether there is any interaction between reliability degradation and irradiation damage. In this paper, the effect of hot-carrier injection (HCI) on γ-ray-irradiated partially depleted (PD) SOI n-MOSFETs with a T-shaped gate structure is investigated. A strange phenomenon that accelerated the annealing effect on irradiation devices caused by HCI in 5 s was observed. That is, HCI has fast recovery ability on the irradiated narrow-channel n-MOSFETs. We explain the physical mechanism of this recovery effect qualitatively. Moreover, we designed a comparable experiment to evaluate the effect on the wide-channel devices. These results show that the narrow-channel devices are more sensitive to irradiation and HCI effects than wide-channel devices.

Keywords: reliability; SOI; MOSFET; radiation; HCI

Citation: Lin, L.; Cong, Z.; Jia, C. Recovery Effect of Hot-Carrier Stress on γ-Irradiated 0.13 µm Partially Depleted SOI n-MOSFETs. *Electronics* 2023, *12*, 4233. https://doi.org/10.3390/electronics12204233

Academic Editor: Paul Leroux

Received: 6 September 2023
Revised: 28 September 2023
Accepted: 28 September 2023
Published: 13 October 2023

Copyright: © 2023 by the authors. Licensee MDPI, Basel, Switzerland. This article is an open access article distributed under the terms and conditions of the Creative Commons Attribution (CC BY) license (https://creativecommons.org/licenses/by/4.0/).

1. Introduction

The development of space electronics technology has traditionally been significantly influenced by the commercial semiconductor industry. The development of metal–oxide–semiconductor (MOS) technology and, in particular, complementary metal–oxide–semiconductor (CMOS) technology, as a dominant commercial technology, have been used to extend the lifespan of devices used in deep space systems [1–5]. Recently, SOI technology has seen widespread applications in the aerospace sector due to its exceptional resistance against transient ionizing radiation, such as single-event effects [6]. However, there is a potential problem associated with the relatively thick buried oxide (BOX), which is sensitive to the total ionizing dose (TID) effect [7–9]. Radiation-induced trapped charges build up in the gate oxide, which causes a shift in the threshold voltage (that is, a change in the voltage that must be applied to turn the device on). In other words, the threshold of the back gate would change with the irradiation dose. Although in practical circuits, the back gate is typically grounded, these devices conduct as the threshold of the back gate drifts below zero, leading to a large channel leakage current [10–12]. If this shift is large enough, the device cannot be turned off, even with zero voltage applied, and the device is suspected to have failed by entering depletion mode.

Furthermore, devices operating in this environment face not only the challenge of the irradiation environment but also issues with conventional reliability, such as hot carrier stress when they are deployed in deep space missions. In many cases, the HCI effect is regarded as one of the most important factors that limit the lifespan of very large-scale integration (VLSI) circuits and maximal devices. Hot carriers may yield interface traps at Si/SiO$_2$ interface, be trapped in the oxide, or generate new oxide traps, resulting in effects such as threshold voltage (V_T) drift, transconductance (G_m) degradation, and an increase in channel leakage current [13]. Both radiation and HCI effects can degrade the device's performance over time and, ultimately, invalidate the device or circuit [14,15]. Devices exposed to deep space environments face the dual challenges of TID irradiation and HCI effects at the same time, both of which degrade the devices' performance by introducing trapped charges into the oxide layer or the oxide/bulk interface. Thus, we wondered whether there is any interaction between reliability degradation and irradiation damage, and whether the lifespan of devices with 0.13 μm PD SOI technology could be further decreased by this interaction. In fact, there are researchers who have explored potential synergistic effects or correlations between hot-carrier effects and TID irradiation in PD SOI MOSFETs [16–20]. Silvestri et al. [21,22] investigated how X-ray exposure impacts the long-term reliability of 130 nm n-MOSFETs as a function of device geometry and irradiation bias conditions. The experimental results presented the opposite effect to the degradation during subsequent hot-carrier injection. Increasing the bias during irradiation slightly reduces the impact on subsequent electrical stress in core MOSFETs. Qi-wen Zheng et al. [23] carried out total-dose irradiation on the hot-carrier reliability of 65 nm n-MOSFETs. The experimental results showed that hot-carrier degradation on irradiated narrow-channel n-MOSFETs are greater than on those without irradiation. Jing-hao Zhao et al. [24] measured the enhancement effect on the degradation of gate voltage, G_m, and I_{Dsat} during hot carrier stress in both T-gate and H-gate SOI p-MOSFETS irradiated by γ-rays. It was found that TID-induced interface states strengthen the process of hot electron injection into the gate oxide, while the radiation-induced weakening of the Si/SiO$_2$ interface aggravates the generation rate of the interface defects. Previous studies have shown that HCI degradation is particularly important in n-MOSFETs, because there are higher electric fields and impact ionization near the drain region as compared to p-MOSFETs [25–27].

However, due to the existence of the BOX layer, the mechanism of irradiation and HCI effects on SOI devices is more complex than on bulk silicon devices. We thought that there may be an interaction between reliability degradation and irradiation damage. So, we carried out HCI tests on γ-ray-irradiated PD SOI n-MOSFETs to verify this idea. During the experiments, a strange phenomenon was observed. The experimental results show that TID leads to a high off-state leakage (I_{off}) current and the obvious negative drift of the threshold voltage. Focusing on the irradiation devices, the parameters—especially the I_{off}—cannot return to their initial values after 190 h of annealing at room temperature (RT). But the results of the HCI experiments show that the I_{off} almost returned to its initial value during the HCI experiment within 5 s. The physical mechanism of this phenomenon is the core content of this paper.

The structure of this paper is organized as follows: Section 1 reviews the background of research on MOSFETs with bulk and SOI processes when they are subjected to radiation and hot carrier stress. Section 2 introduces the device and presents the experimental details. Section 3 discusses the experimental results and analyzes their physical mechanisms. Here, we explain the mechanism of this recovery effect qualitatively. Furthermore, we provide some methods to reduce the value shift of the device characteristics when the devices are used in harsh environments. Finally, Section 4 concludes the whole study.

2. Device and Experimental Details

The I/O n-MOSFETs used in this paper were fabricated using 130 nm PD SOI technology [28] in the Center of Materials Science and Optoelectronics Engineering, University of Chinese Academy of Sciences (in Shanghai, China). Processing was performed on a 200 mm diameter UNIBOND® wafer from SOITEC (in Bernin, France) with a 100 nm top Si film and a 145 nm BOX. The body contacts of all transistors were introduced by a T-shaped gate layout, as shown in Figure 1. The gate oxide thickness is 1.8 nm. In this study, two kinds of n-MOSFETs with different channel widths were used in our experiments. The structure parameter of the narrow channel devices was W/L = 0.15 μm/0.35 μm, and the other structure parameter of the wide channel devices was W/L = 10 μm/0.35 μm. Their working voltage was $V_{DS} = V_{GS}$ = 3.3 V, and their doping concentration in the body was about ~10^{17} cm^{-3}. All the devices were 24-pin DIP-packaged. Three devices were used in our experiments at the same time.

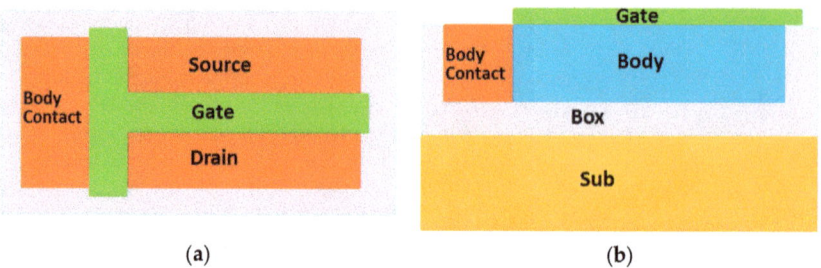

Figure 1. Top and front cross-section diagram of T-gate PD SOI n-MOSFET (not to scale). (**a**) Top diagram. (**b**) Front cross-section diagram.

The γ-ray radiation experiments were carried out at the Xinjiang Technical Institute of Physics and Chemistry, Chinese Academy of Sciences, using ^{60}Co-γ as the radiation source. Before the radiation experiments, we tested the initial parameters of these n-MOSFETS. During the γ-ray radiation process, all devices worked on a bias voltage with V_{GS} = 3.3 V. Other pins were grounded at the same time. The samples were irradiated to 2000 Gy (Si) with a dose rate of 0.5 Gy (Si)/s. After irradiation, the devices were annealed for 190 h at room temperature (25 °C), and they kept the same working conditions as the irradiation process. Then, the devices were sent to hot carrier stress experiments. All the electrical tests were performed using a Keithley 4200 B semiconductor test system at room temperature.

According to Joint Electron Device Engineering Council JESD28-A titled "A Procedure for Measuring N-Channel MOSFET Hot Carrier-Induced Degradation under DC Stress", the bias gate voltage was selected as corresponding to the voltage of the peak substrate current during hot carrier stress experiments. Here, the gate voltage was set to 3.3 V. Additionally, the drain voltage was set to 4.45 V (135% operating voltage) to generate the maximum number of carriers due to impact ionization. Other pins were grounded. Two points including 5 s and 5000 s during hot carrier stress were selected to interrupt electrical stress for the main parameter test.

3. Results and Discussion

The front gate and back gate linear area transfer characteristics of narrow channel devices (W/L = 0.15 μm/0.35 μm) before and after irradiation and 190 h *RT* annealing are shown in Figure 2a,b. We were able to determine that the magnitude of off-state leakage current I_{off} for irradiated devices is about 5~6 orders larger than that of the non-irradiated ones.

Based on the results shown in Figure 2, it is believed that the radiation-induced oxide trap charge in the shallow trench isolation (STI) caused the I_{off}. As reported in Refs. [1,7], parts of the charges in the inversion top Si film are no longer controlled by the main transistor gate, resulting in a negative threshold shift in the main transistor, which can increase the channel current significantly.

Figure 3 shows the electrical equivalent structure activated by irradiation for the PD SOI MOSFET. The primary parasitic element that contributes to the primary MOS transistor is the parasitic bipolar transistor. The floating body node serves as the base of this parasitic bipolar transistor and can be activated by irradiation that forward biases the body–source diode. To prevent its activation, the body region can be connected to the source potential or be grounded. By doing so, the charge generated by radiation in the body is discharged through the "body tie". As a result, the body potential is no longer in a floating state [29,30].

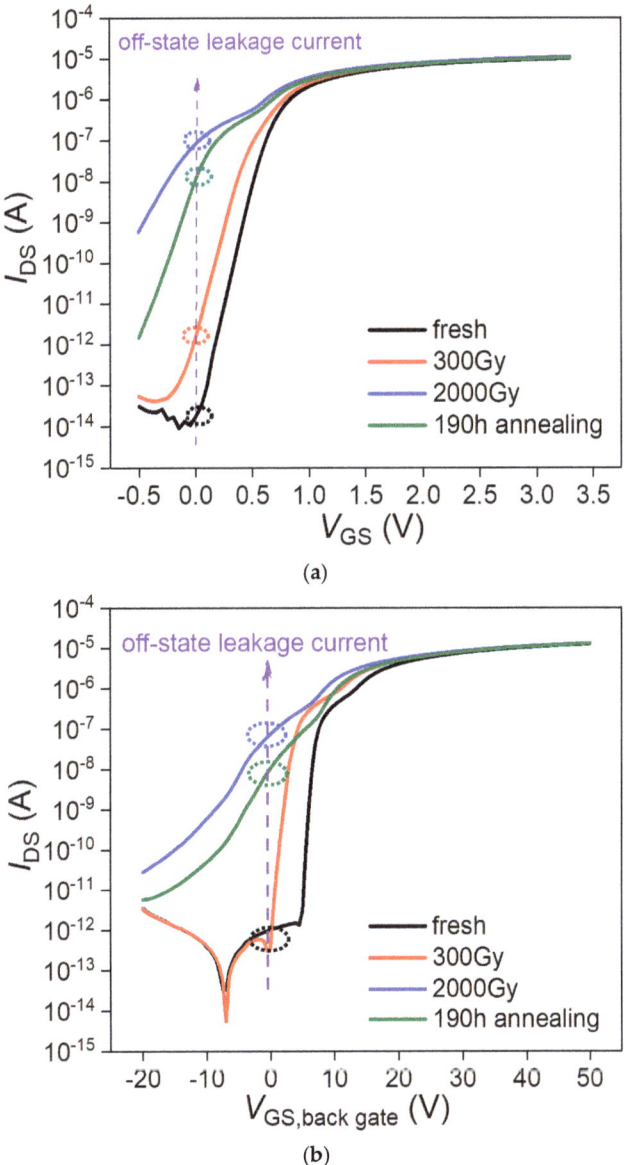

Figure 2. Front gate and back gate linear area transfer characteristics of device (W/L = 10 μm/ 0.35 μm) before and after irradiation and 190 h RT annealing. (**a**) Front gate transistor. (**b**) Back gate transistor.

A previous work [29] demonstrated that nearly all of the radiation-generated holes that manage to avoid immediate recombination become ensnared within the bulk of the oxide, specifically at deep trap sites near their source. Once trapped, a portion of these holes gradually reverts to a neutral state through the thermal emission of electrons from the oxide valence band at room temperature. Besides hole entrapment, electrons are also captured throughout the entirety of the buried oxide. Most of these trapped electrons are thermally released within one second following a radiation pulse. Subsequent to electron release, the resulting charge is predominantly characterized by a high concentration of positively trapped holes, resulting in significant negative shifts in the threshold voltage of the back gate transistors.

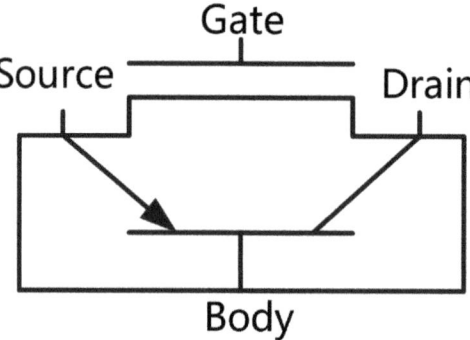

Figure 3. Equivalent electrical structure of PD SOI MOSFET activated by irradiation.

In order to evaluate the impact of the annealing effect, a 190 h *RT* annealing process is applied to irradiated devices under the same bias voltage as irradiation. As shown in Figure 2, the value of I_{off} decreases by about one order after a process of 190 h *RT* annealing. However, the gap is still considerable compared with the initial value. Here, the thermal emission mechanism of annealing at *RT* is explained as follows [7].

$$\Phi_m(t) = \frac{kT}{q} \cdot \ln[AT^2 t] \qquad (1)$$

where $\Phi_m(t)$ presents the energy boundaries of thermal emission, *K* is the Boltzmann constant, *T* is the absolute temperature, *q* is the elementary charge, *t* is the time factor, and *A* presents the constant of the capture cross-section. Since our experiments were performed at room temperature, most thermal emission electrons have energies lower than those required to escape deep energy traps. Consequently, these deep energy traps continue to capture holes, leading to the observed outcome illustrated in Figure 2. In other words, the crucial parameters for irradiated n-MOSFETs cannot be restored to their initial values through annealing at room temperature.

When these devices are utilized in the harsh conditions of deep space, they are simultaneously exposed to both radiation and hot carrier stress. It is crucial to investigate the synergistic effects arising from the combination of radiation and the HCI effect. To explore this proposal, we carried out a series of HCI tests on SOI n-MOSFETs post-irradiation. Based on the experimental results, we also identified the phenomenon of gate-induced drain leakage (GIDL), as shown in Figure 4a. It can be seen that GIDL current experiences significant increases after the 5000 s HCI test. The observed increase in post-stress GIDL current is distinctly different from that induced by oxide traps because oxide traps only induce GIDL current transients over a time scale of seconds. So, this should be dependent on the amount of interface traps created during stress [25,31]. In other words, an additional conduction mechanism involving interface traps should be possible after the hot electron stress. Surprisingly, an interesting phenomenon emerged after a 5 s HCI experiment. In the back gate transistor, the curve of I_{off} nearly returns to its initial value, as shown in Figure 4b.

Focusing on the experiment phenomena, we believe they are a result of the synergistic effect of radiation and the HCI effect. This enhanced synergistic effect has been proved by Hang. Zhou et al. [27] carried out a compared example using irradiated and unirradiated 0.13 μm PD SOI n-MOSFETs and tested front gate I_{ds}–V_{gs} curves before and after 3000 s of hot carrier stress. According to the experimental results, it is obvious that irradiated samples display a larger threshold voltage shift during stress time than un-irradiated samples. The threshold shift of an irradiated device is 108 mV after 3000 s stress, while the threshold shift of an unirradiated device is 46 mV.

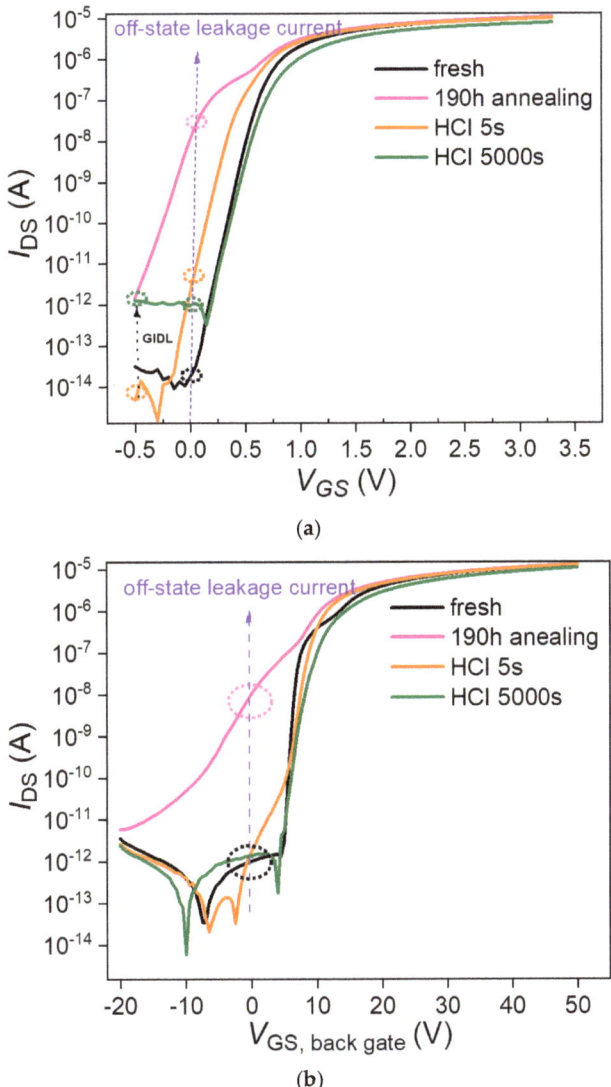

Figure 4. The linear area transfer characteristics of front gate transistor (**a**) and back gate transistor (**b**) after different annealing times.

In this paper, these two effects cause these processes' complexity. The I_{off} caused by radiation induces an oxide trap charge, and the subsequent effects of annealing as a result of the channel-hot electronics leap over the barrier of the Si/SiO$_2$ interface. If electrons collide during travel, they may be incident in the BOX or gate oxide layer. The diagram sketch can be seen in Figure 5a,b.

Moreover, it is important to note that the channel electrons acquire a higher average energy during HCI stress. This increased energy facilitates the annealing of deep-level oxide trap charges, which is shown in Figure 6. The oxide trap charges within the STI region undergo rapid annealing as electrons are injected into them. This electron injection occurs when a high voltage is applied during the HCI test, generating enough hot carriers within the channel. These hot carriers will cross the Si/SiO$_2$ barrier and inject the silicon oxide layer, resulting in the annealing of deep-level oxide trap charges. The corresponding band diagram is depicted in Figure 6, where X_m represents the tunneling front, t presents the time factor, α relates to the attempt frequency for escaping traps, and β is associated with the tunneling barrier [7].

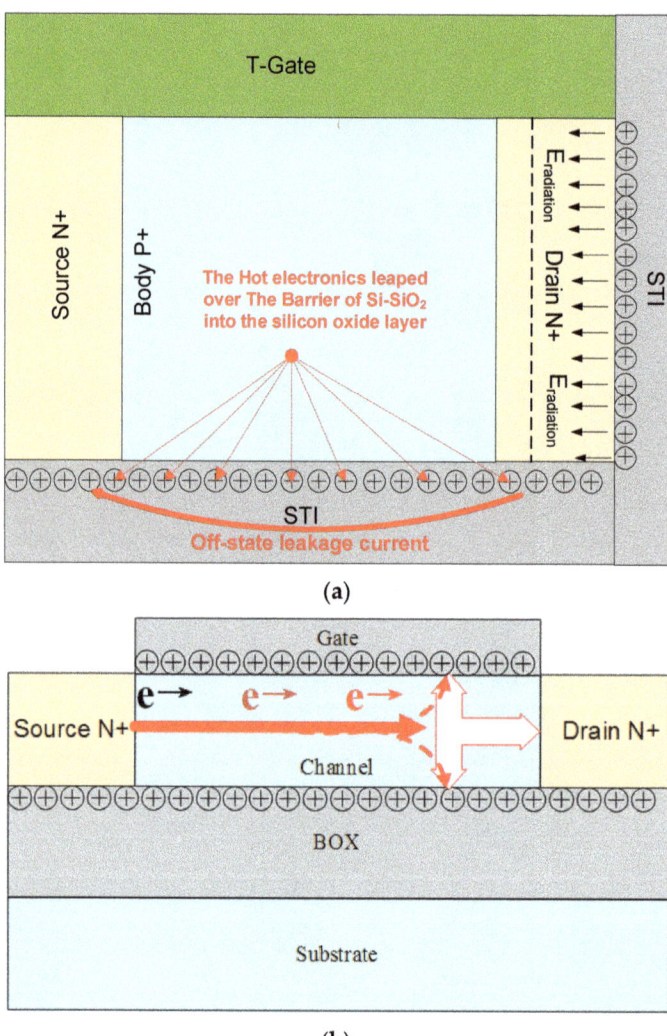

Figure 5. The schematic diagram of the synergistic effect of radiation and HCI effect. (a) Top view. (b) Side view.

Furthermore, it is essential to note that utilizing a simple tunneling front model for cases involving traps distributed in energy due to HCI may not be entirely accurate, as the tunneling barrier β varies with trap depth. However, for the sake of practicality and the development of a simplified predictive methodology, this effect is considered negligible.

Here, we demonstrate that electrons obtain energy W if electron scattering does not occur during the drift from the source to the drain. Additionally, W can be calculated as follows:

$$W = qU \tag{2}$$

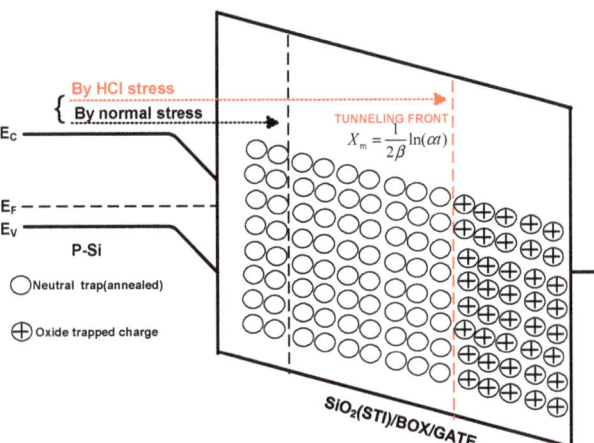

Figure 6. Band diagram of irradiated n-MOSFETs showing tunneling front penetrating into the oxide with HCI stress and normal stress.

When the V_{DS} is set to 4.45 V, the W is equal to 4.45 eV. Because the Si/SiO$_2$ interfacial potential barrier for electrons is 3.15 eV [32,33], high energy-electrons can pass through Si/SiO$_2$ easily. However, it is possible that when the devices work under the normal voltage (3.3 V), the electrons in the channel can still obtain enough energy (3.3 eV) to cross the Si/SiO$_2$ interfacial potential barrier. In this case, we have to calculate the probability of electrons that could pass through Si/SiO$_2$ quantitatively. Here, the probability of electrons in the channel drift distance d without scattering is $e^{-d/\lambda}$, where λ presents the mean free path of electrons. In silicon, λ = 10.5 nm. The channel length of n-MOSFETs in the test is 0.35 μm. We define P_1 as the probability of electrons in the channel drift distance d_1, which obtained 3.15 eV energy, and we define P_2 as the probability of electrons in the channel drift distance d_2 without scattering.

Assuming the electric field in the channel is uniformly distributed, the normal working voltage of the device is set to 3.3 V, and the applied voltage for hot carrier stress experiments is 4.45 V. When the device operates at 4.45 V, the shortest path that electrons need to pass through to obtain energy of 3.15 eV is d_1 = (3.15/4.45) × 0.35 μm = 0.24 μm. When the device operates at working voltage, the shortest path that electrons need to pass through to obtain energy of 3.15 eV is d_2 = (3.15/3.3) × 0.35 μm = 0.334 μm. It is worth noting that in the context mentioned above, the "shortest path" implies that electrons undergo no scattering or collisions along their trajectory.

Based on these calculations, the ratio of P_1/P_2 can be obtained as follows:

$$P_1/P_2 = e^{(-d_1/\lambda + d_2/\lambda)} \approx 8100 \quad (3)$$

Based on this result, the number of hot carrier injections into the STI of devices under HCI stress is 8100 times higher than it is under normal stress. In this case, HCI has a faster recovery capability.

The HCI effect on n-MOSFETs parameters' degradation is a long-term process for the device's reliability. Typically, it causes decreased circuit speed rather than catastrophic failure. In this paper, sensitive parameters such as V_T, G_m, and I_{Dsat} are commonly monitored to identify performance changes. The devices we used in these experiments are ultra-thin gates, so the influence of the oxide trap charge is very small. As shown in Figures 7 and 8, the curve of G_m-V_{GS} and G_m (Max) in different states can be observed.

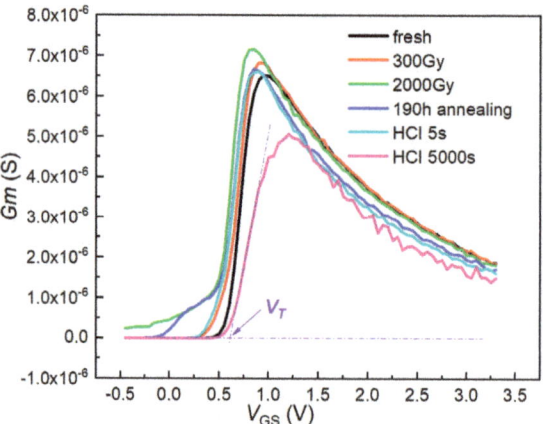

Figure 7. The degradation of transconductance (G_m) in different states.

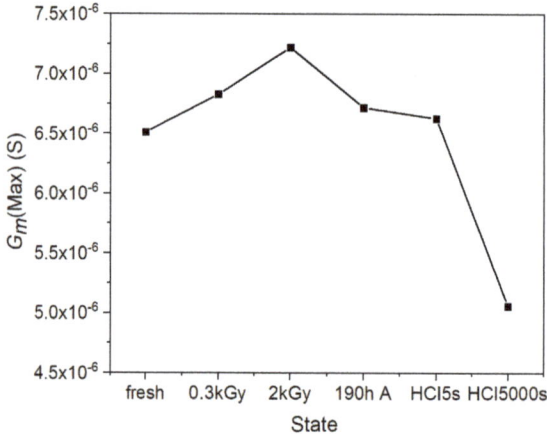

Figure 8. Change trend of the maximum value of transconductance (G_m) in different states.

In order to extract V_T, the G_m extrapolation method in the linear region was used in this study [34,35]. This method suggests that the threshold voltage corresponds to the gate voltage axis intercept of the linear extrapolation of the G_m–V_{GS} characteristics at its maximum first derivative (slope) point. As shown in Figure 9, the extract threshold voltage shift of PD SOI n-MOSFET is −100 mV after 2000 Gy irradiation and is restored to −70 mV after 190 h of annealing. When the devices were tested in the HCI test, the threshold voltage shift was −50 mV after 5 s HCI and 50 mV after 5000 s HCI. The change trend of ΔV_T can be observed in the inline image of Figure 9.

The expression of the threshold voltage shift is shown as follows [7]:

$$\Delta V_T = \Delta V_{Tot} + \Delta V_{Tit} \quad (4)$$

where ΔV_{Tot} is caused by oxide traps and ΔV_{Tit} is caused by the interface state. For n-MOSFETs, ΔV_{Tit} is negative when the interface state level is below the Fermi level under a positive gate voltage. Based on the above formulation, we believe that the degradation of V_T is a result of the oxide trap charge and interface states. Under HCI stress, channel electrons are accelerated to very high energy and lead to an injection into the oxide gate. Some chemical bonds at the Si/SiO$_2$ interface are broken for the hot carrier transfer energy to the lattice through phonon emission. Some carriers are trapped in the SiO$_2$ layer. The trapping or bond breaking creates oxide charge and interface traps that affect the channel carrier mobility and reliability of the gate oxide. The impact on I_{Dsat} is shown in Figure 10.

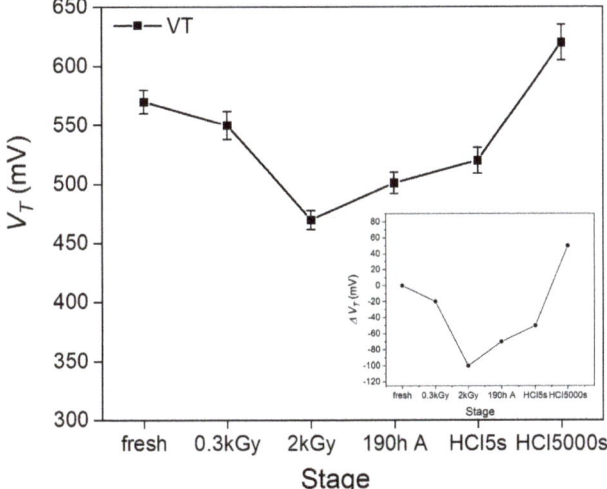

Figure 9. Change trends of V_T and ΔV_T along with different states. Here, the value of V_T (front gate) is extrapolated by calculating the maximum slope of the G_m–V_{GS} curve.

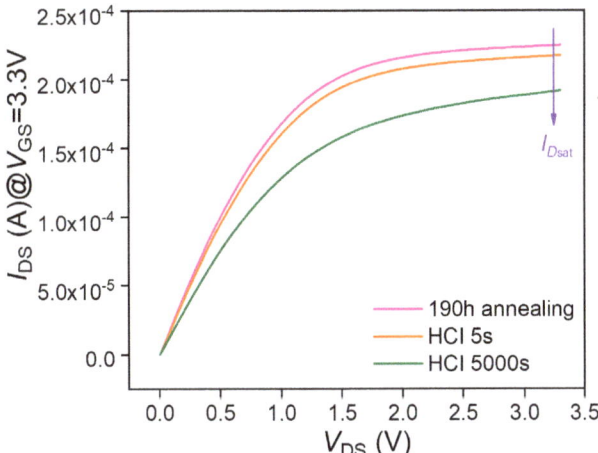

Figure 10. The degradation of I_{Dsat} through the HCI process.

The theme of this paper revolves around the examination of the synergistic effects of narrow-channel transistors. As depicted in Figure 9, the irradiation has a pronounced impact on the threshold voltage of these narrow-channel devices, causing a substantial negative shift. This shift is particularly notable due to the thinness of the top silicon film, which is less than twice the width of the maximum depleted region. This thin film is a consequence of the low doping concentrations in the substrate region.

Consequently, depletion regions emerge independently at both the front and back interfaces. These regions have the potential to interconnect when a sufficient amount of charge becomes trapped in the BOX layer. Similar to fully depleted SOI devices, the radiation-induced charges trapped in the BOX layer alter the electric potential within the substrate region of PD SOI devices. This alteration results in a negative threshold shift at the front gate. Additionally, it is important to note that narrow devices exhibit heightened sensitivity to the charge trapped within the silicon dioxide layer in the shallow trench isolation (STI) along the channel.

To prove this proposal, wide-channel 0.13 μm PD SOI n-MOSFETs (W/L = 10 μm/ 0.35 μm) with the same process are used for γ-ray radiation and HCI experiments. The front gate and back gate I_{DS}–V_{GS} curves of wide channel devices under 2000 Gy irradiation and 3000 s of HCI stress are shown in Figure 11. We found that the I_{off} of the front gate and back gate transistors changed by about three orders after 2000 Gy irradiation, and it almost recovered to its initial value after HCI experiments.

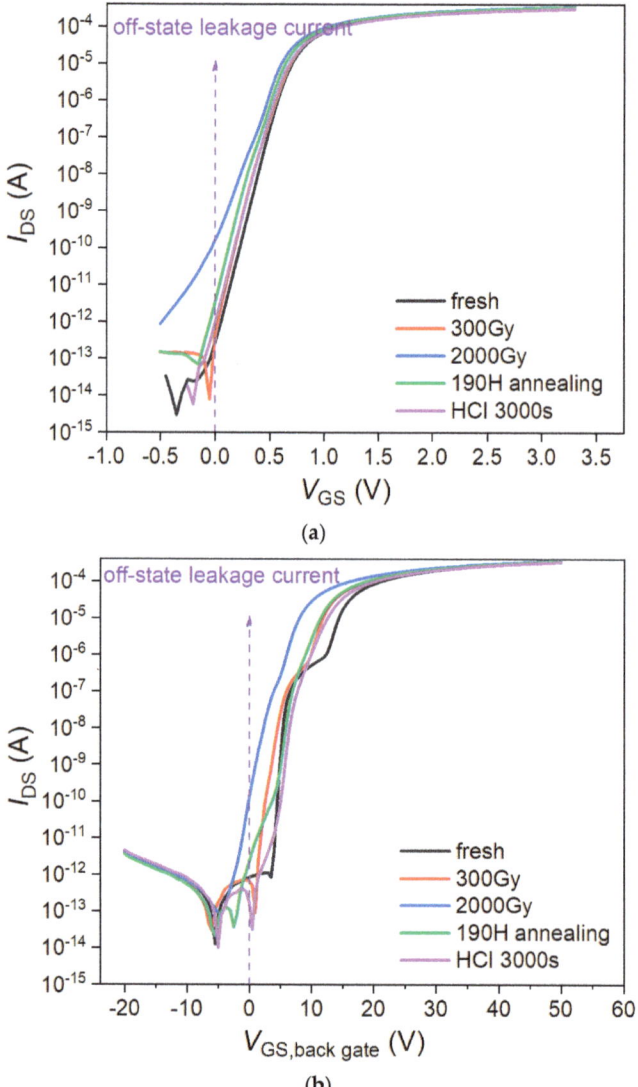

Figure 11. Front gate and back gate linear area transfer characteristics of device (W/L = 10 μm/ 0.35 μm) before and after irradiation and annealing. (**a**) Front gate transistor. (**b**) Back gate transistor.

According to the results shown in Figure 12, the threshold voltage V_T is negatively shifted by radiation-induced positive charges trapped in BOX, while the gate threshold is positively shifted by channel hot carrier stress. The change trend of ΔV_T can be shown in the internal image of Figure 12. Compared to the narrow-channel devices, the V_T of wide-channel devices is insensitive to irradiation and HCI effects.

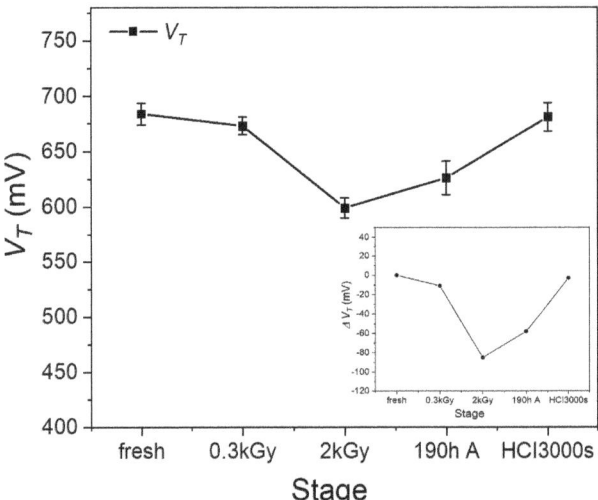

Figure 12. Change trends of V_T and ΔV_T along with different states (W/L = 10 μm/0.35 μm).

To minimize the shift in device characteristics, we think the following methods could be used for combined conditions of radiation and HCI. Faced with the issue of radiation hardness, an H-gate structure design is an effective technology to replace the T-gate design. In the H-gate device, its side wall oxide is completely eliminated, and the radiation resistance performance of the device is greatly improved. However, this design method requires more layout area. For the unique BOX layer of SOI devices, radiation hardness methods such as the Si injection process can be used to suppress or compensate for the effects of radiation-induced positive charges.

For 0.13 μm process technology, a lightly doped drain (LDD) structure design can be used to suppress the HCI effect. That could avoid the design concept of extremely short-channel devices. In addition, some special factories will perform special passivation processes on the Si/SiO$_2$ interface, such as replacing H+ with F+, because the Si–F bond has much stronger bond energy than Si–H. In this case, under the same thermal electron collision, F+-passivated devices will generate fewer dangling bonds, effectively controlling the generation of N_{it}.

4. Conclusions

This paper discusses radiation reliability screens for 0.13 μm PD SOI n-MOSFETs used in applications with HCI environments. The results show that the HCI effect has a recovery effect on the long-term reliability of the n-MOSFETs when applied to a space environment. In our opinion, the physical mechanism of this effect is that the high energy electrons produced by HCI lead to deep-level oxide trap charge annealing, which leads to the almost complete elimination of I_{off} within a few seconds. At the same time, the high-energy electrons injected into the SiO$_2$ layer led to many interface states being produced, which leads to the degradation of the G_m, V_T, and I_{Dsat}. There is a combined effect between HCI and TID. Further, we designed a comparable experiment to evaluate the effect on wide-channel devices. These results show that narrow-channel devices are more sensitive to irradiation and HCI. Based on the results presented in this work, it is useful to place SOI MOSFETs in a biased working state when they are used in space electronic systems, as this will extend their lifespan.

Author Contributions: Conceptualization and writing—original draft preparation, L.L.; methodology, Z.C., formal analysis, C.J. All authors have read and agreed to the published version of the manuscript.

Funding: This research was funded by the Southwest Minzu University Research Startup Fund, grant number RQD2021088, and the Fundamental Research Fund for the Central Universities—Southwest Minzu University, grant number 2021NQNCZ10.

Data Availability Statement: Not applicable.

Acknowledgments: The authors thank Hang Zhou of the Microsystem and Terahertz Research Center, CAEP, for his kind help in data analysis and discussion.

Conflicts of Interest: The authors declare no conflict of interest.

Nomenclature

PD SOI	Partially Depleted Silicon-On-Insulator
MOSFET	Metal–Oxide–Semiconductor Field Effect Transistor
HCI	Hot Carrier Injection
MOS	Metal Oxide Semiconductor
CMOS	Complementary Metal–Oxide–Semiconductor
TID	Total Ionizing Dose
BOX	Buried Oxide
VLSI	Very Large-Scale Integration
RT	Room Temperature
GIDL	Gate-Induced Drain Leakage
STI	Shallow Trench Isolation

References

1. Oldham, T.R.; McLean, F.B. Total ionizing dose effects in MOS oxides and devices. *IEEE Trans. Nucl. Sci.* **2003**, *50*, 483–499. [CrossRef]
2. Brewer, R.M.; Zhang, E.X.; Gorchichko, M.; Wang, P.F.; Cox, J.; Moran, S.L.; Ball, D.R.; Sierawski, B.D.; Fleetwood, D.M.; Schrimpf, R.D.; et al. Total ionizing dose responses of 22-nm FDSOI and 14-nm bulk FinFET charge-trap transistors. *IEEE Trans. Nucl. Sci.* **2021**, *68*, 677–686. [CrossRef]
3. Jiang, R.; Zhang, E.X.; McCurdy, M.W.; Wang, P.; Gong, H.; Yan, D.; Schrimpf, R.D.; Fleetwood, D.M. Dose-rate dependence of the total-ionizing-dose response of GaN-based HEMTs. *IEEE Trans. Nucl. Sci.* **2018**, *66*, 170–176. [CrossRef]
4. Wang, P.F.; Zhang, E.X.; Chuang, K.H.; Liao, W.; Gong, H.; Arutt, C.N.; Ni, K.; Mccurdy, M.W.; Verbauwhede, I.; Bury, E.; et al. X-ray and proton radiation effects on 40 nm CMOS physically unclonable function devices. *IEEE Trans. Nucl. Sci.* **2018**, *65*, 1519–1524. [CrossRef]
5. Wang, P.F.; Zhang, E.X.; Fleetwood, D.M.; McCurdy, M.W.; Lin, J.-T.; Alles, M.L.; Davidson, J.L.; Alphenaar, B.W.; Schrimpf, R.D. Effects of Charge Generation and Trapping on the X-ray Response of Strained AlGaN/GaN HEMTs. In Proceedings of the 2021 IEEE 14th International Conference on ASIC (ASICON), Kunming, China, 26–29 October 2021; IEEE: Piscataway, NJ, USA, 2021; pp. 1–4. [CrossRef]
6. Musseau, O. Single-event effects in SOI technologies and devices. *IEEE Trans. Nucl. Sci.* **1996**, *43*, 603–613. [CrossRef]
7. McWhorter, P.; Miller, S.; Miller, W. Modeling the anneal of radiation-induced trapped holes in a varying thermal environment. *IEEE Trans. Nucl. Sci.* **1990**, *37*, 1682–1689. [CrossRef]
8. Liu, B.; Li, Y.; Wen, L.; Zhang, X.; Guo, Q. Effects of Hot Pixels on Pixel Performance on Backside Illuminated Complementary Metal Oxide Semiconductor (CMOS) Image Sensors. *Sensors* **2023**, *23*, 6159. [CrossRef]
9. Li, X.; Cui, J.; Zheng, Q.; Li, P.; Cui, X.; Li, Y.; Guo, Q. Study of the Within-Batch TID Response Variability on Silicon-Based VDMOS Devices. *Electronics* **2023**, *12*, 1403. [CrossRef]
10. Zhang, Y.W.; Huang, H.X.; Bi, D.W.; Tang, M.; Zhang, Z. Investigation of unique total ionizing dose effects in 0.2 μm partially-depleted silicon-on-insulator technology. *Nucl. Instrum. Methods Phys. Res. Sect. A Accel. Spectrometers Detect. Assoc. Equip.* **2014**, *745*, 128–132. [CrossRef]
11. Fan, S.; Hu, Z.-Y.; Zhang, Z.-X.; Ning, B.-X.; Bi, D.-W.; Dai, L.-H.; Zhang, M.-Y.; Zhang, L.-Q. Total ionizing dose induced single transistor latchup in 130-nm PDSOI input/output NMOSFETs. *Chin. Phys. B* **2017**, *26*, 036103. [CrossRef]
12. Huang, H.-X.; Bi, D.-W.; Peng, C.; Zhang, Y.-W.; Zhang, Z.-X. The Enhanced Role of Shallow-Trench Isolation in Ionizing Radiation Damage of Narrow Width Devices in 0.2 μm Partially-Depleted Silicon-on-Insulator Technology. *Chin. Phys. Lett.* **2013**, *30*, 080701. [CrossRef]

13. Zheng, Q.-W.; Cui, J.-W.; Zhou, H.; Yu, D.-Z.; Yu, X.-F.; Lu, W.; Guo, Q.; Ren, D.-Y. Analysis of functional failure mode of commercial deep sub-micron SRAM induced by total dose irradiation. *Chin. Phys. B* **2015**, *24*, 106106. [CrossRef]
14. Yeh, W.-K.; Wang, W.-H.; Fang, Y.-K.; Chen, M.-C.; Yang, F.-L. Hot-carrier-induced degradation for partially depleted SOI 0.25–0.1/spl mu/m CMOSFET with 2-nm thin gate oxide. *IEEE Trans. Electron. Devices* **2002**, *49*, 2157–2162. [CrossRef]
15. Cui, J.; Xue, Y.; Yu, X.; Ren, D.; Lu, J.; Zhang, X. Total dose irradiation and hot-carrier effects of sub-micro NMOSFETs. *J. Semicond.* **2012**, *33*, 014006. [CrossRef]
16. Rafí, J.M.; Simoen, E.; Mercha, A.; Hayama, K.; Campabadal, F.; Ohyama, H.; Claeys, C. Electrical stress on irradiated thin gate oxide partially depleted SOI nMOSFETs. *Microelectron. Eng.* **2007**, *84*, 2081–2084. [CrossRef]
17. Huang, D.H.; King, E.E.; Wang, J.J.; Ormond, R.; Palkuti, L. Correlation between channel hot-electron degradation and radiation-induced interface trapping in N-channel LDD devices. *IEEE Trans. Nucl. Sci.* **1991**, *38*, 1336–1341. [CrossRef]
18. Dai, L.; Liu, X.; Zhang, M.; Zhang, L.; Hu, Z.; Bi, D.; Zhang, Z.; Zou, S. Degradation induced by TID radiation and hot-carrier stress in 130-nm short channel PDSOI NMOSFETs. *Microelectron. Reliab.* **2017**, *74*, 74–80. [CrossRef]
19. McBrayer, J.D.; Fleetwood, D.M.; Pastorek, R.A.; Jones, R.V. Correlation of hot-carrier and radiation effects in MOS transistors. *IEEE Trans. Nucl. Sci.* **1985**, *32*, 3935–3939. [CrossRef]
20. Palkuti, L.J.; Ormond, R.D.; Hu, C.; Chung, J. Correlation between channel hot-electron degradation and radiation-induced interface trapping in MOS devices. *IEEE Trans. Nucl. Sci.* **1989**, *36*, 2140–2146. [CrossRef]
21. Silvestri, M.; Gerardin, S.; Paccagnella, A.; Faccio, F.; Gonella, L. Channel hot carrier stress on irradiated 130-nm NMOSFETs. *IEEE Trans. Nucl. Sci.* **2008**, *55*, 1960–1967. [CrossRef]
22. Silvestri, M.; Gerardin, S.; Paccagnella, A.; Faccio, F. Degradation induced by X-ray irradiation and channel hot carrier stresses in 130-nm NMOSFETs with enclosed layout. *IEEE Trans. Nucl. Sci.* **2008**, *55*, 3216–3223. [CrossRef]
23. Zheng, Q.-W.; Cui, J.-W.; Zhou, H.; Yu, D.-Z.; Yu, X.-F.; Guo, Q. Hot-carrier effects on total dose irradiated 65 nm n-type metal-oxide-semiconductor field-effect transistors. *Chin. Phys. Lett.* **2016**, *33*, 076102. [CrossRef]
24. Zhao, J.; Zheng, Q.; Cui, J.; Zhou, H.; Liang, X.; Yu, X.; Guo, Q. A study on effects of total ionizing dose on hot carrier effect of PD I/O SOI PMOSFETs. *Results Phys.* **2019**, *13*, 102223. [CrossRef]
25. Lai, P.T.; Xu, J.P.; Wong, W.M.; Lo, H.; Cheng, Y. Correlation between hot-carrier-induced interface states and GIDL current increase in n-MOSFET's. *IEEE Trans. Electron. Devices* **1998**, *45*, 521–528. [CrossRef]
26. Groeseneken, G.; Degraeve, R.; Nigam, T.; Bosch, G.V.D.; Maes, H. Hot carrier degradation and time-dependent dielectric breakdown in oxides. *Microelectron. Eng.* **1999**, *49*, 27–40. [CrossRef]
27. Zhou, H.; Liu, Y.; Zhang, Y. Total-ionizing-dose induced enhanced hot-carrier injection effect in the 130 nm partially depleted SOI I/O nMOSFETs. *Jpn. J. Appl. Phys.* **2020**, *59*, 031001. [CrossRef]
28. Dai, M.; Song, Z.; Lin, C.-H.; Dong, Y.; Wu, T.; Chu, J. Multi-functional multi-gate one-transistor process-in-memory electronics with foundry processing and footprint reduction. *Commun. Mater.* **2022**, *3*, 41. [CrossRef]
29. Schwank, J.R.; Ferlet-Cavrois, V.; Shaneyfelt, M.R.; Paillet, P.; Dodd, P. Radiation effects in SOI technologies. *IEEE Trans. Nucl. Sci.* **2003**, *50*, 522–538. [CrossRef]
30. Paillet, P.; Herve, D.; Leray, J.L.; Devine, R. Evidence of negative charge trapping in high temperature annealed thermal oxide. *IEEE Trans. Nucl. Sci.* **1994**, *41*, 473–478. [CrossRef]
31. Wang, T.; Chang, T.-E.; Huang, C. Interface trap induced thermionic and field emission current in off-state MOSFET's. In Proceedings of the 1994 IEEE International Electron Devices Meeting, San Francisco, CA, USA, 11–14 December 1994; IEEE: Piscataway, NJ, USA, 1994; pp. 161–164. [CrossRef]
32. Majkusiak, B. Gate tunnel current in an MOS transistors. *IEEE Trans. Electron. Devices* **1990**, *37*, 1087–1092. [CrossRef]
33. Tam, S.; Ko, P.-K.; Hu, C. Lucky-electron model of channel hot-electron injection in MOSFET's. *IEEE Trans. Electron. Devices* **1984**, *31*, 1116–1125. [CrossRef]
34. Tsuno, M.; Suga, M.; Tanaka, M.; Shibahara, K.; Miura-Mattausch, M.; Hirose, M. Physically-based threshold voltage determination for MOSFET's of all gate lengths. *IEEE Trans. Electron. Devices* **1999**, *46*, 1429–1434. [CrossRef]
35. Ortiz-Conde, A.; Sánchez, F.J.G.; Liou, J.J.; Cerdeira, A.; Estrada, M.; Yue, Y. A review of recent MOSFET threshold voltage extraction methods. *Microelectron. Reliab.* **2002**, *42*, 583–596. [CrossRef]

Disclaimer/Publisher's Note: The statements, opinions and data contained in all publications are solely those of the individual author(s) and contributor(s) and not of MDPI and/or the editor(s). MDPI and/or the editor(s) disclaim responsibility for any injury to people or property resulting from any ideas, methods, instructions or products referred to in the content.

Article

Analysis of Difference in Areal Density Aluminum Equivalent Method in Ionizing Total Dose Shielding Analysis of Semiconductor Devices

Mingyu Liu [1,2,3], Chengfa He [1,2,*], Jie Feng [1,2,*], Mingzhu Xun [1,2], Jing Sun [1,2], Yudong Li [1,2] and Qi Guo [1,2]

1 Xinjiang Technical Institute of Physics and Chemistry, Chinese Academy of Sciences, Urumqi 830011, China; liumingyu21@mails.ucas.ac.cn (M.L.)
2 Xinjiang Key Laboratory of Electronic Information Material and Device, Urumqi 830011, China
3 University of Chinese Academy of Sciences, Beijing 100049, China
* Correspondence: hecf@ms.xjb.ac.cn (C.H.); fengjie@ms.xjb.ac.cn (J.F.)

Abstract: The space radiation environment has a radiation effect on electronic devices, especially the total ionizing dose effect, which seriously affects the service life of spacecraft on-orbit electronic devices and electronic equipment. Therefore, it is particularly important to enhance the radiation resistance of electronic devices. At present, many scientific research institutions still use the areal density equivalent aluminum method to calculate the shielding dose. This paper sets five common metal materials in aerospace through the GEANT4 Monte-Carlo simulation tool MULASSIS, individually calculating the absorption dose caused by single-energy electrons and protons in the silicon detector after shielding of five different materials, which have the same areal density of 0.8097 g/cm^2. By comparing the above data, it was found that depending on the particle energy, the areal density aluminum equivalent method would overestimate or underestimate the absorbed dose in the shielded silicon detector, especially for the ionization total dose shielding effect of low-energy electrons. The areal density aluminum equivalent method will greatly overestimate the shielding dose, so this difference needs to be taken into account when evaluating the ionizing dose of the electronics on a spacecraft to make the assessment more accurate.

Keywords: Monte-Carlo method; total ionizing dose; radiation shielding; space radiation

Citation: Liu, M.; He, C.; Feng, J.; Xun, M.; Sun, J.; Li, Y.; Guo, Q. Analysis of Difference in Areal Density Aluminum Equivalent Method in Ionizing Total Dose Shielding Analysis of Semiconductor Devices. *Electronics* **2023**, *12*, 4181. https://doi.org/10.3390/electronics12194181

Academic Editor: Francesco Giuseppe Della Corte

Received: 14 August 2023
Revised: 7 October 2023
Accepted: 8 October 2023
Published: 9 October 2023

Copyright: © 2023 by the authors. Licensee MDPI, Basel, Switzerland. This article is an open access article distributed under the terms and conditions of the Creative Commons Attribution (CC BY) license (https://creativecommons.org/licenses/by/4.0/).

1. Introduction

When a spacecraft is in orbit, it inevitably experiences the influence of the space environment, including high-energy electrons, protons, and other heavy ions from the Earth's radiation belts, solar cosmic rays, and galactic cosmic rays. These high-energy particles can significantly impact the performance of semiconductor devices or circuit systems. When these high-energy particles interact with sensitive regions of the devices, they can cause a Total Ionizing Dose effect (TID), Displacement Damage Dose effect (DDD), and Single Event Effect (SEE) [1]. As a result, semiconductor devices may degrade or fail, leading to the potential paralysis of the entire electronic system. Space radiation effects are one of the critical factors contributing to the failure of spacecraft electronic devices and circuit systems, which severely affects the spacecraft's operational life in orbit. This effect is particularly noteworthy, as semiconductor device feature sizes are becoming increasingly smaller, reaching the nanometer scale, and demanding particular attention to the impact of space radiation on semiconductor devices.

Adding shielding to integrated circuits or devices in sensitive areas of spacecraft can effectively mitigate the impact of high-energy particles in space. This passive protection method is widely used, and aluminum is the most common material used due to its excellent metallic properties. The most prevalent shielding method for payload (instruments or equipment carried by satellites or spacecraft) protection is 3 mm aluminum shielding [2]. However, different materials exhibit significant variations in shielding effectiveness under the same thickness due to their density, atomic number (Z), electron density, and other factors. For instance, high atomic number (Z) metals can effectively shield against ionization effects caused by space electrons, but they can also lead to stronger bremsstrahlung [3–6]. On the other hand, for protons, lower atomic number materials provide better shielding effectiveness [7–10]. This is because materials with lower atomic numbers have higher electron density, resulting in greater energy deposition of protons in the material and, consequently, reducing the energy of protons reaching the sensitive layer of semiconductor devices.

In order to accurately assess the potential dose levels that sensitive areas of spacecraft may be exposed to, on-orbit dose simulation is particularly crucial. Currently, commonly used radiation dose simulation methods both domestically and internationally include SHIELDOSE-2 [11] and the Monte-Carlo method [12]. Developed by the National Institute of Standards and Technology (NIST) in the United States, SHIELDOSE-2 can calculate the depth–dose relationship in spacecraft aluminum shielding materials based on electron and proton spectra. Presently, research institutions or entities worldwide predominantly employ this program to calculate shielding doses for spacecraft. However, SHIELDOSE-2 can only compute dose distribution in aluminum material shielding, leading researchers and engineers to primarily employ the areal density aluminum equivalent method. This method transforms other materials into aluminum material thickness using equal areal densities. In this scenario, the dose after aluminum shielding at this thickness is considered to be the dose after shielding with the respective material. For instance, the Space Systems Analysis Tool (SSAT) [13] developed by the European Space Agency (ESA) divides the full solid angle of payload-sensitive regions into several small sectors, traces the ray path through materials using straight lines, and then converts to the geometric thickness, which will be accumulated, of aluminum material based on the areal density aluminum equivalent method. Finally, the dose is calculated based on the depth–dose distribution of particles in the aluminum material. Another approach is using full Monte-Carlo simulations to compute shielding doses of three-dimensional models for the payload [14]. However, this method is computationally intensive, time-consuming, and less commonly used and still in the developmental stage. Common Monte-Carlo simulation software includes EGS*, MNCP**, NOVICE***, Geant4****, etc. (*http://rcwww.kek.jp/research/egs/; **https://mcnp.lanl.gov/; ***http://www.empc.com/novice.php; ****http://geant4.cern.ch/) For instance, the Geant4-based Monte-Carlo simulation tool MULASSIS (The version of MULASSIS used in this article is v1.26) [15] (multi-layered shielding simulation software) can calculate the actual ionizing dose after shielding with any material. Therefore, this study investigates the differences between the areal density aluminum equivalent method and the material Monte-Carlo simulation method in dose calculation. This article utilizes MULASSIS version 1.26, which is built upon the Geant4 toolkit version 4.10.1p3. MULASSIS is an application developed based on the Geant4 toolkit, and it automatically selects the appropriate physics processes. For electrons, it uses the "em_opt3" physics process, while for protons, it employs the "QBBC" physics process. MULASSIS exhibits a statistical error of less than 1% in ionization dosimetry calculations, which is within an acceptable range. Utilizing the Geant4-based Monte-Carlo simulation tool MULASSIS, the study conducts simulation research on the dose of single-energy electrons and protons under the same mass thickness shielding of different shielding materials. It is found that the areal density aluminum equivalent method may overestimate or underestimate ionizing dose for device exposure, depending on the particle energy. This research provides a theoretical basis for payload shielding design optimization.

2. Materials and Methods

MULASSIS is a one-dimensional, multi-layered radiation shielding simulation program based on the Geant4 Monte-Carlo transport software, developed through collaboration between QinetiQ, BiRA, and ESA. It allows for the simulation and calculation of the shielding effectiveness and flux analysis of various shielding materials against space radiation environments. Users can establish models by defining parameters such as particle sources, different shielding materials, and their respective thicknesses.

In this study, the commonly used 3 mm Al equivalent shielding thickness for spacecraft was set as the material constraint. Six materials most commonly used in satellite payloads (aluminum, lead, tantalum, tungsten, molybdenum, and titanium) were chosen as validation targets. The simulation was conducted to calculate the absorbed dose in a silicon detector under equivalent areal density shielding conditions. The material information is presented in Table 1 below.

Table 1. Shielding material information [16].

Material	Atomic Number (Z)	Density (g/cm^3)	Equivalent Areal Density (g/cm^2)	Geometric Thickness (mm)	Electron Density (10^{23} e/g)
Aluminum	13	2.699	0.8097	3.000	2.901
Titanium	22	4.540	0.8097	1.783	2.719
Molybdenum	42	10.220	0.8097	0.792	2.636
Tantalum	73	16.654	0.8097	0.486	2.429
Tungsten	74	19.300	0.8097	0.420	2.424
Lead	82	11.350	0.8097	0.713	2.383

The simulation model established for this study is a planar slab model. The first layer is set as the shielding material with a thickness of the equivalent 3 mm aluminum areal density (0.8097 g/cm^2), using different materials. The second layer represents the absorbing body of the silicon detector, and the third layer is a 5 mm thick layer of aluminum. This setup is designed to simulate the potential backscattering effects from materials such as circuit boards, instrument bases, and outer shells located beneath the device during on-orbit satellite operation. For a 2 MeV electron, the impact of backscattering on the absorbed dose in a silicon detector is significant. When shielded by aluminum, the ionizing dose generated by backscattering accounts for 34.92% of the total dose. It also allows for a convenient comparison with the aluminum shielding material. Regarding the choice of the thickness for the second layer (sensitive area), the MULASSIS internal algorithm computes the average energy deposition for each layer as a whole. Therefore, it is essential to select an appropriate thickness to ensure more accurate dose calculations. As shown in Figure 1, after applying 0.8097 g/cm^2 mass shielding, the dose distribution for 2 MeV electrons with a 200 μm thick silicon detector is displayed. At the interface, there is a significant gradient in the dose distribution, with higher doses closer to the shielding material. Thus, a thinner silicon layer makes the detector more sensitive. Considering that the sensitive region thickness of the silicon detector is on the order of micrometers, a thickness of 20 μm is chosen to ensure the accuracy of the silicon absorption dose.

The schematic diagram of the shielding model is illustrated in Figure 2 (the entire structure needs to be modeled in MULASSIS). In this diagram, the silicon detector represents the sensitive volume for ionizing total dose damage of the electronic device, and we calculated the energy deposition at each layer. However, the absorbed dose at this location is the primary focus of the study. Taking aluminum as an example, the model begins with a 3 mm Al shielding layer, followed by a 20 μm thick silicon detector, and, finally, a 5 mm Al layer. The simulation calculates the absorbed dose in the silicon detector under these conditions. Table 2 presents the parameters of incident particles. To facilitate the comparison of differences between different materials at different energies, the energy spectrum was set as monoenergetic electrons and monoenergetic protons. The energy selection is mainly based on the fact that the energies of electrons and protons in the space environment are less than 7 MeV and 300 MeV, respectively. For low-energy electrons (energy less than 2 MeV) and protons (energy less than 25 MeV), its range in Al is less than 3 mm. Therefore, we chose this energy range.

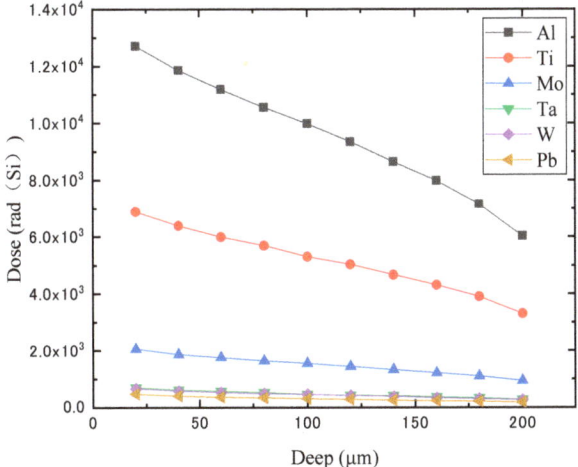

Figure 1. Depth–dose relationship of 2 MeV electron incidence in 200 μm silicon material.

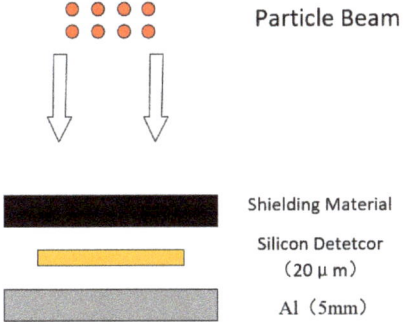

Figure 2. Schematic diagram of the shielding model.

Table 2. Particle parameters.

Particle	Number of Primary Particles	Fluence/(Flux) Intensity (cm^{-2}s^{-1})	Energy (MeV)
Electron	1.0×10^7	1.0×10^{12} cm^{-2}s^{-1}	2–7
Proton	1.0×10^7	7.6×10^9 cm^{-2}s^{-1}	25–300

3. Results

3.1. Monoenergetic Electrons

As shown in Figure 3, the relationship between absorbed dose in the silicon detector and incident electron energy was obtained for different materials under the equivalent areal density shielding (3 mm Al). A smaller absorbed dose in the silicon detector indicates a better shielding effectiveness of the material against the ionizing total dose effect caused by electrons. Overall, within the discussed range of electron energy in Figure 3, high atomic number materials such as tantalum, tungsten, and lead showed a continuous increase in ionizing dose. On the other hand, low atomic number materials such as aluminum, titanium, and molybdenum exhibited a trend of initial increase followed by a decrease in ionizing dose, with a peak point observed. For example, in the case of aluminum shielding, a distinct peak absorbed dose was observed around 4 MeV electron energy, and this peak point shifted to the right with an increase in the atomic number (Z) of the material. Furthermore, there were significant differences in the ionizing dose between the areal density aluminum equivalent method and Monte-Carlo simulation for the same material at the same energy. When the electron energy was less than 4 MeV, the areal density aluminum equivalent method significantly overestimated the absorbed dose in the silicon detector for the other shielding materials. In Figure 4a, for a 2 MeV electron, it can be observed that aluminum shielding resulted in the highest absorbed dose in the silicon detector, while lead shielding resulted in the lowest absorbed dose, with approximately a 96% difference from aluminum. At this point, the shielding effectiveness of lead is 24.8 times that of aluminum. Indeed, for lead, the areal density aluminum equivalent method would overestimate the ionizing dose by a factor of 24.8. It would be 17.8 times for tungsten, 16.8 times for tantalum, 5.9 times for molybdenum, and 1.8 times for titanium. Until 4 MeV, the areal density aluminum equivalent method would begin to underestimate the shielding dose of other materials. In Figure 4b, for 7 MeV electron incidence, the ionizing dose after aluminum shielding is the smallest. At this point, the areal density aluminum equivalent method would underestimate the ionizing dose for tantalum, tungsten, and lead by approximately 72%, for molybdenum by 74%, and for titanium by 88%. Hence, when using the areal density aluminum equivalent method to calculate the total ionizing dose effect of electrons for other materials, there may be differences compared to Monte-Carlo simulations. These differences are closely related to the incident electron energy.

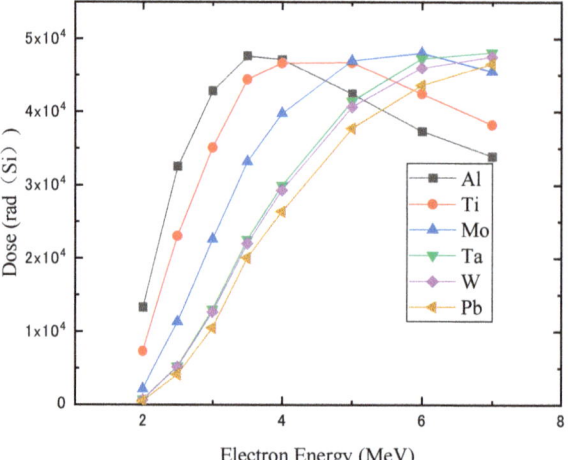

Figure 3. The absorbed dose in the silicon detector for six different materials (aluminum, lead, tungsten, tantalum, molybdenum, and titanium) under the same equivalent areal density shielding (3 mm Al) at various incident electron energies.

To comprehensively evaluate the shielding effectiveness of different materials with an equivalent areal density of 3 mm Al, the ionizing dose at each energy point from Figure 3 was integrated over energy. Based on the results from Table 3, when shielding with 0.8097 g/cm^2 areal density, the differences between the areal density aluminum equivalent method and the Monte-Carlo simulation in ionizing dose calculations was significantly reduced for a wide range of electron energy spectra. For lead, the areal density aluminum equivalent method overestimates the dose by 27.9%, for tungsten by 22.3%, for tantalum by 20.9%, for molybdenum by 7.6%, and there is almost no difference for titanium. Indeed, the areal density aluminum equivalent method for calculating ionization dose possesses certain rationality. However, there is room for further optimization and improvement.

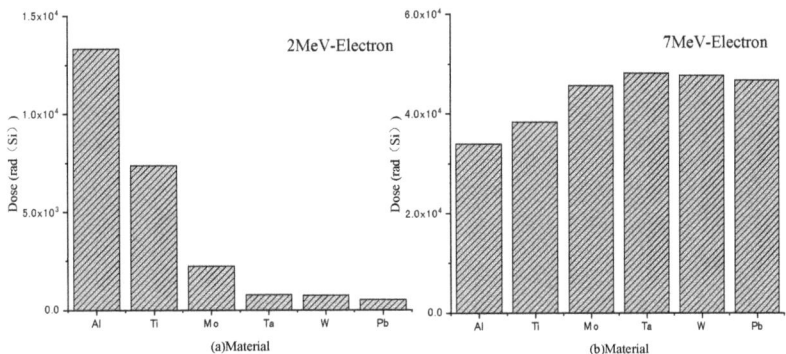

Figure 4. The absorbed dose in the silicon detector for six different materials (aluminum, lead, tungsten, tantalum, molybdenum, and titanium) after equivalent 3 mm Al shielding at an incident electron energy of 2 MeV (**a**) and 7 MeV (**b**).

Table 3. Integration of ionizing dose for 2–7 MeV electrons.

Material	Al	Ti	Mo	Ta	W	Pb
Dose	1.97×10^5	1.97×10^5	1.82×10^5	1.56×10^5	1.53×10^5	1.42×10^5
D_x/D_{Al}	1	0.999	0.924	0.792	0.777	0.721

3.2. Monoenergetic Protons

Figure 5 illustrates the relationship between the absorbed dose in the silicon detector and the proton energy for six different materials after being shielded with an equivalent 3 mm Al thickness. Simultaneously, for ease of observation of the differences, we have inserted ionizing dose graphs for proton energies below 50 MeV. It can be observed that when the proton energy is less than 50 MeV, there are significant differences in the shielding effects among different materials. At this energy range, the areal density aluminum equivalent method significantly deviates from the absorbed dose obtained with actual shielding materials, leading to an overestimation of the shielding dose. The absorbed dose in the silicon detector increases as the atomic number (Z) of the materials decreases. In Figure 6, the absorbed dose in the silicon detector is shown for different materials after shielding with the same areal density at 25 MeV proton energy. Among the various materials, the absorbed dose after shielding with Ta, W, and Pb is approximately 65% higher than that after shielding with Al, while Mo shows a difference of 58%, and Ti has a difference of 43%. The absorbed dose in the silicon detector after the same surface density shielding shows an inverse relationship with the atomic number Z, and there are noticeable differences among different materials. As the proton energy increases beyond 50 MeV, the absorbed dose in the silicon detector gradually decreases and levels off at the same value for all materials. This indicates that at higher energies, the deposited dose in the sensitive area becomes independent of the atomic number of the shielding materials. Both the areal density aluminum equivalent method and the Monte-Carlo method yield similar results for calculating the shielding dose effectiveness at this energy range. Just like in electronic analysis, ionizing dose integrated over 25–300 MeV protons after shielding results in the data shown in Table 4. It can be observed that for a broad energy spectrum, the areal density aluminum equivalent method for calculating ionizing dose exhibits relatively small differences. This also indicates the method's reasonable validity.

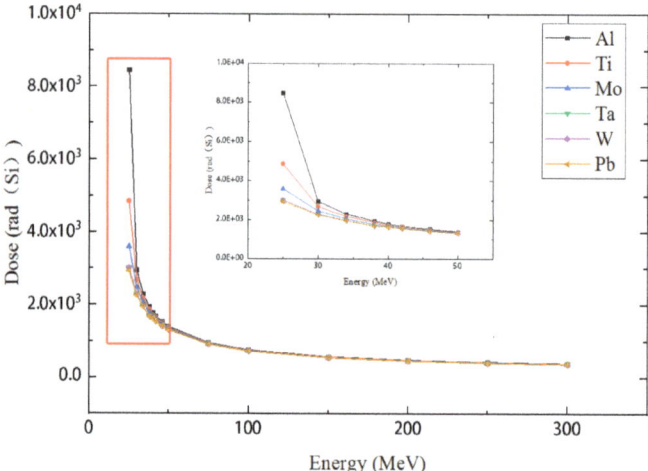

Figure 5. The comparison of absorbed dose in the silicon detector for different materials after single-energy proton irradiation under a shielding thickness of 0.8097 g/cm^2.

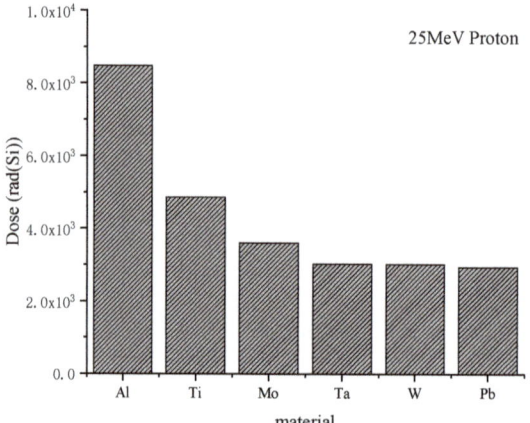

Figure 6. The comparison of absorbed dose in the silicon detector for different materials after shielding with a thickness of 0.8097 g/cm^2 under 25 MeV proton irradiation.

Table 4. Integration of ionizing dose for 25–300 MeV protons.

Material	Al	Ti	Mo	Ta	W	Pb
Dose	2.20×10^5	2.06×10^5	1.98×10^5	1.93×10^5	1.93×10^5	1.92×10^5
D_x/D_{Al}	1	0.936	0.902	0.879	0.878	0.875

4. Discussion

4.1. Monoenergetic Electrons

For electrons, their energy loss within the shielding material is primarily through ionization and bremsstrahlung radiation processes [17]:

$$\left(-\frac{dx}{dt}\right) = \left(-\frac{dx}{dt}\right)_e + \left(-\frac{dx}{dt}\right)_r \quad (1)$$

In the above equation, the first term $\left(-\frac{d_x}{d_t}\right)_e$ represents the ionization energy loss, and the second term $\left(-\frac{d_x}{d_t}\right)_r$ represents the bremsstrahlung energy loss. The sum of these two terms gives the total energy loss of the electron. An increase in ionization energy loss within the shielding material leads to a reduction in the total dose in the silicon detector, while an increase in bremsstrahlung energy loss results in an increase in the total dose in the silicon detector. At lower energies, the contribution of bremsstrahlung is relatively small, and the absorbed dose in silicon is mainly caused by ionization of the residual electrons that penetrate the shielding material. The ionizing dose of the residual electrons in silicon can be calculated using the following formula:

$$D_e = \int_{E_0}^{E_{max}} \varnothing \left(\frac{dE}{dx}\right)_{Si} dE \qquad (2)$$

where $\varnothing$ represents the electron flux, and $\left(\frac{dE}{dx}\right)_{Si}$ denotes the collision stopping power of electrons in silicon, which characterizes the ionization energy loss of electrons in silicon. E_0, E_{max} correspond to the minimum and maximum energy of the electrons, respectively. By utilizing the particle flux calculation function in MULASSIS, we obtained the residual electron energy spectra (Figure 7) for 2 MeV electron incidence on three different materials, Al, Mo, and Pb, after passing through an areal density of 0.8097 g/cm² shielding (as shown in Table 2, the flux of electrons before the shielding is 1.0×10^{12} cm^{-2}s^{-1}). It is evident from Figure 7 that the remaining electron flux in Al is significantly higher than in Mo and Pb, with Pb exhibiting the lowest flux. This trend is inversely proportional to the atomic number (Z) of the materials. Consequently, at lower energies, the areal density aluminum equivalent method tends to overestimate the ionizing dose.

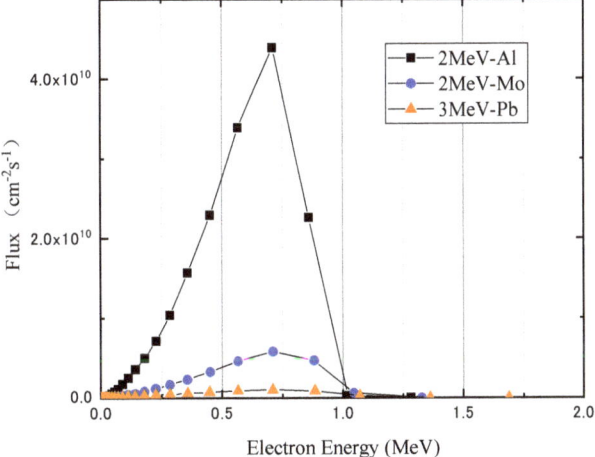

Figure 7. The remaining electron energy spectra after 2 MeV electron incidence on three different materials.

As the electron energy increases, the bremsstrahlung becomes more pronounced within the shielding material. In Figure 8, the radiative stopping power of 2 MeV to 7 MeV electrons in Al, Mo, and Pb materials was calculated using the ESTAR tool developed by the National Institute of Standards and Technology (NIST). It is evident that as the incident electron energy increases, the radiative stopping power also increases, with higher atomic number materials exhibiting a faster increase. The bremsstrahlung X-ray photons generated within the shielding material transfer energy to electrons in silicon through processes such as the photoelectric effect, Compton scattering, and electron–positron pair production, resulting in ionizing dose. As the radiation loss within the shielding material increases, the absorbed dose in silicon also increases. The ionizing dose caused by X-ray photons can be calculated using the following formula:

$$D_r = \int_0^{E_{max}} \Psi \frac{\mu_{en}}{\rho} dE, \quad (3)$$

where Ψ represents the photon energy flux, $\frac{\mu_{en}}{\rho}$ denotes the mass energy absorption coefficient of photons in Si material (μ_{en} is the coefficient of linear energy absorption, and ρ is the density of the material through which the rays pass), and D_r stands for dose, which represents the dose deposited in silicon by secondary bremsstrahlung produced by electrons in the shielding material. This dose can be obtained by integrating $\Psi \frac{\mu_{en}}{\rho}$ over the energy range.

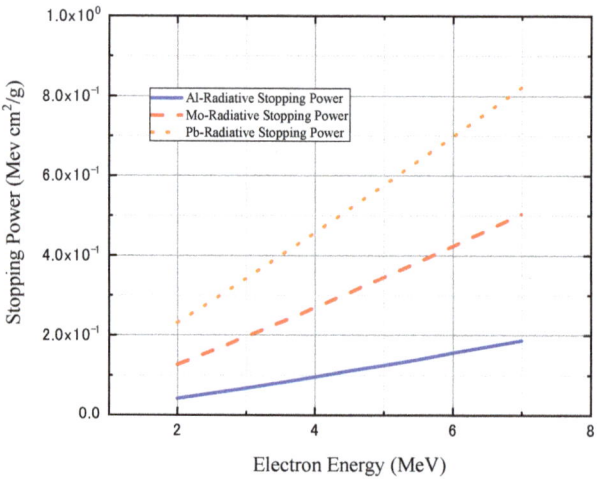

Figure 8. The radiative stopping power of 2 MeV to 7 MeV electrons in Al, Mo, and Pb materials.

Through the MULASSIS particle flux calculation function, with other parameters held constant, simulations were conducted for electron incident energies of 2 MeV and 7 MeV, obtaining the X-ray energy spectra after passing through Al, Mo, and Pb shielding materials, as shown in Figure 9. At both energy levels, high atomic number shielding materials produce more secondary X-rays. Integrating the energy E over the energy spectrum yields the photon energy fluence spectra (Figure 10), and integrating $\Psi \frac{\mu_{en}}{\rho}$ over the X-ray photon energy E provides the dose deposited by X-rays in Si material (Figure 11). Notably, the dose from bremsstrahlung increases by two orders of magnitude for Pb shielding material from 2 MeV to 7 MeV (from 3.07×10^2 rad(Si) to 1.328×10^4 rad(Si)), whereas for Al, it only increases by one order of magnitude (from 1.313×10^2 rad(Si) to 2.584×10^3 rad(Si)). The difference in bremsstrahlung dose between these two materials increases by 59.1 times. Mo shielding material lies between the two (increasing from 2.573×10^2 rad (Si) to 8.041×10^3 rad (Si)). Thus, for different metal materials, higher atomic numbers lead to a greater increase in secondary X-ray ionizing dose. This is due to the fact that cross-sections for the photoelectric effect ($\sigma_k \propto Z^5$), Compton scattering ($\sigma_c \propto Z$), and electron–positron pair production ($\sigma_p \propto Z^2$) are proportional to the atomic number Z raised to the fifth power, first power, and square, respectively [17]. Larger atomic numbers result in higher probabilities for these three energy transfer processes, leading to stronger bremsstrahlung. The subsequent decrease in total dose in the peak region may be attributed to the increase in electron velocity to a certain value, reducing the number of collisions with outer-shell electrons of target atoms in the shielding material. Additionally, the influence of the sensitive region's thickness contributes to the reduction in ionizing dose in the 20 μm Si detector.

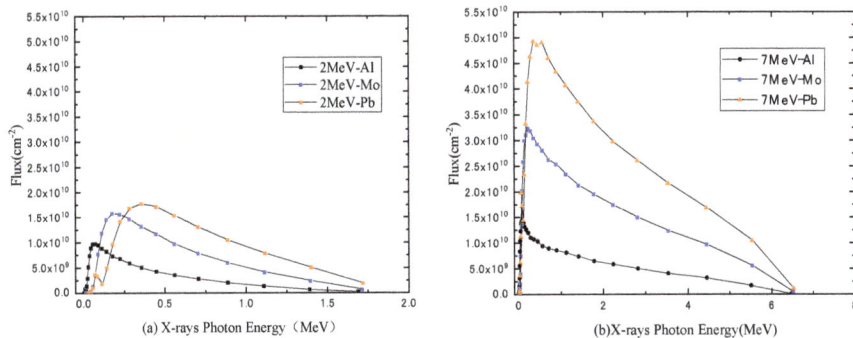

Figure 9. X-ray fluence spectra for 2 MeV (**a**) and 7 MeV (**b**) electrons after passing through Al, Mo, and Pb materials with the same mass shielding.

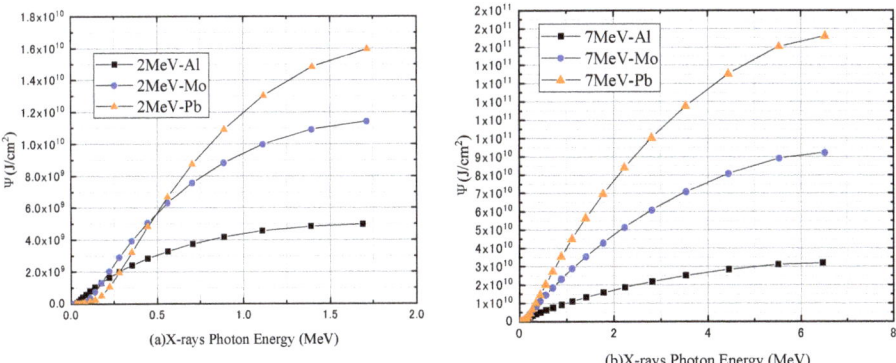

Figure 10. X-ray photon energy-fluence spectra for 2 MeV and 7 MeV electrons after passing through Al, Mo, and Pb materials.

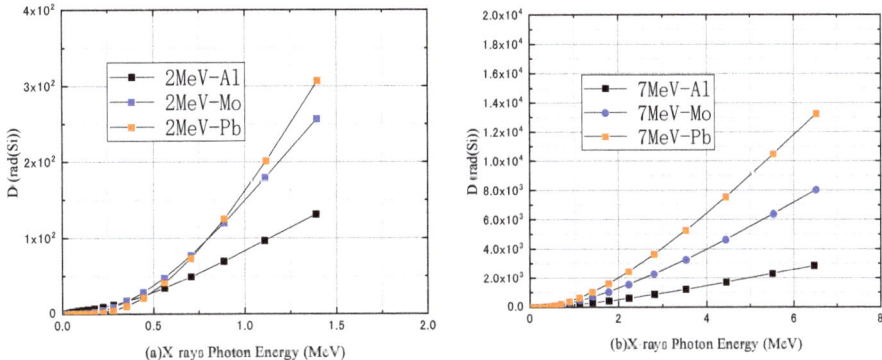

Figure 11. The ionizing dose produced by X-ray photons in Si material.

4.2. Monoenergetic Protons

For protons, their interaction with the target material involves a process of energy loss and deceleration. Upon entering the target material, protons undergo collisions with the atomic nuclei or the outer-shell electrons of the target material, leading to a continuous loss of energy and a gradual reduction in their velocity. This process continues until the proton's energy is reduced to zero, and it comes to a stop, becoming stationary within the material. The primary mechanism responsible for proton ionization effects is the energy loss resulting from collisions between the protons and the outer-shell electrons of the target material. As protons lose energy through these collisions, they transfer energy to the electrons, leading to the ionization of atoms in the material. Figure 12 shows the depth–dose curves of 25 MeV protons in Al, Mo, and Pb metals, with the areal density used as the horizontal axis for comparison. In Figure 13, all three materials exhibit a distinct peak in the depth–dose distribution, known as the Bragg peak. The formation of the Bragg peak is due to the reduction in proton velocity as it penetrates deeper into the target material. As the proton's velocity decreases, its energy loss through collisions with the outer-shell electrons of the target material increases, leading to an increase in ionizing dose. At the end of the proton's trajectory, it comes to a stop, resulting in the maximum deposition of energy and the highest ionizing dose, forming the peak. The ionizing dose is larger on the left side of the Bragg peak and decreases with an increase in atomic number Z. For 0.8097 g/cm^2 Al shielding, it is closest to the Bragg peak, resulting in the highest ionizing dose in the Si material downstream of the shielding. As the proton energy increases, the 0.8097 g/cm^2 shielding thickness moves further away from the Bragg peak. Additionally, due to the increasing proton energy, its energy loss within the 0.8097 g/cm^2 thickness decreases. Figure 13 shows the depth–dose distribution after 100 MeV proton irradiation in Al, Mo, and Pb materials. At this energy, the 0.8097 g/cm^2 shielding for all three materials is located in the plateau region, far from the Bragg peak. Consequently, in Figure 10, with increasing energy, the dose after 3 mm equivalent Al shielding gradually decreases and eventually levels off to the same level.

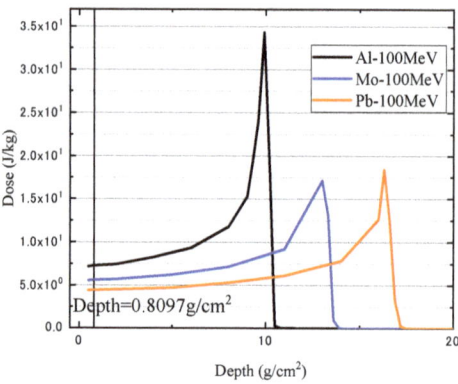

Figure 12. The depth–dose distribution of 25 MeV protons in three different materials.

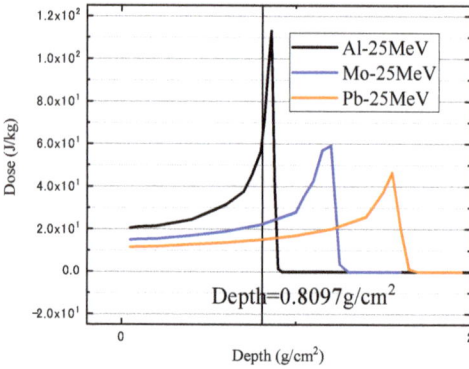

Figure 13. The depth–dose distribution of 100 MeV protons in three different materials.

5. Conclusions

In this study, we utilized the Monte-Carlo simulation tool MULASSIS in Geant4 to investigate the shielding effects on ionizing total dose in microelectronic devices for single-energy electrons or protons after shielding with lead, tungsten, tantalum, molybdenum, and titanium materials. We compared the absorbed doses and the areal density aluminum equivalent method shielding dose calculations at areal density for the materials. Our findings are as follows:

1. For single-energy electrons (2–7 MeV), significant differences exist between the areal density aluminum equivalent method and the Monte-Carlo (MULASSIS) method when calculating the absorbed dose in the silicon detector after shielding with different materials at an equivalent 3 mm Al areal density. At lower energies, the areal density aluminum equivalent method severely underestimates the shielding effectiveness of the other five materials against total ionizing dose, and the difference in absorbed dose in the silicon detector increases with larger atomic number differences. For instance, at 2 MeV electron energy, the absorbed dose in the silicon detector after lead shielding differs by 96% compared to aluminum shielding. For electron space environments with energies below 5 MeV, materials with higher atomic numbers seem to provide better shielding effects against ionization in microelectronic devices under the same areal density shielding. However, for electrons with energies above 5 MeV, the absorbed dose in the silicon detector is lower after aluminum shielding compared to other materials. Therefore, when evaluating the total dose effects caused by electrons, using the areal density aluminum equivalent method for dose assessment will overestimate the shielding effectiveness of other materials.

2. For protons, under a 0.8097 g/cm^2 areal density shielding, the areal density aluminum equivalent method overestimates the shielding effectiveness of the other five materials when proton energy is below 50 MeV. For example, after shielding with materials such as Pb, W, and Ta, the absorbed dose is around 35% of the dose obtained with aluminum shielding. However, as the proton energy increases, when the proton energy is greater than 50 MeV, the absorbed dose gradually converges to the same level. Therefore, for the ionizing effects caused by protons in microelectronic devices, the areal density aluminum equivalent method is not sufficiently accurate when the proton energy is below 50 MeV.

In conclusion, in radiation shielding design for payload protection, using only the areal density aluminum equivalent method to evaluate the total dose effects caused by single-energy electrons and protons may lead to inaccuracies in dose assessment. Radiation shielding and passive protection must consider the differences in dose among different materials. This approach helps accurately evaluate the radiation dose in sensitive areas of the payload and allows for targeted radiation protection designs for different regions. Strengthening the radiation resistance of devices and electronic equipment in orbit can increase the on-orbit lifetime of satellite space missions. Furthermore, in this study, single-energy electrons and protons were chosen as the subjects for comparing the differences in the ionizing total dose effects among different metal materials. However, for continuous spectrum radiation in space with various types of particles, further research is needed to investigate the variability in shielding effectiveness.

Author Contributions: Conceptualization, M.L. and C.H.; methodology, M.L., C.H. and J.F.; software, M.L. and M.X.; validation, M.L. and C.H.; formal analysis, C.H. and M.X.; investigation, J.F.; data curation, M.L.; writing—original draft preparation, M.L.; writing—review and editing, J.F. and C.H.; visualization, M.L.; supervision, C.H.; project administration, J.S., Y.L. and Q.G.; funding acquisition, J.F. and C.H. All authors have read and agreed to the published version of the manuscript.

Funding: This research was funded by the National Natural Science Foundation of China under grant No. 12175307, the National Natural Science Foundation of China under grant No. 11975305, the West Light Talent Training Plan of the Chinese Academy of Sciences under grant No. 2022-XBQNXZ-010, the West Light Talent Training Plan of the Chinese Academy of Sciences under grant No. 2021-XBQNXZ-020, and the Youth Science and Technology Talents Project of Xinjiang Uygur

Autonomous Region No. 2022TSYCCX0094, Fund of Robot Technology Used for Special Environment Key Laboratory of Sichuan Province No. 21kftk03.

Data Availability Statement: Not applicable.

Conflicts of Interest: The authors declare no conflict of interest.

References

1. Wang, H.; Hu, Y.; Zheng, Y.; Wang, H. The present situation of technologies for analysis of space radiation effects to spacecraft and some retrospection. *Spacecr. Environ. Eng.* **2022**, *39*, 4.
2. Janet, L.B. Space weather effects on spacecraft systems. In Proceedings of the American Meteorological Society Annual Meeting Space Weather Symposium, Seattle, WA, USA, 13 January 2004.
3. Brown, L.M. Bethe-Bloch formula with density effect correction. *Nucl. Instrum. Methods* **1979**, *164*, 15–21.
4. Fujimoto, T.; Monzen, H.; Nakata, M.; Okada, T.; Yano, S.; Takakura, T.; Kuwahara, J.; Sasaki, M.; Higashimura, K.; Hiraoka, M. Dosimetric shield evaluation with tungsten sheet in 4, 6, and 9 MeV electron beams. *Phys. Medica* **2014**, *37*, 838–842. [CrossRef] [PubMed]
5. Fujita, Y.; Myojoyama, A.; Saitoh, H. Bremsstrahlung and photoneutron production in a steel shield for 15-22-MeV clinical electron beams. *Radiat. Prot. Dosim.* **2014**, *163*, 149–159. [CrossRef] [PubMed]
6. Hu, J.; Feng, Y.; Han, J.; Cai, M.; Yang, T. Simulation and validation of composite shielding for total ionizing dose. *Chin. J. Space Sci.* **2014**, *34*, 180–185. [CrossRef]
7. Xun, M.; He, C.; Zheng, Y. Analysis of shielding ability of different materials for 100 MeV proton incident. *Manned Spacefl.* **2018**, *24*, 740–744. [CrossRef]
8. Wang, C.; Luo, W.; Zha, Y.; Wang, C. Monte-Carlo simulation of optimization choice of shielding materials for proton radiation in space. *Radiat. Prot.* **2007**, *27*, 79–86. [CrossRef]
9. Waterman, G.; Kase, K.; Orion, I.; Broisman, A.; Milstein, O. Selective shielding of bone marrow: An approach to protecting humans from external gamma radiation. *Health Phys.* **2017**, *113*, 195–208. [CrossRef] [PubMed]
10. Wu, Z.-X.; Sun, H.-B.; He, C.-F.; Tong, Y.-P.; Ma, Y.-G.; Lu, J.-B.; Xun, M.-Z.; Hu, Y.-Q.; Cai, Z.-B.; Zhu, Z.-P. Analysis of radiation environment and its dose within sealed cabin of manned spacecraft. *Spacecr. Environ. Eng.* **2016**, *33*, 154–157.
11. Seltzer, S.M. *Updated Calculations for Routine Space-Shielding Radiation Dose Estimates: SHIELDOSE-2*; NIST Publication NISTIR 5477; NIST Publication: Gaithersburg, MD, USA, 1994.
12. Xu, S. *Application of Monte-Carlo Methods in Experimental Physics*; Atomic Energy Press: Beijing, China, 2006.
13. Santin, G. New Geant 4 based simulation tools for space radiation shielding and effects analysis. *Nucl. Phys. B* **2003**, *125*, 69–74. [CrossRef]
14. Cai, M.; Han, J. Method for Evaluating Shielding Thicknesses and Radiation Dose Inside Spacecraft Based on ProE. *J. Astronaut.* **2012**, *33*, 6. [CrossRef]
15. Lei, F.; Truscott, R.R.; Dyer, C.S.; Quaghebeur, B.; Heynderickx, D.; Nieminen, R.; Evans, H.; Daly, E. MULASSIS: A Geant4-based multilayered shielding simulation tool. *IEEE Trans. Nucl. Sci.* **2002**, *49*, 2788–2793. [CrossRef]
16. Attix, F.H. *Introduction to Radiological Physics and Radiation Dosimetry*; John Wiley & Sons: Hoboken, NJ, USA, 1986. [CrossRef]
17. Fudan University; Tsinghua University; Peking University (Eds.) *Experimental Methods in Nuclear Physics*; Atomic Energy Press: Beijing, China, 1997.

Disclaimer/Publisher's Note: The statements, opinions and data contained in all publications are solely those of the individual author(s) and contributor(s) and not of MDPI and/or the editor(s). MDPI and/or the editor(s) disclaim responsibility for any injury to people or property resulting from any ideas, methods, instructions or products referred to in the content.

Communication

Research on High-Dose-Rate Transient Ionizing Radiation Effect in Nano-Scale FDSOI Flip-Flops

Tongde Li [1,2], Jingshuang Yuan [1,2], Yang Bai [1,2], Chunqing Yu [1,2], Chunliang Gou [1,2], Lei Shu [1,2], Liang Wang [1,2] and Yuanfu Zhao [1,2,3,*]

1. Beijing Microelectronics Technology Institute, Beijing 100076, China; ltdeam@163.com (T.L.); yuanjsh_9826@163.com (J.Y.)
2. Laboratory of Science and Technology on Radiation-Hardened Integrated Circuits in CASC, Beijing 100076, China
3. China Academy of Aerospace Electronics Technology, Beijing 100094, China
* Correspondence: zhaoyf@vip.163.com

Abstract: This paper presents an experimental study on the high-dose-rate transient ionizing radiation response and influencing factors of a Nano-Scale Fully Depleted Silicon-On-Insulator (FDSOI) D flip-flops (DFFs) circuit. Results indicate that data errors occur in DFFs at the lowest dose rate of 4.70×10^{11} rad(Si)/s in experiments, and the number of data errors shows a nonlinear increasing trend with the increase in dose rate and supply voltage. Three-dimensional technology computer-aided design (TCAD) simulations were conducted to analyze the transient photocurrent and charge collection mechanism at advanced process. The simulation results indicated that the charge collection efficiency is heightened with an increase in supply voltage, resulting in the higher photocurrent. This plays a major role in the process of charge collection for Ultra-Thin Body and Buried oxide (UTBB) FDSOI technology. The investigation into the high-dose-rate transient ionizing radiation effect (HDR-TIRE) in Nano-Scale FDSOI DFFs will aid in the assessment and application of advanced integrated circuits in aerospace.

Keywords: high-dose-rate transient ionizing effect; FDSOI; TCAD simulation; supply voltage

1. Introduction

Exposure to various radiation environments may result in diverse radiation effects on Integrated Circuits (ICs), such as high-dose-rate transient ionizing radiation effect (HDR-TIRE), total ionizing dose effect (TID) and single-event effect (SEE). HDR-TIRE is a phenomenon in which circuits experience signal upset and strong disturbance when exposed to high-dose-rate radiation environments. This effect is primarily caused by the generation of transient photocurrents in semiconductor devices due to ionizing effects. The stability of the supply voltage is crucial for the proper functioning of ICs. Unfortunately, HDR-TIRE causes significant disruptions to the supply voltage, which directly affects the normal operation of circuits [1–5]. Currently, experimental studies of HDR-TIRE depend on large ground-based devices to build up the radiation environment, while relevant research is carried out by using "Qiangguang I" accelerator.

As the feature size of Nano-Scale ICs continues to scale down, short-channel effects (SCEs) limit the performance of traditional bulk planar technology. To extend CMOS scaling beyond the sub-28 nm node, Fin Field-Effect Transistor (FinFET) and FDSOI technology routes have been proposed. Nano-Scale FDSOI technology has overcome the performance of traditional planar bulk transistors due to improved electrostatic control, and it is scalable according to Moore's Law. In addition, it provides best the performance-power trade-offs using the body biasing [6]. Moreover, FDSOI has great potential for military and space applications.

Citation: Li, T.; Yuan, J.; Bai, Y.; Yu, C.; Gou, C.; Shu, L.; Wang, L.; Zhao, Y. Research on High-Dose-Rate Transient Ionizing Radiation Effect in Nano-Scale FDSOI Flip-Flops. *Electronics* **2023**, *12*, 3149. https://doi.org/10.3390/electronics12143149

Academic Editor: Lodovico Ratti

Received: 9 June 2023
Revised: 6 July 2023
Accepted: 7 July 2023
Published: 20 July 2023

Copyright: © 2023 by the authors. Licensee MDPI, Basel, Switzerland. This article is an open access article distributed under the terms and conditions of the Creative Commons Attribution (CC BY) license (https://creativecommons.org/licenses/by/4.0/).

SOI, as a modern type of semiconductor device structure, has also attracted considerable attention in the field of radiation hardening [7–10]. To achieve a complete dielectric isolation structure, a thin layer of SiO$_2$ is inserted into the substrate of SOI devices, which eliminates the latch-up effect commonly found in traditional bulk devices [11]. Additionally, the buried oxide (BOX) layer prevents a significant amount of charge generated in the substrate from being collected by the top silicon film, leading to a lower intensity of photocurrent generated by SOI circuits compared to bulk circuits in a high dose rate irradiation environment [12–14]. A previous study showed that SOI presents a bipolar parasitic transistor, which may amplify the charge injected by irradiation [15]. Under the same process conditions, the tolerance to HDR-TIRE of an SOI device is substantially improved [16]. SOI devices can be divided into PDSOI (Partially Depleted) and FDSOI (Fully Depleted), in which the FDSOI structure gains high tolerance against radiation [17]. The structure comparison between FDSOI and a planar bulk device is shown in Figure 1.

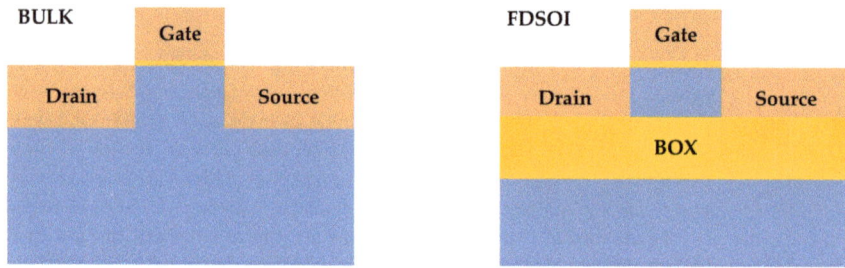

Figure 1. Schematic diagram of the device structure of bulk and FDSOI processes.

A study has investigated the HDR-TIRE of SOI CMOS RF ICs [18]. A radiation-hardened FPGA chip based on a 0.5 μm SOI process has been shown to tolerate a dose rate of over 1.5×10^{11} rad(Si)/s [19]. For a radiation-hardened SRAM chip with a 0.8 μm SOI process, although the stored data remained intact and the read/write function worked normally after being exposed to a dose rate of 2.45×10^{11} rad(Si)/s, there was an increase in current after irradiation from the experiment results [20]. However, related research into advanced integrated circuits based on Ultra-Thin Body and Buried oxide (UTBB) FDSOI technology has not been reported.

Flip-flops are one of the fundamental and widely used components in digital circuits. Therefore, it is essential to discover the radiation response of flip-flops. This paper focuses on an experimental and simulation study to investigate the impact of supply voltage and dose rate on the high-dose-rate transient ionizing radiation response of FDSOI-based D Flip-Flops (DFFs). In the following Sections 2 and 3, the test circuit based on Nano-Scale FDSOI and experiment setups is described. Additionally, experimental and simulation results are discussed in Section 4. Finally, the conclusion of the analyzed data is provided in Section 5.

2. Circuit Samples

The test sample was a customized Nano-Scale UTBB FDSOI DFFs circuit, which was developed to test and verify the radiation tolerance performance of FDSOI technology. Figure 2 shows the structure schematic of conventional DFF [21]; it is composed of two latches, each designed with back-to-back connect inverters, where D is the input, clk is the clock signal, and Q is the output.

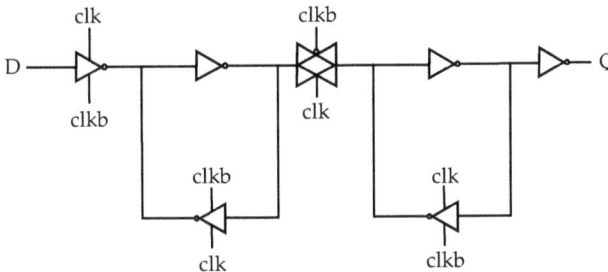

Figure 2. Schematic of Conventional DFF.

3. Radiation Experiments

The high-dose-rate transient ionizing radiation experiments were carried on the "Qiangguang-I" accelerator at Northwest Institute of Nuclear Technology. This accelerator is capable of simulating a variety of pulsed radiation environments formed by nuclear explosions. Experiments were performed by simultaneously bombarding multiple samples. In order to shield electromagnetic interference, the devices under the test board and configuration board were placed in aluminum boxes. The dose rate values required were obtained by placing the sample at a certain position from the gamma ray radiation source. Additionally, two power supply voltages (0.8 V and 0.88 V) were set up, respectively, in this experiment for observing the radiation characteristics of supply voltage effect.

4. Experimental Analysis and Discussion

The relationship between data errors of FFs and dose rate values for different supply voltages is represented in Figure 3, where the number of data errors is normalized to the value at 0.88 V supply voltage and 1.80×10^{10} rad(Si)/s dose rate. The supply voltage was set to 0.8 V in Figure 3a, and it can be calculated that the normalized error increased by 4.2% as the dose rate was raised from 4.70×10^9 rad(Si)/s to 5.90×10^{10} rad(Si)/s; the dose rate is increased by a factor of 12.6. It increased by 16.8% as the dose rate was raised from 4.70×10^9 rad(Si)/s to 4.10×10^{11} rad(Si)/s, and the dose rate increased by a factor of 87.2. Additionally, the normalized error was increased by 12.1%, with the dose rate increasing by a factor of 6.9 from 5.90×10^{10} rad(Si)/s to 4.10×10^{11} rad(Si)/s. When the supply voltage was 0.88 V, as shown in Figure 3b, the normalized error was raised by 13.0% and 99.0% with the dose rate increased from 1.80×10^{10} rad(Si)/s to 1.60×10^{11} rad(Si)/s and 1.80×10^{10} rad(Si)/s to 6.90×10^{11} rad(Si)/s, respectively; the dose rates increased by a factor of 8.9 and 38.3. Test results have confirmed a nonlinear increase in data errors induced by transient ionizing radiation as the dose rate increased. It is noted that though the dose rate value could be roughly expected according to the distance from the radiation source, the same order of magnitude of dose rate values could not be completely equivalent during multiple trials.

Supply voltage acted as a global influence factor on the high-dose-rate transient ionizing radiation response in the circuit. As illustrated in Figure 4, the number of the normalized error was lower for the 0.8 V and 4.10×10^{11} rad(Si)/s combination than the 0.88 V and 1.60×10^{11} rad(Si)/s combination, which indicates that even with a lower dose rate a higher supply voltage results in a higher number of errors in DFFs.

The impact of the supply voltage on the high-dose-rate transient ionizing radiation response of FDSOI circuits is closely related to the circuit structure and power supply network. This complexity is reflected in the behavior of the electron-hole pairs. Some work has been based on circuit level simulation [22], but this is not adequate for research on the microphysical mechanism. Therefore, this paper presents TCAD simulation in order to analyze the HDR-TIRE on Nano-Scale FDSOI devices for different supply voltages. The width of the transistor was 324 nm and 396 nm, corresponding to NMOS and PMOS, respectively. The buried oxide layer thickness was 20 nm. The top silicon film thickness was 6 nm. Three-dimensional TCAD models were developed using the Synopsys Sentaurus suit of TCAD tools.

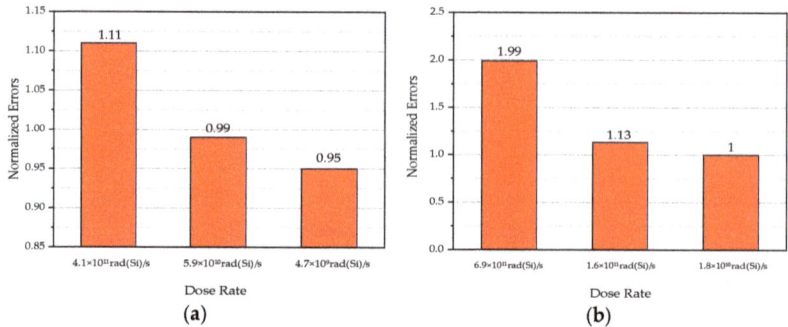

Figure 3. The relationship between normalized data errors and dose rate for different supply voltages. (**a**) Supply voltage is 0.8 V; (**b**) supply voltage is 0.88 V.

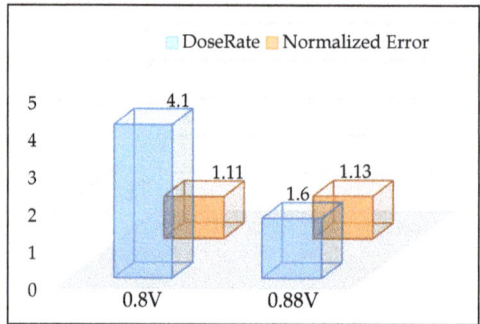

Figure 4. Comparison of normalized errors with different dose rates and supply voltages. The unit of DoseRate is 1×10^{11} rad(Si)/s.

The initial device models of NMOS and PMOS were established and the electrical characteristic curves were simulated. The gate voltage was swept from 0 V to 0.8 V for NMOS and from −0.8 V to 0 V for PMOS. Additionally, the electrical characteristic curves of NMOS and PMOS in the SPICE model were obtained. The results of the electrical characteristics calibration are shown in Figure 5. It can be seen that the electrical characteristic curves of the two models could be matched nicely with an average error of less than 5%, indicating that the established FDSOI device models can be used for HDR-TIRE simulations.

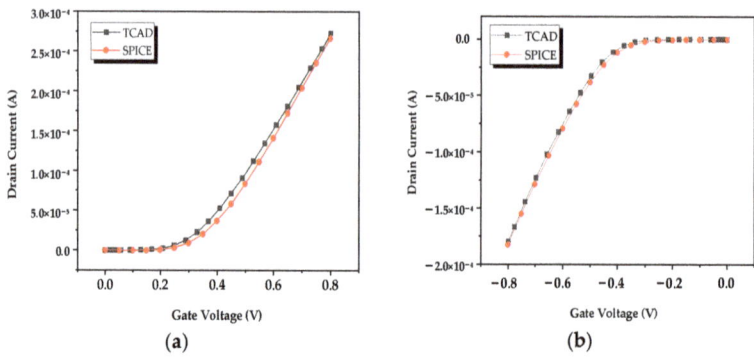

Figure 5. FDSOI device model calibration results. (**a**) NMOS; (**b**) PMOS.

As depicted in Figure 6, the 3D model of the Nano-Scale FDSOI inverter was established based on NMOS and PMOS devices. The logic function of the inverter was correct and the inverter gate switching time was less than 60 ps. Therefore, it could be operated normally at the frequency of 15 GHz.

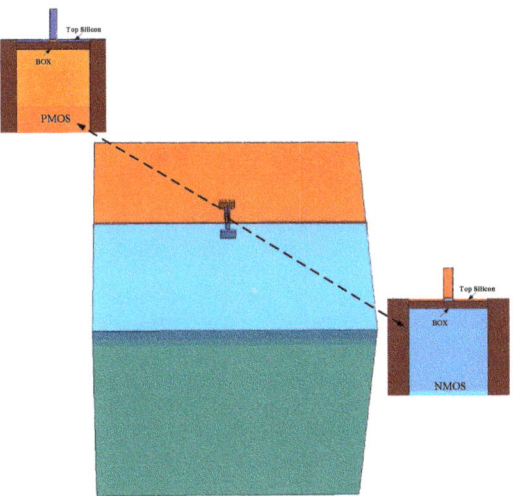

Figure 6. TCAD model of nano-scale FDSOI inverter.

Supply voltage is a key factor for both SEE [23] and HDR-TIRE [24]. As ICs technology advances, Moore's law describes ICs development accurately, namely, performance improvement and power consumption reduction. For Nano-Scale FDSOI, the standard core supply voltage has been reduced to 0.8 V. Additionally, since the thickness of the top silicon film is on the order of a few nanometers, the mechanism of HDR-TIRE for different supply voltages is quite complicated. Therefore, this paper presents a simulation study of the FDSOI inverter in order to explore the effect of power supply voltage on high-dose-rate transient ionizing radiation response and charge collection mechanism. The simulations were carried out for two supply voltages (0.8 V, 1.2 V) and two dose rates (1×10^{12} rad(Si)/s, 1×10^{11} rad(Si)/s).

HDR-TIRE results in the appearance of photocurrent pulses in the device. To facilitate the observation of the leakage current variation, the ΔI_{peak} for different supply voltages is compared in Figure 7a. ΔI_{peak} is defined as the difference between photocurrent peak and initial leakage current since raising the supply voltage can increase the initial static leakage current of the device. The ΔI_{peak} was normalized with a 0.8 V supply voltage and a dose rate of 1×10^{12} rad(Si)/s condition, which was 8.24×10^{-8} A. It can be illustrated that an increase in supply voltage leads to an increase in ΔI_{peak}, as the dose rate is fixed. For instance, the ΔI_{peak} increased by a factor of 5.5 when the supply voltage was increased from 0.8 V to 1.2 V at the dose rate of 1×10^{12} rad(Si)/s. And the ΔI_{peak} at a dose rate of 1×10^{12} rad(Si)/s was 5× larger than that of 1×10^{11} rad(Si)/s for the 0.8 V supply voltage. Moreover, compared with the 0.8 V@1×10^{12} rad(Si)/s case in Figure 7a, the supply voltage increased to 1.2 V, although the dose rate reduced to 1×10^{11} rad(Si)/s. The ΔI_{peak} increased by a factor of 2.2, which indicates the dominant effect of supply voltage on the photocurrent.

Coupled with the photocurrent integral, the charge collection (Q_{col}) was obtained and the comparison is shown in Figure 7b. It can be concluded that the supply voltage increased from 0.8 V to 1.2 V, and Q_{col} increased by a factor of 13.8 at a dose rate of 1×10^{12} rad(Si)/s and increased by a factor of 30.3 at a dose rate of 1×10^{11} rad(Si)/s. This suggests that the increase in supply voltage leads to a broadening of the photocurrent pulse width, which results in a greater increase in Q_{col} than ΔI_{peak}.

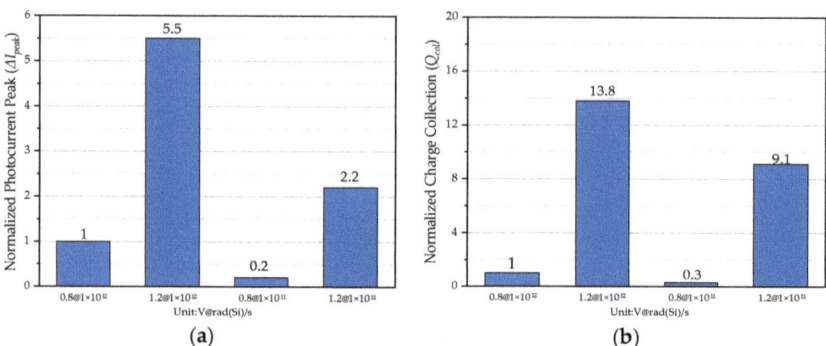

Figure 7. Comparison of normalized photocurrent peak and charge collection with different supply voltages and dose rates of inverter. (**a**) ΔI_{peak}; (**b**) Q_{col}.

Increasing the supply voltage resulted in an enhanced internal electric field, which led to an increase in the amount of charge collection and a rise in the body potential, as shown in Figure 8. As a result, the parasitic bipolar amplification became more efficient, leading to an increase in the photocurrent peak. It should be noted that increasing the supply voltage can contribute to the increase in the restore current and the competitive factor may mitigate the photocurrent, while from the simulation results the bipolar effect is the dominant mechanism.

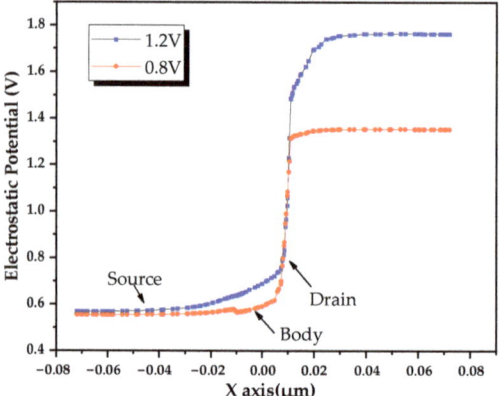

Figure 8. Effect of supply voltage on electrostatic potential distribution. The dose rate is 1×10^{12} rad(Si)/s.

5. Conclusions

In order to investigate the high-dose-rate transient ionizing radiation response of a Nano-Scale FDSOI circuit with an ultra-thin top silicon film under different dose rates and supply voltages, this paper presents an experimental study of a Nano-Scale FDSOI DFFs circuit. The results indicated that HDR-TIRE causes a data error at the dose rate of 4.70×10^9 rad(Si)/s, and the number of errors increases nonlinearly with the increase in dose rate value.

For the 0.8 V supply voltage, the dose rate increased 6.9-fold and 12.6-fold, from 5.9×10^{10} rad(Si)/s to 4.1×10^{11} rad(Si)/s and 4.7×10^9 rad(Si)/s to 5.9×10^{10} rad(Si)/s; the normalized error increased by 12.1% and 4.2%, respectively. An increase in the dose rate led to a higher number of generated electron-hole pairs. This impacted the electrostatic potential and charge collection mechanisms such as drift process and parasitic bipolar amplification. There was a synergistic effect, resulting in the nonlinear relationship between dose rate value and data errors. The experimental results showed that even with a lower dose rate, a higher supply voltage results in a higher number of data errors in DFFs. Three-dimensional TCAD simulations of the devices were performed. It was found that the number of charges collected increases with the increase in supply voltage for the enhancement of the parasitic bipolar amplification effect in the Nano-Scale FDSOI device. Understanding the high-dose-rate transient ionizing radiation response of the Nano-Scale FDSOI circuit and the impact of factors such as supply voltage is important when considering radiation-hardening techniques.

Author Contributions: Conceptualization and methodology, T.L.; validation, L.W.; formal analysis, T.L.; writing—original draft preparation, T.L.; writing—review and editing, Y.B. and L.S.; visualization, J.Y., C.Y. and C.G.; supervision, Y.Z.; project administration, Y.Z. All authors have read and agreed to the published version of the manuscript.

Funding: This research was supported by the Qian Xuesen Youth Innovation Fund. NO. 2021.01.

Data Availability Statement: Not applicable.

Acknowledgments: The authors would like to thank Chenhui Wang at the Northwest Institute of Nuclear Technology for the help and support with the chip test.

Conflicts of Interest: The authors declare no conflict of interest.

References

1. Hu, C.; He, C. Numerical Simulation of Neutron Irradiation in Ultra-Deep Submicron SOI NMOSFETs. *At. Energy Sci. Technol.* **2011**, *45*, 456–460.
2. Xue, H.; Zhang, M. Numerical Simulation of Transient Dose Rate Effects in NMOS Devices with Deep Submicron SOI process. *Appl. Electron. Technol.* **2019**, *45*, 59–61.
3. Niu, Z.; Tu, Y. Research on Neutron Irradiation Effect of Computer Based on DSP. In Proceedings of the 10th Annual National Conference on Radiation Resistant Electronics and Electromagnetic Pulse, Shenyang, China, 1 July 2009; pp. 97–102.
4. Liu, W. *Radiation Effects of Silicon Semiconductor Devices and Reinforcement Techniques*, 1st ed.; Beijing Science Press: Beijing, China, 2013; pp. 132–136.
5. Ellis, T.D.; Kim, Y.D. Use of a Pulsed Laser as an Aid to Transient Upset Testing of I2L LSI Microcircuits. *IEEE Trans. Nucl. Sci.* **1978**, *25*, 1489–1493. [CrossRef]
6. Weber, O. FDSOI vs FinFET: Differentiating Device Features for Ultra Low Power & IoT applications. In Proceedings of the 2017 IEEE International Conference on IC Design and Technology (ICICDT), Austin, TX, USA, 23–25 May 2017; pp. 1–3.
7. Luo, H. SOI Devices and Applications. *Electron. Packag.* **2007**, *7*, 41–45.
8. Perin, L.; Pereira, A. SOI Stacked Transistors Tolerance to Single-Event Effects. *IEEE Trans. Device Mater. Reliab.* **2019**, *19*, 393–401. [CrossRef]
9. Heidel, A. Single-Event Upsets and Multiple-Bit Upsets on a 45 nm SOI SRAM. *IEEE Trans. Nucl. Sci.* **2009**, *56*, 3499–3504. [CrossRef]
10. Elesin, V.; Nazarova, G. Investigation of the Possibility to Develop Radiation-Hardness LSIs for Navigational Purposes According to the 0.35-μm Domestic CMOS SOI Technology. *Russ. Microelectron.* **2012**, *41*, 266–277. [CrossRef]
11. Zhang, Z.; Liu, J. Investigation of Threshold Ion Range for Accurate Single Event Upset Measurements in Both SOI and Bulk Technologies. *IEEE Trans. Nucl. Sci.* **2014**, *61*, 1459–1467. [CrossRef]
12. Raine, M.; Gaillardin, M. Experimental Evidence of Large Dispersion of Deposited Energy in Thin Active Layer Devices. *IEEE Trans. Nucl. Sci.* **2011**, *58*, 2664–2672. [CrossRef]
13. Wirth, J.; Rogers, S. The Transient Response of Transistors and Diodes to Ionizing Radiation. *IEEE Trans. Nucl. Sci.* **1964**, *11*, 24–38. [CrossRef]
14. Palkuti, L.; Alles, M. The Role of Radiation Effects in SOI Technology Development. In Proceedings of the 2014 SOI-3D-Subthreshold Microelectronics Technology Unified Conference (S3S), Millbrae, CA, USA, 6–9 October 2014; pp. 1–2.
15. Schwank, J.; Ferlet, V. Radiation Effects in SOI Technologies. *IEEE Trans. Nucl. Sci.* **2003**, *50*, 522–538. [CrossRef]
16. Zhang, Z.; Zou, S. Radiation Hardening Technology of SOI Materials and Devices. *Chin. Sci. Bull.* **2017**, *62*, 1004–1017. [CrossRef]
17. Ferlet, V.; Gasiot, G. Insights on the Transient Response of Fully and Partially Depleted SOI Technologies Under Heavy-ion and Dose-Rate Irradiations. *IEEE Trans. Nucl. Sci.* **2002**, *49*, 2948–2956. [CrossRef]
18. Nazarova, G.; Elesin, V. Long-Term Transient Radiation Effects in SOI CMOS RF ICs. In Proceedings of the 2015 15th European Conference on Radiation and Its Effects on Components and Systems (RADECS), Moscow, Russia, 14–18 September 2015; pp. 1–4.
19. Wu, L.; Han, X. Radiation-hardened SOI Process Design and FPGA. *Inf. Electron. Eng.* **2012**, *10*, 627–632.

20. Zhao, K.; Liu, Z. Resist Radiation 128 KB PDSO1 Static Random Access Memory. *J. Semicond.* **2007**, *7*, 1139–1143.
21. Ball, D.; Alles, M. The Impact of Charge Collection Volume and Parasitic Capacitance on SEUs in SOI- and Bulk-FinFET D Flip-Flops. *IEEE Trans. Nucl. Sci.* **2018**, *65*, 326–330. [CrossRef]
22. Liu, H.; Golke, K. A New Dose Rate Model for SOI MOSFETs and Its Implementation in SPICE. In Proceedings of the 2005 IEEE International SOI Conference Proceedings, Honolulu, HI, USA, 3–6 October 2005; pp. 112–113.
23. Kauppila, J.; Kay, W. Single-Event Upset Characterization Across Temperature and Supply Voltage for a 20-nm Bulk Planar CMOS Technology. *IEEE Trans. Nucl. Sci.* **2015**, *62*, 2613–2619. [CrossRef]
24. Li, T.; Zhao, Y. Investigation on Transient Ionizing Radiation Effects in a 4-Mb SRAM with Dual Supply Voltages. *IEEE Trans. Nucl. Sci.* **2022**, *69*, 340–348. [CrossRef]

Disclaimer/Publisher's Note: The statements, opinions and data contained in all publications are solely those of the individual author(s) and contributor(s) and not of MDPI and/or the editor(s). MDPI and/or the editor(s) disclaim responsibility for any injury to people or property resulting from any ideas, methods, instructions or products referred to in the content.

Article

Oxide Electric Field-Induced Degradation of SiC MOSFET for Heavy-Ion Irradiation

Xiaowen Liang [1], Haonan Feng [1,2], Yutang Xiang [1,2], Jing Sun [1], Ying Wei [1], Dan Zhang [1,2], Yudong Li [1], Jie Feng [1,*], Xuefeng Yu [1,*] and Qi Guo [1]

[1] Key Laboratory of Functional Materials and Devices for Special Environments, Xinjiang Technical Institute of Physics and Chemistry, Chinese Academy of Sciences, Urumqi 830011, China
[2] University of Chinese Academy of Sciences, Beijing 100049, China
* Correspondence: fengjie@ms.xjb.ac.cn (J.F.); yuxf@ms.xjb.ac.cn (X.Y.)

Abstract: This work presents an experimental study of heavy-ion irradiation with different particle linear energy transfer (LET), gate biases, and drain biases. The results reveal that when the irradiation biases are low, the SiC MOSFET does not experience single event effect (SEE) and the electrical properties remain unchanged (the devices are in the safe operating area (SOA)). However, the oxide breakdown voltage of the device is significantly decreased due to the latent damage generated by the irradiation. The experimental results, along with TCAD simulations, suggest that the latent damage induced by the irradiation in the gate oxide is closely related to the peak electric field in the gate oxide at the time of particle incidence. This peak electric field is determined by the potential difference between the two sides of the gate oxide, which is affected by the particle LET, gate biases, and drain biases together. The high potential is determined by the combined effect of the LET and the drain-source voltage. The impact ionization of the particle by the applied electric field causes the accumulation of holes in the JFET oxide, which leads to a decrease in the doping of the N^- epitaxial layer and eventually causes a rise in the high potential near the JFET oxide. The low potential is determined by the gate bias, and the negative bias applied to the gate can further increase the potential difference between the two sides of the oxide, causing an increase in the peak electric field in the gate oxide and aggravating the gate oxide damage.

Keywords: SiC MOSFET; heavy-ion irradiation; oxide reliability; TCAD

1. Introduction

Silicon carbide (SiC) MOSFETs are a new generation of power devices based on wide bandgap semiconductor materials with excellent high-voltage, high-temperature, and high-frequency characteristics. SiC MOSFETs can significantly improve system efficiency and power density while reducing system size and weight and have gained widespread attention and application in various fields, including for electric vehicles, high-voltage power grids, photovoltaic inverters, and railroad traction [1–3]. Furthermore, there is a growing demand for new power devices with high performance and high reliability in aerospace, nuclear power, and other radiation fields [4,5]. However, the performance of Si-based power devices has reached its physical limit and cannot be further enhanced; thus, researchers have also turned their attention to SiC power devices. Rays and particles in the radiation environment can affect the characteristics of SiC power devices and pose a threat to the performance and reliability of the devices; therefore, there is an urgent need to investigate the radiation effect and reliability degradation mechanisms of SiC MOSFETs in the radiation environment [6].

Because of the thick gate oxide, power MOSFETs are susceptible to the total ionizing dose (TID) effects in the space radiation environment, causing the degradation of electrical parameters such as the threshold and blocking voltages [7,8]. In addition, the power devices are sensitive to single-event effects due to the high-rated drain voltage and ease of particles passing through the sensitive region of the device. The most severe challenges currently faced by SiC MOSFETs in the space radiation environment are single-event burnout (SEB) and single-event gate rupture (SEGR) caused by high-energy particles [9–16]. Single-event effects can lead to the instantaneous catastrophic failure of the device, impacting the reliability of the spacecraft. To ensure the reliability of SiC MOSFETs in space applications, high-energy particle irradiation experiments are conducted on the devices to determine their failure threshold voltage and establish the SEE safe operating area for devices [6,17]. It has been demonstrated that the safe operating area of SiC MOSFETs under heavy-ion irradiation is closely correlated to V_{DS}, as illustrated in Figure 1 [18,19]. In Region 3, high V_{DS} was applied during heavy-ion irradiation, and SEB occurred in the device [11]. At this stage, the PN junction between the source and drain burnt out, and the device lost the blocking characteristics. In Region 2, with medium V_{DS}, damage or latent damage occurred in the source-drain PN junction and gate oxide of the SiC MOSFET. These damages caused a significant increase in the leakage current of the device, resulting in the degradation of the characteristics. It is worth noting that the increased gate-source leakage current I_{GSS} in this region indicated that the device had undergone the SEGR effect. It was concluded that the oxide damage induced by heavy-ion irradiation could be attributed to multiple particle impacts [14,20], high electric fields generated by accumulated holes [21], or localized high-power density [15,19,22]. In Region 1, devices were irradiated at low V_{DS}, the electrical parameters of the device did not change significantly, and the devices were considered to be in the safe operating area [6].

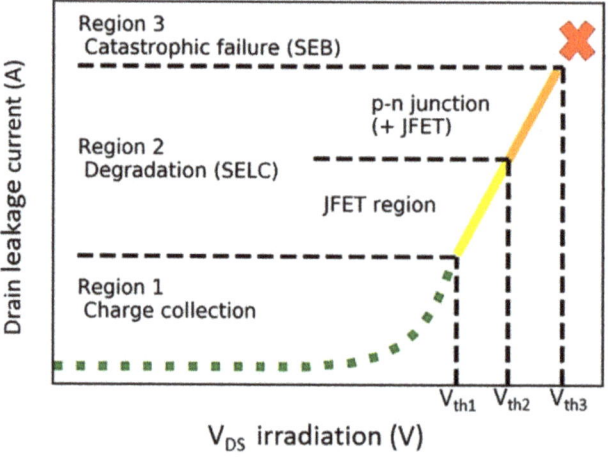

Figure 1. Current response at different V_{DS} during heavy-ion irradiation.

The above study found that the determination of the safe operating area for SiC MOSFETs in the space environment pertains to the effect of drain bias. However, SiC MOSFETs are high-voltage, high-power devices that typically require the application of negative gate voltage to prevent improper conductivity during operation. Therefore, the effect of gate bias (V_{GS}) must also be considered when determining the safe operating area of the device through experiments. Moreover, the long-term reliability of devices in the safe operating area must be taken into account. The gate oxide reliability of SiC MOSFETs has been a significant concern [23,24]. Although the gate oxide reliability of SiC MOSFETs in conventional environments has been largely solved [25], the damage in the gate oxide caused by heavy-ion irradiation in region 2 has raised concerns regarding the long-term reliability of SiC MOSFETs in space radiation environments. Therefore, the variation of oxide reliability in the safe operating region must be further investigated.

In this study, heavy-ion irradiation experiments were conducted at low biases (both the V_{DS} and V_{GS}) to ensure that the devices operated within the safe operating area. The irradiated devices were subjected to accelerated stress experiments on gate oxide to obtain the changes in oxide reliability. Based on the experimental results, the effect of gate and drain bias on the oxide reliability is summarized, and the mechanism of the oxide damage is analyzed by combining the experimental results and TCAD simulation.

2. Experimental Setup

The devices under test (DUT) used in this study were commercial 1200 V, 60 mΩ N-channel SiC MOSFETs (CGE1M120060). The recommended gate-source voltage of the device was −5/+20 V and the measured drain-source breakdown voltage $V_{(BR)DSS}$ was approximately 1500 V (test condition was I_{DS} = 1 mA). The thickness of the oxide layer was approximately 50 nm, and the thickness of the epitaxial layer was approximately 10 μm. The die of this device was selected to allow heavy ions to penetrate the sensitive region of the device. The die was encapsulated for the bias experiments. Heavy-ion irradiation experiments were conducted at the China Institute of Atomic Energy and the Lanzhou Heavy Ion Accelerator National Laboratory. The irradiated heavy ions were ^{35}Cl, ^{73}Ge, and ^{181}Ta, and the corresponding LETs in SiC were 15.89, 39.6, and 78.7 MeV/(mg/cm^2), respectively. The flux in the experiments was 1×10^4 ions/cm^2/s, and the fluence was 1×10^6 ions/cm^2. Low drain and gate biases were used to ensure that no single-event effect occurred in the device. The experimental configurations are shown in Table 1. The incidence depths of three ions in the device were greater than 20 μm, indicating that all could pass through the sensitive region (10 μm epitaxial layer); thus, the effect of incidence depth could be excluded in the subsequent analysis.

Table 1. Particle information and irradiation bias.

Ions	Energy MeV	LET in SiC MeV/(mg/cm^2)	Depth in SiC μm	Flux ions/cm^2/s	Fluence ions/cm^2	Irradiation Bias
Cl	110	15.89	20.31			V_{GS} = 0 V, V_{DS} = 60 V
Ge	210	39.6	20.19			V_{GS} = 0 V, V_{DS} = 60 V
^{181}Ta	2005.5	78.7	79.29	10^4	10^6	V_{GS} = 0 V, V_{DS} = 60 V
						V_{GS} = 0 V, V_{DS} = 30 V
						V_{GS} = −3 V, V_{DS} = 30 V
						V_{GS} = −5 V, V_{DS} = 30 V

The Keithley Source Measure Unit models 2636 and 2410 were connected to a PC to record the changes in the gate-source leakage current (I_{GS}) and drain-source leakage current (I_{DS}) during irradiation in real time, and the connection is shown in Figure 2. In the test, the high-voltage Source Measure Unit model 2410 applied a drain-source voltage and monitored the leakage current at this voltage; the Source Measure Unit model 2636 applied a negative gate-source voltage to ensure channel shutdown and monitored the leakage current corresponding to this negative gate voltage in real-time. In order to ensure that the source meter would not be damaged during the experiment, the current limit of the source meter was set to 1 μA. Only one device could be irradiated at a time with this system, and to ensure the accuracy of the experimental results, three devices were irradiated for each bias condition (see Table 1).

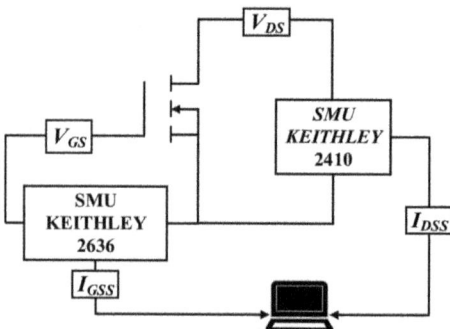

Figure 2. Online current monitoring system for SiC MOSFETs during irradiation.

The change in the subthreshold transfer characteristic curve of the SiC MOSFET was measured before and after irradiation using a Keithley 4200-SCS semiconductor parametric instrument. The I–V curve was tested with the drain voltage V_{DS} = 50 mV, and the gate voltage swept from −5 V to 10 V. The blocking voltage of the device was measured using BC3193 at I_{DS} = 1 mA. Heavy-ion irradiation can induce latent damage in the gate oxide as a precursor to oxide breakdown [14]. Some of the latent damage in oxide can be easily activated by the applied gate stress and cause oxide breakdown. Therefore, post-irradiation gate stress (PIGS) tests were conducted on the devices using Keithley 4200-SCS after irradiation. In the PIGS test, the gate-source voltage V_{GS} was scanned from 0 V to 20 V, and the variation of the gate oxide leakage current I_{GSS} was monitored.

Since the activation energy of the latent damage in the oxide is unknown, the PIGS test from 0 to 20 V did not guarantee the activation of the gate oxide latent damage to fully characterize the change in gate oxide reliability. Therefore, a ramp voltage stress (RVS) test with higher gate voltage was performed on the device. In the RVS test, the gate voltage started from 20 V and increased by 200 mV every 20 s until oxide breakdown occurred, and the gate voltage at the time of oxide breakdown was recorded. Both irradiation and tests were performed at room temperature.

3. Results and Analysis

3.1. Heavy-Ion Experiment Results

During the heavy-ion irradiation test, biases were applied to the gate-source and drain-source terminals simultaneously. The variations of the gate-source leakage current (corresponding to V_{GS} = 0 V, −3 V, and −5 V) and drain-source leakage current (corresponding to V_{DS} = 30 V and 60 V) of the devices are shown in Figure 3. During the irradiation period, the leakage current jumped significantly, while after the irradiation ceased, the current no longer displayed significant jumps, and only minor fluctuations were observed. The analysis suggested that the jump in leakage current during irradiation was caused by the interaction between heavy ions and the extranuclear electron of the material. The high flux of heavy ions during irradiation generated a large number of electron-hole pairs within the material, leading to carrier fluctuations inside the material, which were ultimately manifested as jumps in the leakage current. This effect ceased after the irradiation was terminated, resulting in the current returning to its initial value.

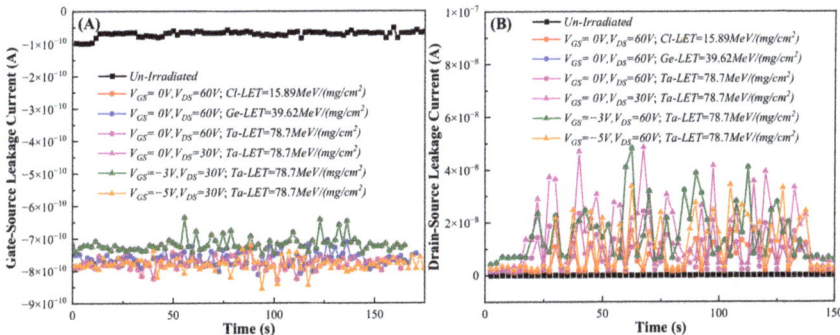

Figure 3. Variation of SiC MOSFET leakage current during irradiation: (**A**) I_{GSS}; (**B**) I_{DSS}.

The transfer characteristic curves of the SiC MOSFETs did not exhibit significant drift after irradiation, as depicted in Figure 4A. According to the maximum transconductance method, the threshold voltage of the device was extracted in the I–V curve, and the V_{TH} of the device was approximately 2.4 V before and after irradiation, without significant changes. This indicated that the equivalent total ionizing dose produced by heavy-ion irradiation was low and did not accumulate trapped charges in the oxide of the device, causing a change in the threshold voltage. The change in the blocking voltage of the device before and after irradiation is shown in Figure 4B. The blocking voltage was approximately 1500 V, and no degradation occurred. This result indicates that heavy-ion irradiation does not produce significant defects in the SiC material that causes degradation of the reverse blocking characteristics of the PN junction composed of N-epitaxy and P-well.

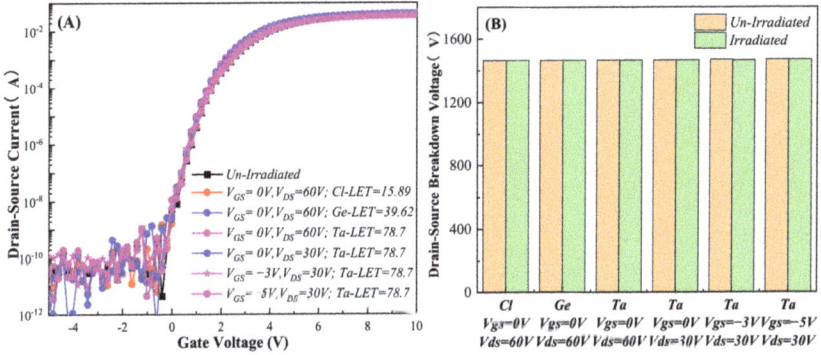

Figure 4. Changes in the electrical parameters of SiC MOSFETs after heavy-ion irradiation: (**A**) sub-threshold transfer characteristic curves; (**B**) blocking voltage (BV_{DSS}).

The results of the PIGS test (0–20 V) of the irradiated devices under different conditions are depicted in Figure 5. The oxide leakage current of the irradiated device did not exhibit significant changes in comparison to the unirradiated device. This suggests that the irradiation biases used in the test were within the SEE SOA of the device. Although no SEGR occurred in the gate oxide of the device under these irradiation biases, the heavy-ion irradiation may have produced latent damage in the gate oxide that was difficult to activate. Therefore, the RVS experiments with higher gate voltage were continued for the device.

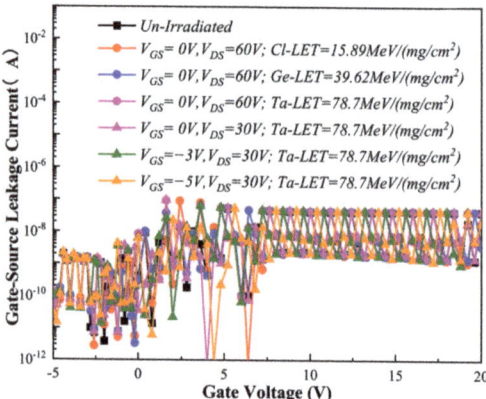

Figure 5. Experimental results of post-irradiation gate stress.

Figure 6A illustrates the changes in the oxide breakdown voltage of the device after heavy-ion irradiation with different LET at V_{GS} = 0 V and V_{DS} = 60 V. Figure 6B presents the degradation of the oxide breakdown voltage of SiC MOSFETs after irradiation with different gate and drain biases. It can be observed that as the LET, drain voltage, and gate voltage increased, the oxide breakdown voltage of the device degraded drastically.

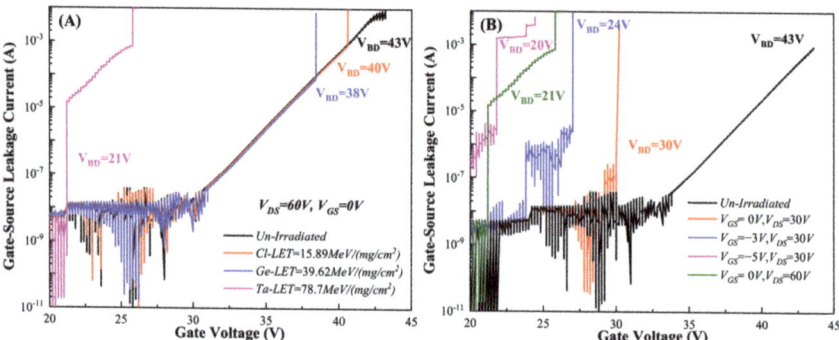

Figure 6. Variation of oxide breakdown voltage after heavy-ion irradiation: (**A**) Different LETs; (**B**) Different irradiation biases.

The degradation of the oxide breakdown voltage in irradiated devices is suggested to be closely associated with the transient high electric field in the oxide during irradiation. The high electric field can cause a rapid rise in the generation rate of defects in SiO_2 [26], resulting in a higher density of latent damage in the oxide. The latent damage does not affect the characteristics when it is not activated but can seriously affect the oxide reliability once activated.

3.2. Degradation Mechanism of Oxide Reliability

The experimental results discussed above reveal that heavy-ion irradiation leads to a decrease in oxide breakdown voltage, influenced by the LET, gate bias, and drain bias. When heavy ions penetrate the device, a significant number of electron-hole pairs can be generated along its traces, affecting the carrier concentration in the device and, consequently, altering the potential distribution. Due to the extremely short response time of heavy-ion incidence, the instantaneous potential and field changes in the oxide could not be monitored experimentally using the equipment in the experiment. Therefore, TCAD simulations were used to analyze the electrical parameters at the moment of particle incidence.

The irradiated device used in this study was a planar gate SiC MOSFET, and the cross-section of the devices is depicted in Figure 7. Based on this structure, a two-dimensional model was constructed in TCAD. The simulation parameters were obtained from previously published papers [11,12,27,28], as shown in Table 2. In the simulation, the incident position of the heavy ions was located above the JFET region, which is the most sensitive area of the oxide. The heavy-ion incidence path is shown as the red dashed line in Figure 7, which penetrated through the gate oxide and epitaxial layer. The number of electron-hole pairs generated by the particle along its incident path was related to the LET, and unit conversion was required in the simulation: 1 pC/μm(SiC) = 151 MeV/mg/cm^2. The electric field in the oxide reached its peak value of approximately 10 ps of ion incidence, followed by a rapid decrease in the electric field in the gate oxide. Hence, the distribution of potential barriers and electric fields at 10 ps for the irradiated device under different conditions was extracted during the simulation. The models used in the simulations included the drift-diffusion model for transport, the Shockley–Read–Hall model for generation-recombination, the doping dependence model, and a high field saturation model for mobility.

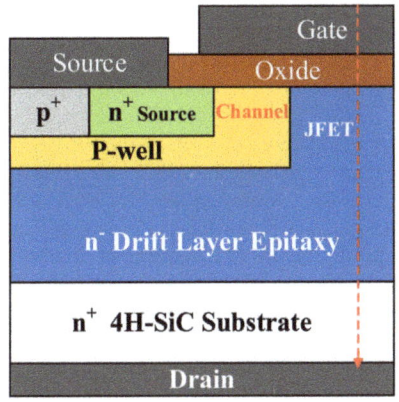

Figure 7. Cross-sectional view of the device.

Table 2. Parameters used in TCAD simulations.

Parameter	Value
N-Epi Doping/Depth	1×10^{16} cm^{-3}, 10 μm
N+ Substrate	1×10^{19} cm^{-3}
Body Doping/Depth	2×10^{17} cm^{-3}, 1.5 μm
N+ Drain Doping	10^{19} cm^{-3}
Oxide Thickness	50 nm
Ion Track Radius/Length	50 nm, 15 μm

Heavy ions generate electron-hole pairs along their incident traces. The applied voltage during irradiation induces an electric field in the device, and the collisional ionization of carriers under the electric field produces a lot of electron-hole pairs. At the same time, the electrons and holes move in opposite directions under the action of the electric field. The electrons are more mobile than the holes in SiC [29]; this causes holes to move and accumulate at the SiC/SiO$_2$ interface. As shown in Figure 8, the concentration of holes inside the device varies with LETs. Heavy ions with high LET result in a much higher hole concentration inside the device than the low LET particle. The analysis suggested that the irradiation bias was the same in both figures, indicating that the electric field inside the device was the same. However, the incident particle with high LET produced a higher number of electron-hole pairs along its path, leading to more intense carrier collision ionization, which, ultimately, makes a higher density of accumulated holes.

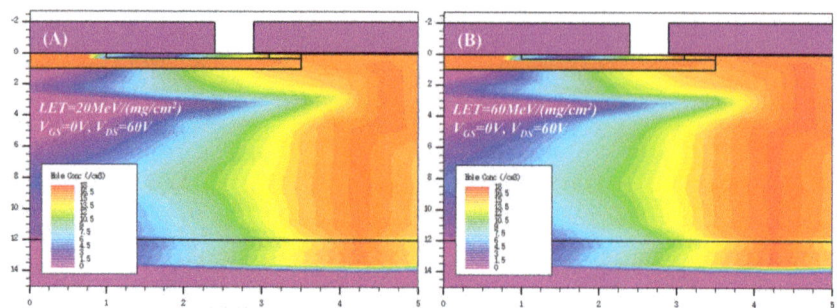

Figure 8. Concentration of holes in the device after the incidence of heavy ions with different LETs: (**A**) LET = 20 MeV/(mg/cm^2); (**B**) LET = 60 MeV/(mg/cm^2).

In general, SiC MOSFETs have low doping in the N-type epitaxial region to increase the blocking voltage. Therefore, the accumulated holes from heavy-ion irradiation can severely reduce the doping in the N-type epitaxial region near the incident path. The change in doping concentration further affects the potential distribution inside the device.

The simulations of potential distributions inside the device for different LETs are given in Figure 9A,B. The potential distribution near the JFET oxide is more intensive with high LET heavy ions. The analysis suggested that this was due to the high density of holes generated by the high LET heavy ions causing a reduction in N$^-$ epitaxial doping near the JFET oxide. The doping of the N$^-$ epitaxial region is closely related to the blocking characteristics of the device, and the drain voltage drops mainly in the depletion layer of the lower-doped N$^-$ epitaxial region during the blocking state. Therefore, the reduced N$^-$ doping near the JFET oxide led to a more intensive potential distribution near it.

Figure 9B,C show the simulated potential distribution in the device when irradiated at different V_{DS}. The potential near the JFET oxide increased remarkably when irradiated with heavy ions at high V_{DS}. The analysis suggested that the increase in high potential was closely related to the high electric field generated by the V_{DS}. At the high electric field, the impact generation rate of the carriers increased, which generated more electron-hole pairs in the device. The simulations of the impact generation rate in irradiated devices at different V_{DS} are depicted in Figure 10. The impact generation rate near the JFET oxide of the devices irradiated at high V_{DS} was significantly higher than those irradiated at low V_{DS}. The high impact generation rate could generate more electrons and holes. The accumulation of holes further decreased N$^-$ doping near the JFET oxide, which also increased the potential.

A comprehensive analysis of the influence mechanisms of LET and V_{DS} suggested that they jointly determine the high potential value on the SiC side of the JFET gate oxide layer at heavy-ion incidence.

Comparing Figure 9C,D, it is found that the potential distribution near the JFET oxide was approximately the same when irradiated at the same LET and drain bias. However, the negative gate bias could affect the low potential of the metal oxide. As the negative gate bias voltage increased, the low potential value on the metal side of the gate oxide decreased. The difference between the high and low potential values on both sides of the gate oxide determined the peak electric field in the oxide.

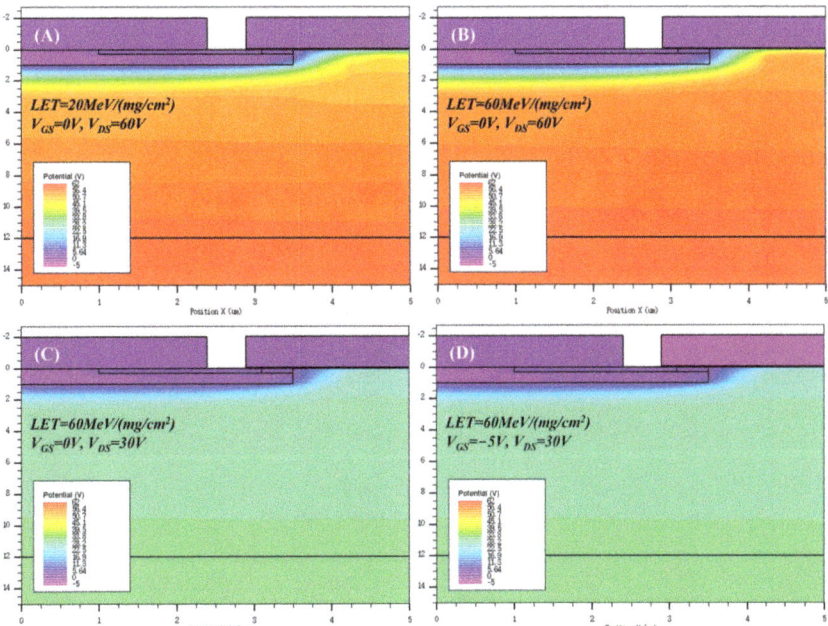

Figure 9. The potential distribution in the device when irradiated under different conditions: (**A**) LET = 20 MeV/(mg/cm^2), V_{DS} = 60 V, V_{GS} = 0 V; (**B**) LET = 60 MeV/(mg/cm^2), V_{DS} = 60 V, V_{GS} = 0 V; (**C**) LET = 60 MeV/(mg/cm^2), V_{DS} = 30 V, V_{GS} = 0 V; (**D**) LET = 60 MeV/(mg/cm^2), V_{DS} = 30 V, V_{GS} = −5 V.

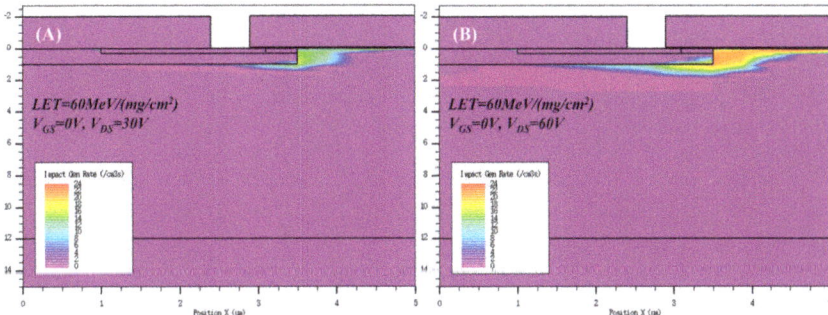

Figure 10. Variation of impact generation rate due to particle incidence at different V_{DS}: (**A**) V_{DS} = 30 V; (**B**) V_{DS} = 60 V.

In order to provide a more intuitive analysis of the effects of LET, V_{DS}, and V_{GS} on the peak electric field in the gate oxide, the potential distribution along the particle trace was extracted in the simulation, and the results are shown in Figure 10. For the unirradiated device, the high voltage applied at the drain uniformly dropped in the epitaxial layer of approximately 3 µm. However, for the irradiated device, the drain bias experienced a significant drop near the gate oxide. By examining the enlarged plot in Figure 11A, it is evident that the LET mainly affected the high potential value on the SiC side of the gate oxide. The proportion of the drain voltage coupled to the oxide increased continuously with the increase in the LET. The simulation of the potential distributions in the irradiated device at different V_{DS} and V_{GS} are given in Figure 11B. The high potential of the gate oxide in the irradiated device at different V_{DS} was significantly different, and the potential in the irradiated device at high V_{DS} was much higher than that at low V_{DS}. This discrepancy is mainly attributed to the intensified collisional ionization resulting from the high electric field at high V_{DS}. The V_{DS} and LET jointly affected the high potential of the gate oxide. There was no significant difference in the high potential of irradiated devices under different V_{GS}, but the low potential varied with V_{GS}. The potential difference between the high and low potentials determined the peak electric field in the gate oxide.

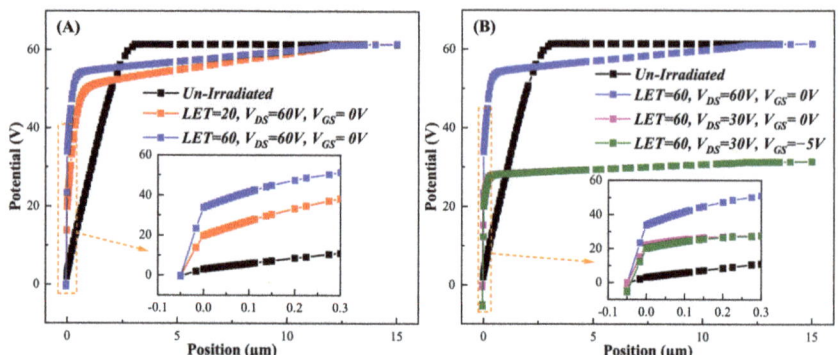

Figure 11. Changes in potential along the particle traces when irradiated with heavy ions under different conditions: (**A**) Different LETs; (**B**) Different irradiation biases.

The gate oxide thickness of the device in the simulation was 50 nm, and, as can be concluded from Figure 11, the potential difference between the two sides of the gate oxide reached more than 20 V, even at a lower bias. Therefore, it can be calculated that the instantaneous peak electric field in the gate oxide was greater than 4 MV/cm, which was even close to the critical breakdown electric field of SiO_2 (10 MV/cm) at higher V_{DS}. The increase in the transient peak electric field greatly increased the defect generation rate, thus creating more defects in the gate oxide and affecting the reliability of the gate oxide.

4. Conclusions

This study investigated the effect of the peak electric field in the gate oxide on the generation of oxide latent damage during heavy-ion irradiation. The irradiated SiC MOSFETs were found to have no single-event effect and good functional characteristics, but the oxide reliability was degraded. The experimental results show that the degree of oxide reliability degradation is affected by a combination of LET, V_{GS}, and, V_{DS}.

The analysis and simulation concluded that the degradation of the oxide reliability of SiC MOSFETs is caused by the defects generated by the peak electric field during heavy-ion irradiation. The particle LET and V_{DS} can affect the high potential coupled to one side of the gate oxide during irradiation, while the applied VGS affects the low potential on the other side of the oxide. The potential difference between the two determines the peak electric field in the gate oxide.

The results of this study suggest that even if irradiation biases are in the SEE safe operating area, heavy-ion irradiation can severely limit the reliability and lifetime of the device, which poses a new challenge for the space application of SiC MOSFETs. In summary, for the study of the adaptability of SiC MOSFETs in the space environment, in addition to the tricky SEE study of SiC MOSFETs, the gate oxide reliability must be considered.

Author Contributions: Methodology, X.L. and J.F.; Software, Y.W.; Validation, Y.L.; Investigation, D.Z.; Resources, X.Y. and Q.G.; Data curation, H.F. and Y.X.; Funding acquisition, J.S., J.F. and Y.W. All authors have read and agreed to the published version of the manuscript.

Funding: This work was partially funded by the National Natural Science Foundation of China (Grant no. 11975305 and 12175307) and Young Scholars in Western China, the Chinese Academy of Sciences (Grant no. 2021-XBQNXZ-021).

Data Availability Statement: Not applicable.

Acknowledgments: This work acknowledges the support of the Institute of modern physics, Chinese Academy of Sciences and the China institute of atomic energy in the heavy ion experiment.

Conflicts of Interest: The authors declare no conflict of interest.

References

1. She, X.; Huang, A.Q.; Lucia, O.; Ozpineci, B. Review of Silicon Carbide Power Devices and Their Applications. *IEEE Trans. Ind. Electron.* **2017**, *64*, 8193–8205. [CrossRef]
2. Roccaforte, F.; Fiorenza, P.; Greco, G.; Nigro, R.L.; Giannazzo, F.; Iucolano, F.; Saggio, M. Emerging Trends in Wide Band Gap Semiconductors (SiC and GaN) Technology for Power Devices. *Microelectron. Eng.* **2018**, *187–188*, 66–77. [CrossRef]
3. Li, J.; Igarashi, S.; Fujishima, N. SiC Power Devices and Application to Power Electronics. In Proceedings of the PCIM Asia 2022, International Exhibition and Conference for Power Electronics, Intelligent Motion, Renewable Energy and Energy Management, Shanghai, China, 26–27 October 2022.
4. Boomer, K.; Lauenstein, J.M.; Hammoud, A. *Body of Knowledge for Silicon Carbide Power Electronics*; NASA: Washington, DC, USA, 2016.
5. Jie, X.; Qing, K.; Xuan, Z.; Feng, L. Application Prospect of SiC Power Semiconductor Devices in Spacecraft Power Systems. In Proceedings of the 2017 IEEE 13th International Conference on Electronic Measurement & Instruments (ICEMI), Yangzhou, China, 20–22 October 2017; IEEE: Piscataway, NJ, USA, 2017.
6. Lauenstein, J.M. Wide Bandgap Power SiC, GaN Radiation Reliability. In Proceedings of the IEEE Nuclear and Space Radiation Effects Conference (NSREC), Santa Fe, NM, USA, 30 November 2020; IEEE: Piscataway, NJ, USA, 2020.
7. Akturk, A.; McGarrity, J.M.; Potbhare, S.; Goldsman, N. Radiation Effects in Commercial 1200 V 24 A Silicon Carbide Power MOSFETs. *IEEE Trans. Nucl. Sci.* **2012**, *59*, 3258–3264. [CrossRef]
8. Schwank, J.R.; Shaneyfelt, M.R.; Fleetwood, D.M.; Felix, J.A.; Dodd, P.E.; Paillet, P.; Ferlet-Cavrois, V. Radiation Effects in MOS Oxides. *IEEE Trans. Nucl. Sci.* **2008**, *55*, 1833–1853. [CrossRef]
9. Asai, H.; Nashiyama, I.; Sugimoto, K.; Shiba, K.; Sakaide, Y.; Ishimaru, Y.; Okazaki, Y.; Noguchi, K.; Morimura, T. Tolerance Against Terrestrial Neutron-Induced Single-Event Burnout in SiC MOSFETs. *IEEE Trans. Nucl. Sci.* **2014**, *61*, 3109–3114. [CrossRef]
10. Shoji, T.; Nishida, S.; Hamada, K.; Tadano, H. Analysis of neutron-induced single-event burnout in SiC power MOSFETs. *Microelectron. Reliab.* **2015**, *55*, 1517–1521. [CrossRef]
11. Witulski, A.F.; Ball, D.R.; Galloway, K.F.; Javanainen, A.; Lauenstein, J.-M.; Sternberg, A.L.; Schrimpf, R.D. Single-Event Burnout Mechanisms in SiC Power MOSFETs. *IEEE Trans. Nucl. Sci.* **2018**, *65*, 1951–1955. [CrossRef]
12. Ball, D.R.; Galloway, K.F.; Johnson, R.A.; Alles, M.L.; Sternberg, A.L.; Sierawski, B.D.; Witulski, A.F.; Reed, R.A.; Schrimpf, R.D.; Hutson, J.M.; et al. Ion-Induced Energy Pulse Mechanism for Single-Event Burnout in High-Voltage SiC Power MOSFETs and Junction Barrier Schottky Diodes. *IEEE Trans. Nucl. Sci.* **2020**, *67*, 22–28. [CrossRef]
13. Peng, C.; Lei, Z.; Chen, Z.; Yue, S.; Zhang, Z.; He, Y.; Huang, Y. Experimental and Simulation Studies of Radiation-Induced Single Event Burnout in SiC-Based Power MOSFETs. *IET Power Electron.* **2020**, *14*, 1700–1712. [CrossRef]
14. Abbate, C.; Busatto, G.; Tedesco, D.; Sanseverino, A.; Silvestrin, L.; Velardi, F.; Wyss, J. Gate Damages Induced in SiC Power MOSFETs During Heavy-Ion Irradiation—Part I. *IEEE Trans. Electron Devices* **2019**, *66*, 4235–4242. [CrossRef]
15. Busatto, G.; Di Pasquale, A.; Marciano, D.; Palazzo, S.; Sanseverino, A.; Velardi, F. Physical mechanisms for gate damage induced by heavy ions in SiC power MOSFET. *Microelectron. Reliab.* **2020**, *114*, 113903. [CrossRef]
16. Lauenstein, J.M.; Casey, M.C.; Ladbury, R.L.; Kim, H.S.; Phan, A.M.; Topper, A.D. Space Radiation Effects on SiC Power Device Reliability. In Proceedings of the 2021 IEEE International Reliability Physics Symposium (IRPS), Monterey, CA, USA, 21–25 March 2021; IEEE: Piscataway, NJ, USA, 2021.
17. Single-Event Burnout and Single-Event Gate Rupture, Mil-Std-750e, Method 1080 [S]. 2006. Available online: http://everyspec.com/MIL-STD/MIL-STD-0700-0799/MIL-STD-750E_15413/ (accessed on 18 June 2023).
18. Martinella, C.; Ziemann, T.; Stark, R.; Tsibizov, A.; Voss, K.O.; Alia, R.G.; Kadi, Y.; Grossner, U.; Javanainen, A. Heavy-Ion Microbeam Studies of Single-Event Leakage Current Mechanism in SiC VD-MOSFETs. *IEEE Trans. Nucl. Sci.* **2020**, *67*, 1381–1389. [CrossRef]
19. Martinella, C.; Natzke, P.; Alia, R.; Kadi, Y.; Niskanen, K.; Rossi, M.; Jaatinen, J.; Kettunen, H.; Tsibizov, A.; Grossner, U.; et al. Heavy-ion induced single event effects and latent damages in SiC power MOSFETs. *Microelectron. Reliab.* **2022**, *128*, 114423. [CrossRef]
20. Abbate, C.; Busatto, G.; Tedesco, D.; Sanseverino, A.; Silvestrin, L.; Velardi, F.; Wyss, J. Gate Damages Induced in SiC Power MOSFETs During Heavy-Ion Irradiation—Part II. *IEEE Trans. Electron. Devices* **2019**, *66*, 4243–4250. [CrossRef]
21. Zhou, X.; Pang, H.; Jia, Y.; Hu, D.; Wu, Y.; Zhang, S.; Li, Y.; Li, X.; Wang, L.; Fang, X.; et al. Gate Oxide Damage of SiC MOSFETs Induced by Heavy-Ion Strike. *IEEE Trans. Electron. Devices* **2021**, *68*, 4010–4015. [CrossRef]
22. Pintacuda, F.; Massett, S.; Vitanza, E.; Muschitiello, M.; Cantarella, V. SEGR and PIGS Failure Analysis of SiC MOSFET. In Proceedings of the IEEE 2019 European Space Power Conference (ESPC), Juan-les-Pins, France, 30 September–4 October 2019; IEEE: Piscataway, NJ, USA, 2019.

23. Cheung, K.P. SiC Power MOSFET Gate Oxide Breakdown Reliability—Current Status. In Proceedings of the 2018 IEEE International Reliability Physics Symposium (IRPS), Burlingame, CA, USA, 11–15 March 2018; IEEE: Piscataway, NJ, USA, 2018.
24. Matocha, K.; Ji, I.H.; Zhang, X.; Chowdhury, S. SiC Power MOSFETs: Designing for Reliability in Wide-Bandgap Semiconductors. In Proceedings of the 2019 IEEE International Reliability Physics Symposium (IRPS), Monterey, CA, USA, 31 March–4 April 2019; IEEE: Piscataway, NJ, USA, 2019.
25. Lichtenwalner, D.J.; Hull, B.; Van Brunt, E.; Sabri, S.; Gajewski, D.A.; Grider, D.; Allen, S.; Palmour, J.W.; Akturk, A.; McGarrity, J. Reliability Studies of SiC Vertical Power MOSFETs. In Proceedings of the 2018 IEEE International Reliability Physics Symposium (IRPS), Burlingame, CA, USA, 11–15 March 2018; IEEE: Piscataway, NJ, USA, 2018.
26. Stathis, J.H. Physical and Predictive Models of Ultra Thin Oxide Reliability in CMOS Devices Circuits. *IEEE Trans. Device Mater. Reliab.* **2001**, *1*, 43–59. [CrossRef]
27. Zhou, X.; Jia, Y.; Hu, D.; Wu, Y. A Simulation-Based Comparison between Si and SiC MOSFETs on Single-Event Burnout Susceptibility. *IEEE Trans. Electron. Devices* **2019**, *66*, 2551–2556. [CrossRef]
28. Li, Q.; Chen, X.; Luo, H.; Li, X.; Ma, X.; Tao, L.; Qian, J.; Tan, C. Study on Single-Event Burnout of SiC VDMOSFET Failure Mechanism and Influence Factors. In Proceedings of the 2019 20th International Conference on Electronic Packaging Technology (ICEPT), Hong Kong, China, 12–15 August 2019; IEEE: Piscataway, NJ, USA, 2019.
29. Kimoto, T.; Cooper, J.A. *Fundamentals of Silicon Carbide Technology: Growth, Characterization, Devices and Applications*; Wiley: Hoboken, NJ, USA, 2014.

Disclaimer/Publisher's Note: The statements, opinions and data contained in all publications are solely those of the individual author(s) and contributor(s) and not of MDPI and/or the editor(s). MDPI and/or the editor(s) disclaim responsibility for any injury to people or property resulting from any ideas, methods, instructions or products referred to in the content.

Article

Heavy Ion Single Event Effects in CMOS Image Sensors: SET and SEU

Zhikang Yang [1,2], Lin Wen [1,*], Yudong Li [1,*], Jie Feng [1], Dong Zhou [1], Bingkai Liu [1], Zitao Zhao [1,2] and Qi Guo [1]

[1] Xinjiang Technical Institute of Physics and Chemistry, Chinese Academy of Sciences, Urumqi 830011, China; yangzhikang19@mails.ucas.ac.cn (Z.Y.)
[2] University of Chinese Academy of Sciences, Beijing 100049, China
* Correspondence: wenlin@ms.xjb.ac.cn (L.W.); lydong@ms.xjb.ac.cn (Y.L.)

Abstract: High-energy particles in space often induce single event effects in CMOS image sensors, resulting in performance degradation and functional failure. This paper focuses on the formation and morphology of transient bright spots in CMOS image sensors and analyzes the formation process of transient bright spots by conducting heavy ion irradiation experiments to obtain the variation law of transient bright spots with heavy ion linear energy transfer values and background gray values; in addition, we classify the single event upset that occurred in the experiments according to the state of transient bright spots and extract the characteristics of different single event upsets. The failure mechanisms of different single event upsets are analyzed according to their characteristics and are combined with the information given by transient bright spots. This provides an essential reference for rapidly evaluating single event effects and the reinforcement design of CMOS image sensors.

Keywords: single event effects; CMOS image sensor; transient bright spot; single event upset

1. Introduction

Complementary metal oxide semiconductor (CMOS) image sensors (CISs) are widely used in spacecraft imaging systems such as remote sensing imaging and star-sensitive vehicles due to their excellent system performance in terms of power consumption, size, and quality [1,2]. However, the high-energy particle radiation environment in space, such as high-energy protons and heavy ions, can induce single event effects (SEEs) in CISs, which can affect the performance of CISs and even cause functional failure [3–5].

Due to the frequent occurrence of SEEs in CISs in the space environment, a series of studies have been carried out in China and abroad about single event transient (SET) and single event upset (SEU) [6–14]. Hopkinson et al. found SET in the pixel arrays during heavy ion evaluation of the radiation-resistant STAR-250, which appear as transient white bright spots that disappear in the following image [6]. Lalucaa et al. investigated the charge collection process of bright spots by conducting heavy ion irradiation experiments and discussed the effect of blooming, whereby the additional charges of the saturated diode diffuse into neighboring diodes [7]. Yang et al. discovered that the distribution of the total collected charge of each bright spot could be well-fitted by Landau distribution [8]. SEU is a single high-energy particle incident on a semiconductor device that causes a flip in the logic state of that sensitive unit [9]. Beaumel et al. conducted heavy ion irradiation experiments on the HAS2 CIS, which illustrated potential SEU-sensitive cells in the readout circuit of this CIS [10]. Virmontois et al. tested more than 30 registers in the readout circuit in a heavy ion experiment. They found that not all registers triggered image anomalies, and only a few registers (e.g., gain or integration time registers) corrupted the image [11].

The main focus of the current study is on the description and failure explanation of single transient phenomena (SET or SEU). However, SETs and SEUs often occur simultaneously in the experiment, but these two anomalies have yet to be linked and analyzed simultaneously in previous studies. In this paper, by analyzing the morphological size

of the transient bright spots, the SEU phenomena that appeared in the experiment are classified into bright spots that disappeared, were unaffected, and were affected. The failure mechanism of the SEU is analyzed accordingly, and the anomalous phenomena that appeared in this experiment are elaborated on and classified, providing an experimental basis and a theoretical foundation for the systematic study of SEUs in CISs.

2. Samples and Irradiation Conditions

The sample in this experiment is a four-transistor active pixel sensor (4T-APS) with a resolution of 2048 × 2048 and a pixel size of 5.5 µm × 5.5 µm. Figure 1 shows the block diagram of the selected CIS, whose readout circuit includes the analog front-end, addressing circuit, Serial Peripheral Interface (SPI) registers, and LVDS block. The driver board has a 10-bit pixel depth and eight data output channels.

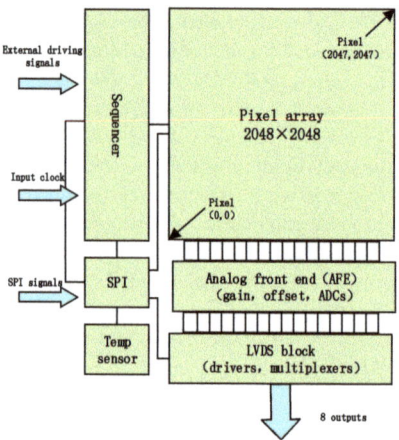

Figure 1. CIS architecture block diagram.

The experiment setup is shown in Figure 2a. The CIS is mounted on a test board and fixed to the irradiation board, which is moved to the particle beam irradiation position by laser positioning and guide rails during the experiment. The rest of the test board was protected with a shield layer, except for the exposed irradiated CIS. The CIS is remotely controlled by a signal line, such as a camera-link cable, and the online test operation is performed outside the irradiation room. During the test, the CIS was in global exposure mode with a frame rate of 180 frames/s and an exposure time set to 1000 lines. Figure 2b shows the schematic diagram of 4T-APS, whose pixel unit mainly consists of a pinned photodiode (PPD), a reset transistor (Trst), a source follower (SF) transistor, a selector transistor (Tsel), a transmission gate (TG) transistor, and a floating diffusion (FD) area.

The heavy ion irradiation experiments were conducted at the HI-13 Tandem Accelerator at the China Institute of Atomic Energy Science [15]. The encapsulated optical glass window was removed in advance. The heavy ions were incident vertically, and the heavy ion beam spot was 30 mm × 30 mm, which could simultaneously cover the entire surface of the CIS chip. Table 1 shows the ion species, range, and linear energy transfer (LET) values used in the experiment.

Table 1. Experimental heavy ion types and energies.

Accelerator	Ion Species	Initial Energy (MeV)	LET (MeV cm^2 mg^{-1})	Range (µm (Si))
HI-13	^{16}O	100	3.01	95.2
	^{28}Si	135	9.3	50.7
	^{35}Cl	150	13.4	42.8
	^{48}Ti	160	22.2	32.9
	^{74}Ge	205	37.37	29.95

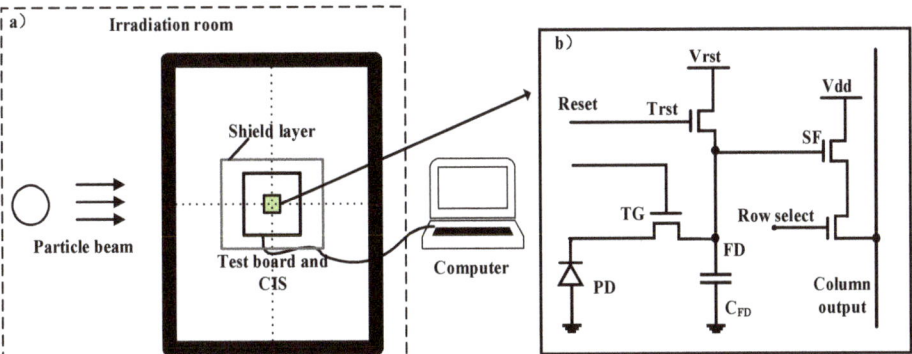

Figure 2. (a) Experimental setup; (b) 4T-APS circuit schematic diagram.

3. Results and Discussion

3.1. SET Bright Spots

The SET bright spot is a transient phenomenon unique to optoelectronic imaging devices. It manifests as a bright white spot at the particle incidence location of the acquired image, and the bright spot at that location disappears in the following image.

In order to understand the formation process of the SET bright spot, it is necessary first to understand the signal readout process of the pixel, as shown in Figure 3. Before exposure, Trst and TG are on to reset the PPD so that the PPD is depleted and in reverse bias; during exposure, Trst and TG are off and the light signal is irradiated at the PPD to generate a photogenerated charge, which is collected by the depletion zone; after a period of integration, Trst is turned on to reset the FD to clear the residual charge before the photogenerated charge is transferred to the FD via TG; immediately afterwards, TG is turned on and the photogenerated charge accumulated on PD is transferred to FD, and TG is turned off after the charge transfer is completed; then, the photogenerated charge transferred to FD is converted into a voltage by the parasitic capacitor of FD, which is amplified by SF and output to the column output bus; the output reset voltage and the optical signal voltage are passed through the CDS circuit to eliminate the noise, and then through the amplifier for signal amplification; the amplified voltage signal is turned into a digital signal by the AD converter, which is processed by a specific image signal inside the sensor and finally output to the external [16].

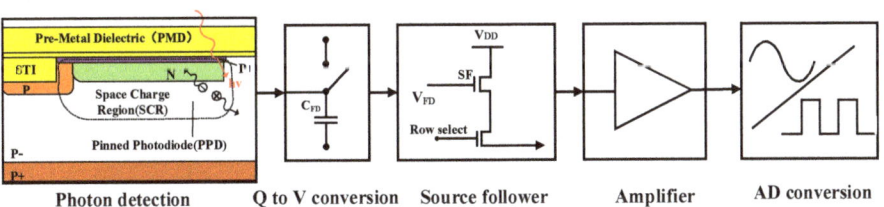

Figure 3. The basic process of pixel readout.

The appearance of SET bright spots occurs during the exposure period. The electron-hole pairs generated by heavy ions in the sensitive layer of the sample are collected by the photodiode depletion region, causing a change in the potential in the PPD region, followed by a readout of the changed potential through the transistor, which is expressed in the image as a bright spot with a gray value greater than the background value. One particle incidence changes the current potential in the PPD. After the current potential change in the PPD is read out, the potential in the PPD is reset to a high level before the subsequent integration. Therefore, the bright spots in the current image will disappear in the following image.

As shown in Figure 4, the SET bright spot is not a dot, but a circular-like cluster formed at the center of the incident point, with the maximum gray value at the center and the decreasing gray value at the edges as the distance from the center increases. The reason is that the neighbor pixels in the pixel array are interconnected and isolated only by a shallow trench (STI). Therefore, the electron–hole pairs incident on the particle traces in the sensitive layer of a pixel, in addition to being collected by the PPD of that pixel unit, will also move to neighbor pixels by drifting and diffusion and thus appear on the image as a bright spot instead of a dot. A single-pixel unit charge collection sensitive body is a region that collects charge into an integrating capacitor. The sensitive body of the photodetection region consists of the depletion region of the p-n junction and the epitaxial layer part. The PPD collects the electron–hole pairs generated in the depletion layer by drifting under the action of the electric field. The electron–hole pairs generated in the epitaxial layer outside the depletion region are partially collected by random diffusion to the adjacent PPD through thermal diffusion, and the rest are compounded or captured. Moreover, at the edge of the bright spot, due to the blooming effect, the diffusion of charges is not collected in the saturated photodiode into the neighboring photodiodes.

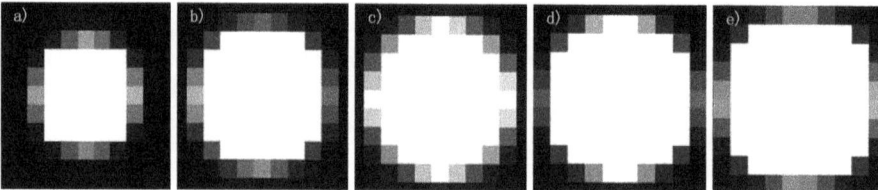

Figure 4. Transient bright spot at different LET values: (**a**) ^{16}O; (**b**) ^{28}Si; (**c**) ^{35}Cl; (**d**) ^{48}Ti; (**e**) ^{74}Ge.

3.2. Effect of Different Conditions on SET Bright Spots

Previous works have described the SET bright spot size concerning the conditions of integration time, process, and incident angle [8–10], and this paper focuses on the size of the SET bright spot concerning the LET value of the incident heavy ions and the background gray value.

The collected charge increment and size are two critical characteristic parameters of SET bright spots. The collection charge increment is obtained by subtracting the background gray value from the total gray value of all covered pixel cells of the SET bright spot. The size refers to the total number of covered pixels. During the test, the collected image signal value is the gray value, and its unit is DN (digital number), which indicates the digital signal value obtained directly by AD conversion.

$$\Delta N_e = (\mu_{DN} - \mu_{DN.dark})/CVG \tag{1}$$

The formula ΔN_e denotes the collected charge increment, $\mu_{DN.dark}$ denotes the total gray value of the background, and μ_{DN} denotes the total gray value of an SET bright spot. CVG (charge voltage gain) indicates the gain of the charge collected in the photodiode into a digital signal through the readout circuit.

3.2.1. LET Value

To reasonably characterize this energy transfer, the physical quantity LET (linear energy transfer) is introduced [17], which is the energy transferred per unit distance of the incident particle in the target material,

$$\text{LET} = -\frac{1}{\rho}dE/dx \tag{2}$$

where ρ is the density of the incident material, E is the energy, and x is the transmission distance. Different LET values were obtained by changing the heavy ion species in the experiment. In Si materials, the ionization energy of Si, the energy required for an electron outside the nucleus to leave the nucleus of a silicon atom to be a free electron, E_{Si} = 3.6 eV/pair, so that the number of electrons and holes produced by an incident particle per unit path is expressed as:

$$\frac{dN_e}{dx} = \frac{dN_h}{dx} = \frac{1}{E_{Si}}\frac{dE}{dx} = LET \cdot \frac{\rho}{3.6\text{eV}} \qquad (3)$$

N_e and N_h are the numbers of electrons and holes produced by ionization, respectively. From Equations (2) and (3), it can be seen that as the value of LET is larger, more electron–hole pairs are generated per unit distance.

As shown in Figure 4, the size of the SET bright spot increases as the LET value increases. Figure 5 shows that the size and charge collected by the SET bright spot increase rapidly and then slowly with the LET value of the heavy ions. The more extensive the LET value, the larger the energy loss per unit distance of the incident heavy ions. Therefore, the larger the number of charges generated, the increasing size and charge collected by the SET bright spot. However, it eventually saturates because it is limited by the carriers' diffusion length and the pixel's full well charge (FWC).

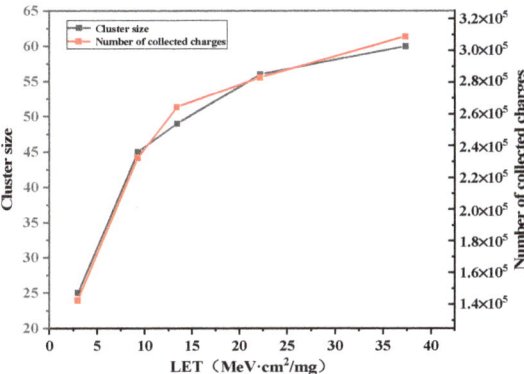

Figure 5. Correlation curve of LET value and bright spot parameters.

3.2.2. Background Gray Value

Figures 6 and 7 show the variation of the bright spot size and the collected charge for different background gray values under ^{35}Cl irradiation. It can be seen that the size of the SET bright spot increases with the increase of the background gray value in a nearly proportional relationship. In contrast, the total collected charge increases first and then decreases.

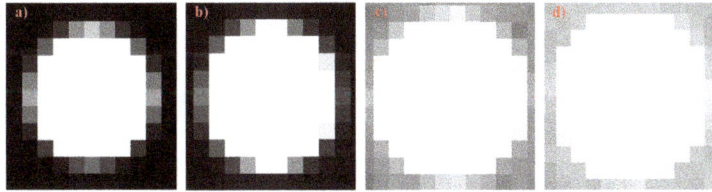

Figure 6. Transient bright spots at different background gray values: (a) 80 DN; (b) 170 DN; (c) 630 DN; (d) 800 DN.

By the Fick law of diffusion [18]:

$$\left.\frac{d\Delta n}{dt}\right|_x = -D_n \frac{d\Delta n}{dx} \qquad (4)$$

$$L_d = \sqrt{D\tau} \tag{5}$$

For large injections, the carrier lifetime is:

$$\tau_n = \tau_p = \frac{1}{(G_{th}/n_i^2) \cdot \Delta n} \tag{6}$$

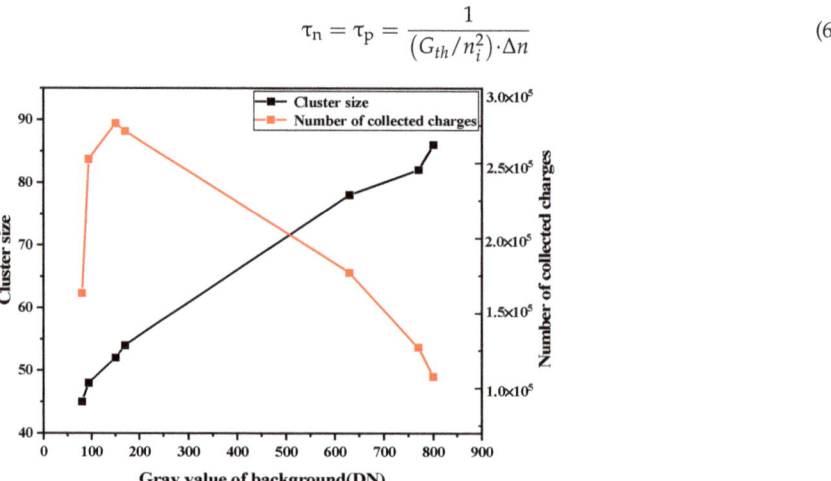

Figure 7. Correlation curve of background gray value and bright spot parameters.

In Equations (4)–(6), Δn is the injected electron concentration, D_n is the electron diffusion coefficient, L_d is the diffusion length, τ_n is the electron lifetime, G_{th} is the heat generation rate, and n_i is the intrinsic electron concentration.

Since the charge generated by the incident heavy ions are much larger than the number of photogenerated charges in the current environment, it meets the large injection condition ($\Delta n = \Delta p > n$ and p). As the background gray value increases, the number of charges required to saturate the pixel unit decreases, and the excess charge density becomes larger. Since the diffusion coefficient D is proportional to the concentration gradient, D becomes larger; since Δn remains constant, the carrier lifetime remains almost unchanged. In summary, the diffusion length becomes larger. Therefore, the bright spot size keeps increasing with the background gray value.

The total collected charge is affected by the amount of charge generated by the incident heavy ions and the lifetime of a few carriers. When the background gray value is small, it is limited by the diffusion length, resulting in the diffused charge being compounded and challenging to be collected effectively; when the background gray value is large enough, the saturation level of the readout circuit is smaller than the saturation level of the PPD, resulting in the calculated charge on the image being smaller than the actual amount, so the total number of collected charges increases and then decreases with the background gray value.

3.3. Classification of SEU Phenomenon by SET Bright Spot

As the most common type of SEE, SET bright spots often coexist with other anomaly phenomena, such as SEUs. Since the features of SET bright spots are easy to extract and the morphology of transient bright spots is correlated with readout circuits, evaluating the SEU events that occur in experiments is meaningful.

According to the description above, the transient white bright spots appearing on the image are due to electron–hole pairs generated by the ionization of heavy ions across the pixel units. The pixel unit collects and spreads these charges to form circular bright spots of similar size. Its size is related to the exposure time, background brightness, and heavy ion LET value. Several SEUs, such as the row, column, and output anomalies, were seen during the experiment. We classify some of the SEU phenomena according to the morphological size of the transient bright spots on the images and discuss each type of SEU failure mechanism based on the bright spot information.

3.3.1. Disappeared Bright Spot

The disappeared bright spot means that the transient bright spots in the image disappear and can be restored only after power-off and restart. As shown in Figure 8, pixel outputs on the image were from 135 DN (gray value) to zero, and all bright spots disappeared in three consecutive pictures.

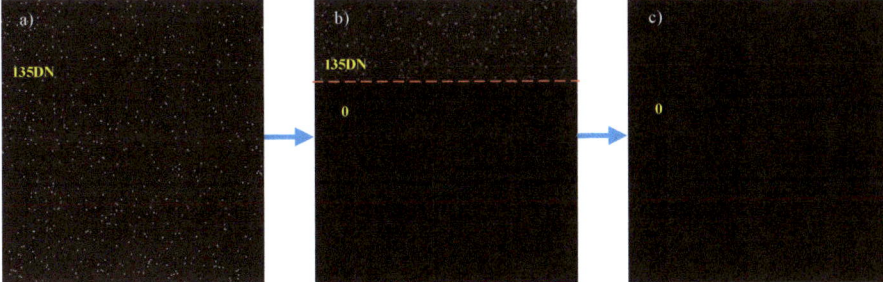

Figure 8. The gray value from 135 DN to 0 and bright spots disappeared (three consecutive pictures): (**a**) Normal image; (**b**) Next frame image; (**c**) Global gray value reduced to 0.

The global disappearance of the transient bright spot in the figure and the fact that the outputs were zero indicated an output anomaly. The reason may arise from the process of pixel cell signal readout. According to the process of transient change in Figure 8b, it can be seen that the row readout mainly causes the anomaly. Hence, this phenomenon is induced by heavy ion bombardment of the row address decoder within the CIS. When a row of pixels in the array is selected, the row address decoder can control the pixel operation of the row by controlling the selection of the transfer gate, reset transistor, and selector transistor. However, when a heavy ion hits the row address decoder and SEU occurs, it may cause the transfer gate, reset transistor, and selector transistor to be generally on or off. Therefore, the pixels of that row each output the same signal during the signal sampling and reset phases. After the associated double-sampling process, the output value after subtracting the two is zero. Therefore, no transient bright spots appear on these images, and the global gray value is zero.

3.3.2. Unaffected Bright Spot

The unaffected bright spot means that after the SEU phenomenon, the size and shape of the bright spot in the image remain the same as usual, and there is no change. As shown in Figure 9, the global gray value of the image dropped from the original 170 to about 30, and the bright spots did not disappear. Comparing the bright spots in the before and after images, as shown in Figure 9c,d, it can be found that the size of the bright spots does not change, having sixty saturated pixels. Since the size of the transient bright spot is in connection with the average gray value of the background, but the figure shows the parallel size of the bright spot, it means that the cause of the abnormality has nothing to do with the row or column readout circuit. Therefore, the gray value of the background dropped due to a flip in one of the registers in the analog front end, causing a change in the global gray value, not due to external light leakage or other reasons.

According to the failure characteristics, this global gray value drop was caused by the failure of the offset register. The function of the offset register in CISs is to calibrate the output to compensate for the output signal dark level value [19], as shown in Equation (7).

$$\text{Dark-level output} = 70 + \text{setting} - 16{,}383 \tag{7}$$

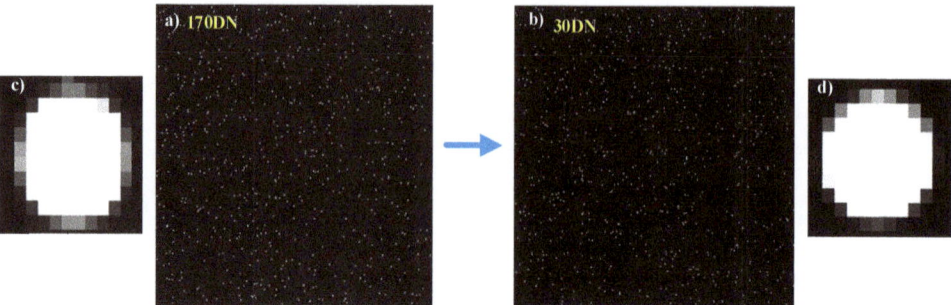

Figure 9. Global gray value drop (two consecutive images): (**a**) Normal image; (**b**) Global gray value drop; (**c**) Bright spot pattern in normal image; (**d**) Bright spot pattern in anomalous images.

When heavy ions bombard the register and SEU occurs, the offset register setting will no longer be the correct value. When the offset register setting value decreases from the original 16,323, it causes the gray values of all pixels to decrease, as shown in Figure 9. There are still bright spots on them because heavy ions generate a large amount of ionized charge on the incident traces to be collected, increasing the pixel output dark level. Because enough charge is collected, the pixel output reaches 1023 DN (maximum value) at the bright spot location. The ionizing charge formed by a single heavy ion can affect multiple pixels near the incident location by diffusion, drift, and the funnel effect. Hence, the bright spot still exists under this anomaly.

3.3.3. Affected Bright Spot

The affected bright spot means that the size of the bright spot in the image is affected, mainly as the bright spot becomes larger or smaller. The impact on the bright spot mainly occurs in the row and column anomalies, as shown in Figure 10; we can observe two white vertical stripes with a constant gray value of 934 DN, much higher than the average gray value of 159 DN in the background. The positions of the two anomalous vertical stripes are X127 and X255. We can get more information through the transient bright spot situation, such as the misalignment problem at the anomalous column. The X127 column is down one wrong, and the X255 column is up one wrong, precisely equivalent to the two columns in the middle of the data line signal value read out in the following line. Secondly, comparing the bright spot size, it is found that the cut bright spots on the left and right add up to the same size as the typical bright spot, which indicates that the X127 column and X255 column data are invalid data, and the value is locked. Zooming in at Y = 0, as shown in Figure 11, verifies that the signal value in the middle of the abnormal column is panned down one row.

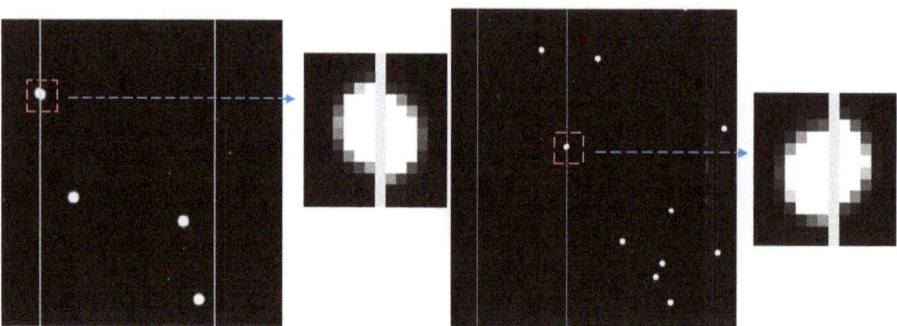

Figure 10. Column anomalies and transient bright spots on anomalous columns.

With the abnormal characteristics described above, it can be inferred that the abnormality is due to an abnormality in the signal output. This CIS has 18 LVDS output channels, including 16-pixel output channels, one clock channel, and one control channel. The driver board used in the test had eight output channels. The anomalous column in Figure 4 is located in the last of the two parallel segments of 128 pixels on LVDS channel one in columns 127 and 255. This anomaly may be due to SEU occurring on LVDS channel one during pixel data transfer, causing an anomaly in the row addressing register and resulting in a misalignment of the output values between them.

Figure 11. Complete image of the misaligned row anomaly in Figure 10.

Figure 12 illustrates one type of row anomaly; the gray value on the anomalous row (90 DN) is smaller than the regular row (125 DN), as shown in Figure 12b. The transient bright spot in the anomalous row is smaller than the bright spot on the normal background outside the row, caused by the lower gray value of the background.

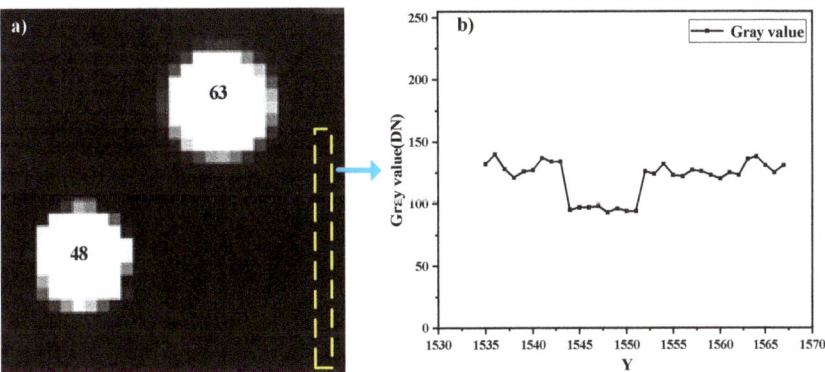

Figure 12. Transient bright spots on row anomalies: (**a**) row anomaly image; (**b**) variation curve of column gray value.

This anomaly is similar to one SEU phenomenon reported by Beaumel in 2013 [10], which is thought to be a flip in the row reset register or row read register. The read or reset register pointer jumped to a non-corresponding line, resulting in visible disturbances in the integration time of some lines on the image. As a result, the gray values at the disturbed rows are abnormal, as shown in Figure 12.

In conclusion, a summary of the classification of bright spots in SEU phenomena is shown in Table 2.

Table 2. The classification of SEU events with the bright spots.

Category	Characteristic	SEU Phenomenon	Failure Localization
Disappeared	Bright spot which disappeared	Global gray value drop to 0	Row address decoder Offset register
Unaffected	No change in bright spot size	Global gray value drop	
Affected	Change in bright spot size	Column anomalies, row anomalies	LVDS block, row reset register, or row read register

4. Conclusions

This paper investigated the relationship between the variation of SET bright spots with different experimental conditions by conducting CIS heavy ion irradiation experiments. The size of the SET bright spot increases with the heavy ion LET value and then tends to saturate; it gradually increases with the increase of the background gray value. Meanwhile, a variety of SEU events were observed in the experiment, and the SEU events were classified into three categories: disappeared bright spot, affected bright spot, and unaffected bright spot, by feature extraction of the SET bright spots in the SEU events. The above classification narrowed down the failure location of SEU. The disappearance of the bright spot indicates a failure at the readout circuit of the CIS; the unaffected bright spot indicates a failure at the analog front-end; and the change of the bright spot indicates an abnormality in the readout circuit of the CIS or the LVDS block.

In summary, this paper proposes to identify and classify SEUs using SET bright spot characteristics and establishes a fast identification method to analyze SEU patterns and sensitive areas based on transient bright spot size, background gray value, and other parameters. It provides an essential reference for the rapid evaluation and reinforcement design of the SEE of CISs. In the future, this analysis method can be combined with machine learning methods to collect relevant SEE experiment data, which can be used for online identification analysis and failure localization of SEEs in optical imaging devices.

Author Contributions: Conceptualization, Z.Y. and L.W.; methodology, Y.L. and L.W.; software, J.F. and Z.Z.; validation, Z.Y., L.W. and Y.L.; formal analysis, Z.Y.; investigation, L.W.; resources, L.W.; data curation, Z.Y.; writing—original draft preparation, Z.Y.; writing—review and editing, L.W.; visualization, Z.Y.; supervision, D.Z. and B.L.; project administration, Q.G.; funding acquisition, L.W. and B.L. All authors have read and agreed to the published version of the manuscript.

Funding: This research was funded by the CAS "Light of West China" Program, grant number 2020-XBQNXZ-004; the National Natural Science Foundation of China, grant number NSFC 12175308, and the Natural Science Foundation of Xinjiang Uygur Autonomous Region, grant number 2022D01B205.

Data Availability Statement: Not applicable.

Acknowledgments: The authors would like to thank the teachers at the Heavy Ion Research Facility in Lanzhou, at the HI-13 Tandem Accelerator the China Institute of Atomic Energy.

Conflicts of Interest: The authors declare no conflict of interest.

References

1. Velichko, S.; Hynecek, J.J.; Johnson, R.S.; Lenchenkov, V.; Komori, H.; Lee, H.-W.; Chen, F.Y.J. CMOS Global Shutter Charge Storage Pixels with Improved Performance. *IEEE Trans. Electron Devices* **2016**, *63*, 106–112. [CrossRef]
2. Oike, Y.; Akiyama, K.; Hung, L.D.; Niitsuma, W.; Kato, A.; Sato, M.; Kato, Y.; Nakamura, W.; Shiroshita, H.; Sakano, Y.; et al. An 8.3 M-pixel 480 fps global-shutter CMOS image sensor with gain-adaptive column ADCs and 2-on-1 stacked device structure. *IEEE J. Solid-State Circuits* **2016**, *52*, 985–993. [CrossRef]
3. Rolando, S.; Goiffon, V.; Magnan, P.; Corbière, F.; Molina, R.; Tulet, M.; Bréart-De-Boisanger, M.; Saint-Pé, O.; Guiry, S.; Larnaudie, F.; et al. Smart CMOS image sensor for lightning detection and imaging. *Appl. Opt.* **2013**, *52*, C16–C23. [CrossRef]
4. Belredon, X.; David, J.-P.; Lewis, D.; Beauchene, T.; Pouget, V.; Barde, S.; Magnan, P. Heavy ion-induced charge collection mechanisms in CMOS active pixel sensor. *IEEE Trans. Nucl. Sci.* **2002**, *49*, 2836–2843. [CrossRef]

5. Lalucaa, V.; Goiffon, V.; Magnan, P.; Virmontois, C.; Rolland, G.; Petit, S. Single Event Effects in 4T Pinned Photodiode Image Sensors. *IEEE Trans. Nucl. Sci.* **2004**, *51*, 2753–2762. [CrossRef]
6. Hopkinson, G.; Mohammadzadeh, A.; Harboe-Sorensen, R. Radiation effects on a radiation-tolerant CMOS active pixel sensor. *IEEE Trans. Nucl. Sci.* **2004**, *51*, 2753–2762. [CrossRef]
7. Lalucaa, V.; Goiffon, V.; Magnan, P.; Rolland, G.; Petit, S. Single-Event Effects in CMOS Image Sensors. *IEEE Trans. Nucl. Sci.* **2013**, *60*, 2494–2502. [CrossRef]
8. Yang, H.B.; Mai, F.T.; Liao, J.W.; Zhang, H.; Ma, X.; Gao, C.; Ren, W.; Zhou, W.; Sun, X.; Liu, J.; et al. Heavy-ion beam test of a monolithic silicon pixel sensor with a new 130 nm High-Resistivity CMOS process. *Nucl. Instrum. Methods Phys. Res. Sect. A* **2022**, *1039*, 167049. [CrossRef]
9. Baumann, R. Radiation-induced soft errors in advanced semiconductor technologies. *IEEE Trans. Device Mater. Reliab.* **2005**, *5*, 305–316. [CrossRef]
10. Beaumel, M.; Herve, D.; Van Aken, D.; Pourrouquet, P.; Poizat, M. Proton, Electron, and Heavy Ion Single Event Effects on the HAS2 CMOS Image Sensor. *IEEE Trans. Nucl. Sci.* **2013**, *61*, 1909–1917. [CrossRef]
11. Virmontois, C.; Toulemont, A.; Rolland, G.; Materne, A.; Lalucaa, V.; Goiffon, V.; Codreanu, C.; Durnez, C.; Bardoux, A. Radiation-Induced Dose and Single Event Effects in Digital CMOS Image Sensors. *IEEE Trans. Nucl. Sci.* **2014**, *61*, 3331–3340. [CrossRef]
12. Cai, Y.L.; Li, Y.D.; Wen, L.; Wen, L.; Zhou, D.; Feng, J.; Ma, L.-D.; Zhang, X.; Wang, T.-H. Heavy ion-induced single event effects in active pixel sensor array. *Solid-State Electron.* **2008**, *152*, 93–99. [CrossRef]
13. Virmontois, C.; Lalucaa, V.; Belloir, J.-M.; Bascoul, G.; Bardoux, A. Single Event Effect Similarities between Heavy Ions and LASER Tests in Advanced CMOS Image Sensors. In Proceedings of the 2019 19th European Conference on Radiation and Its Effects on Components and Systems (RADECS), Montpellier, France, 16–20 September 2019. [CrossRef]
14. Van Aken, D.; Hervé, D.; Beaumel, M. Total Dose, Displacement Damage and Single Event Effects in the Radiation Hardened CMOS APS HAS2. *Proc. SPIE* **2009**, *7474*, 1–12.
15. Zhu, S.Y.; Guo, G.; He, M.; Wu, Z.; Yuan, D.; Sui, L.; Jiao, X.; Chang, H.; Zuo, Y.; Fan, P.; et al. Present Status and Future Prospect of Applied Nuclear Physics Research at HI-13 Tandem Accelerator. *At. Energy Sci. Technol.* **2016**, *54*, 1–16.
16. Kuroda, T. *Essential Principles of Image Sensors*, 1st ed.; CRC Press: New York, NY, USA, 2015; pp. 29–34.
17. Massengill, L. SEU Modeling and Prediction Techniques. In Proceedings of the 30th International IEEE Nuclear and Space Radiation Conference (NSREC), Snowbird, UT, USA, 19 July 1993.
18. Sze, S.M.; Ng, K.K. *Physics of Semiconductor Devices*, 3rd ed.; John Wiley & Sons: Hoboken, NJ, USA, 2008; pp. 34–35. [CrossRef]
19. AMS OSRAM Group. *Global Shutter CMOS Image Sensor for Machine Vision*, 4th ed.; AMS AG: Premstätten, Austria, 2021; pp. 58–59.

Disclaimer/Publisher's Note: The statements, opinions and data contained in all publications are solely those of the individual author(s) and contributor(s) and not of MDPI and/or the editor(s). MDPI and/or the editor(s) disclaim responsibility for any injury to people or property resulting from any ideas, methods, instructions or products referred to in the content.

Article

Mechanism of Total Ionizing Dose Effects of CMOS Image Sensors on Camera Resolution

Jie Feng [1,2,*], Hai-Chuan Wang [1,2,3], Yu-Dong Li [1,2,*], Lin Wen [1,2] and Qi Guo [1,2]

1. Xinjiang Technical Institute of Physics and Chemistry, Chinese Academy of Sciences, Urumqi 830011, China
2. Xinjiang Key Laboratory of Electronic Information Material and Device, Urumqi 830011, China
3. University of Chinese Academy of Sciences, Beijing 100049, China
* Correspondence: fengjie@ms.xjb.ac.cn (J.F.); lydong@ms.xjb.ac.cn (Y.-D.L.)

Abstract: The nuclear industry and other high-radiation environments often need remote monitoring equipment with advanced cameras to achieve precise remote control operations. CMOS image sensors, as a critical component of these cameras, get exposed to γ-ray irradiation while operating in such environments, which causes performance degradation that adversely affects camera resolution. This study conducted total ionizing dose experiments on CMOS image sensors and camera systems and thoroughly analyzed the impact mechanisms of the dark current, Full Well Capacity, and quantum efficiency of CMOS image sensors on camera resolution. A quantitative evaluation formula was established to evaluate the impact of Full Well Capacity and quantum efficiency of the CMOS image sensor on camera resolution. This study provides a theoretical basis for the evaluation of the radiation resistance of cameras in environments with strong nuclear radiation and the development of radiation-resistant cameras.

Keywords: CMOS image sensor; TID; radiation effects; camera resolution

Citation: Feng, J.; Wang, H.-C.; Li, Y.-D.; Wen, L.; Guo, Q. Mechanism of Total Ionizing Dose Effects of CMOS Image Sensors on Camera Resolution. *Electronics* **2023**, *12*, 2667. https://doi.org/10.3390/electronics12122667

Academic Editors: Gian-Franco Dalla Betta, Yaqing Chi, Li Cai and Chang Cai

Received: 9 May 2023
Revised: 8 June 2023
Accepted: 12 June 2023
Published: 14 June 2023

Copyright: © 2023 by the authors. Licensee MDPI, Basel, Switzerland. This article is an open access article distributed under the terms and conditions of the Creative Commons Attribution (CC BY) license (https://creativecommons.org/licenses/by/4.0/).

1. Introduction

The nuclear industry is crucial to national security, but the presence of strong nuclear radiation in the environment poses significant risks with regard to the operation, maintenance, and emergency response of nuclear facilities. These radiation environments are extremely harmful to human health; therefore, to ensure the safety of staff and facilities, it is necessary to use remote monitoring equipment and radiation-resistant robots for refined remote control operations [1,2]. However, both remote monitoring equipment and radiation-resistant robots rely on cameras to acquire target information, and the environment contains high levels of neutrons and α-, β-, and γ-rays with high dose rates and total doses. The neutron dose levels are generally low outside the operating reactor, and α and β radiation can be effectively shielded by relatively thin housing. γ-rays have strong penetration ability, and their impact on cameras cannot be ignored. The effects of strong nuclear radiation on electronic systems can cause significant camera performance degradation. The degradation of the camera resolution will lead to the loss of information with potentially disastrous consequences. As such, mitigating the impact of γ-rays on cameras is critical for effective remote monitoring and control in nuclear facilities.

Several studies have investigated the impact of radiation on camera resolution in the past. These include studies by KIM et al. in 2004 and 2007, who evaluated the resolution of scintillator-coupled CMOS sensors under X-rays based on the Modulation Transfer Function (MTF) and its sensitivity to dark signals [3,4]. In 2010, Jie Yu et al. conducted an analysis of the effect of CCD (Charge-Coupled Device) camera transient noise on imaging resolution in neutron photography, taking into account specific shielding requirements [5]. However, these studies mainly focused on the relationship between the radiation dose, ray type, and CMOS image sensor (Complementary Metal Oxide Semiconductor Image Sensor, CIS) noise or total ionizing dose (TID) effects on the CCD system, without a detailed analysis of the mechanism by which radiation-sensitive parameters in the CIS affect camera resolution. Furthermore, these studies did not establish quantitative relationships between CIS radiation-sensitive parameters and camera resolution.

This paper focuses on investigating the degradation mechanism of camera resolution under a γ-ray radiation environment. Specifically, it establishes a quantitative evaluation formula for the impact of the CMOS image sensor's Full Well Capacity (FWC) and quantum efficiency (QE) on camera resolution. This study provides a theoretical foundation for evaluating camera radiation resistance in strong nuclear radiation environments and developing radiation-resistant cameras.

2. Materials and Methods

The test camera follows a modular design and comprises three main components: an optical lens, an image sensor, and a peripheral circuit. The optical lens is connected to the CIS, while the peripheral circuit is linked to the CIS device through a flexible cable. For this test, an ON Semiconductor AR series commercial image sensor with 2.1 million pixels and a single pixel size of 3 μm × 3 μm utilizing RGB Bayer array color filters was used as the image sensor. The image sensor utilized rolling exposure mode, while a commercial automatic zoom lens was employed to capture high-quality images. Additionally, the camera features a self-designed anti-radiation circuit encompassing a power supply module, a digital signal processing module, and a network transmission module.

The irradiation test was carried out on the ^{60}Co-γ radiation source of the Xinjiang Institute of Physics and Chemistry, Chinese Academy of Sciences. The camera system irradiation test is shown in Figure 1. The camera system was connected to the PC outside the irradiation room via a network cable.

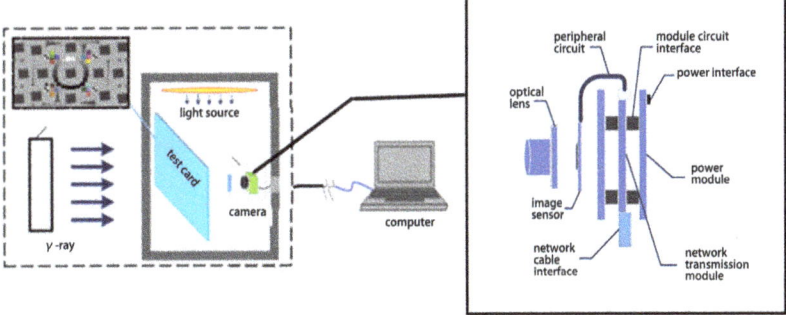

Figure 1. Schematic diagram of the camera irradiation test.

The irradiation test involved two parts: the camera system irradiation test and the CIS irradiation test. Firstly, the camera system was subjected to irradiation at a dose rate of 28 rad(Si)/s while in an online working state. After reaching the dose points of 70, 110, 140, 180, 210, and 280 krad(Si), the resolution of the camera was tested. When the radiation dose of the camera system exceeded 280 krad(Si), the performance index of the camera dropped significantly, and the working state became abnormal, so the irradiation test was stopped. As the TID of the camera increased, the light transmittance of the camera lens decreased. Thus, during the displacement test of a portion of the dose of the camera system, a supplementary test was conducted by replacing the unirradiated lens. In the CIS irradiation test, only the CIS was irradiated while the peripheral circuit was shielded and protected. The device worked normally during the irradiation process at a dose rate of 28 rad(Si)/s, with irradiation doses of 70, 110, 140, 180, 210, and 280 krad(Si). After reaching the corresponding dose, the key parameters of the CIS and the combined camera system were tested using the photoelectric imaging device radiation damage test system at the Xinjiang Institute of Physics and Chemistry, Chinese Academy of Sciences.

3. Results

3.1. Camera System Resolution Degradation

Camera resolution is the ability to distinguish the number of line pairs per unit length, and it serves as an important parameter to determine the clarity of camera imaging. This is crucial for the human eye to discern whether the actual image is clear and effective. MTF is a function that varies with spatial frequency, and its value ranges from 0 to 1 with constant spatial frequency. A higher MTF value indicates better imaging quality, as it represents higher restoration of the contrast between the object and the image [6]. In this study, the imaging system resolution has been evaluated by calculating the MTF value, which possesses the property of being cascadable. The formula for calculating the MTF in this experiment is given in Equation (1).

$$MTF = MTF_{cis} * MTF_{\text{Peripheral Circuits}} * MTF_{\text{Optical Lens}} = \frac{I_{max} - I_{min}}{I_{max} + I_{min}} \quad (1)$$

When image noise is low, the Spatial Frequency Response (SFR) test method recommended by ISO 12233 yields stable results. However, if the image noise surpasses the algorithm's threshold value, the test outcomes will change dramatically. This discrepancy with human perception during actual use can lead to inaccurate assessment of the impact of radiation-induced noise on the camera system's performance. To precisely assess noise's effects on the camera's actual performance in high-radiation environments, this study employed wedge diagrams to evaluate the camera's resolution, along with Imatest Master to determine the value of Aliasing onset and MTF10.

MTF10 is a classical theoretical value used to describe the resolution of an optical system, while Aliasing onset is the spatial frequency at which the number of bars detected by the software is lower than the total number of wedges. The results obtained from Aliasing onset are not affected by signal processing, such as sharpening or noise reduction, making it suitable for evaluating the resolution of different types of cameras. In practical camera usage, Aliasing onset is more in line with manual subjective discrimination than the theoretical limit value of MTF10. Therefore, it can better solve the problem of evaluating camera resolution in a strong nuclear radiation environment. When pictures taken by the camera have noise, the MTF calculation formula can be deduced from Equation (1) [5].

$$MTF = \frac{(I_{max} + I_{\text{light-noise}}) - (I_{min} + I_{\text{dark-noise}})}{(I_{max} + I_{\text{light-noise}}) + (I_{min} + I_{\text{dark-noise}})} = \frac{(I_{max} + \sigma_{light}ng_2) - (I_{min} + \sigma_{dark}ng_1)}{(I_{max} + \sigma_{light}ng_2) + (I_{min} + \sigma_{dark}ng_1)} \quad (2)$$

The maximum and minimum grayscale values of the image target region under ideal conditions are represented by I_{max} and I_{min}, respectively. The noise captured by the camera after irradiation can significantly impact the resolution. Thus, the ratio of the number of noise to the total number of pixels in the maximum gray value region of the image target area is σ_{light}, the ratio of the number of noise to the total number of pixels in the minimum gray value region of the image target area is σ_{dark}, the average gray value of the noise in the minimum gray value region is g_1, and the number of pixels is n. Then, the increment of the gray value of the image in the region of maximum gray value where noise exists is represented by $I_{\text{light-noise}}$, and the average gray value of the noise in this region of maximum gray value is g_2, while its value is $\sigma_{light}ng_2$. Similarly, the increment of the gray value of the image in the region of minimum gray value where noise exists is represented by $I_{\text{dark-noise}}$, and its value is $\sigma_{dark}ng_1$. Finally, the $MTF_{camera-radiation}$ formula in the size of $I_{max} + \sigma_{light}ng_2$ and $I_{min} + \sigma_{dark}ng_1$ has been obtained, which represents the maximum and minimum gray values of the image measured.

Figure 2a shows the change in the calculated MTF value of the camera with the TID. As the irradiation dose increases, the calculated MTF value of the camera decreases. The trend of the camera system's resolution after irradiation with the TID is presented in Figure 2b. The measured values of the camera resolution are consistent with the trend of decreasing

MTF calculated values with the increase of the TID. Moreover, after the irradiation dose reaches 210 krad(Si), the rate of decrease in the camera resolution becomes faster.

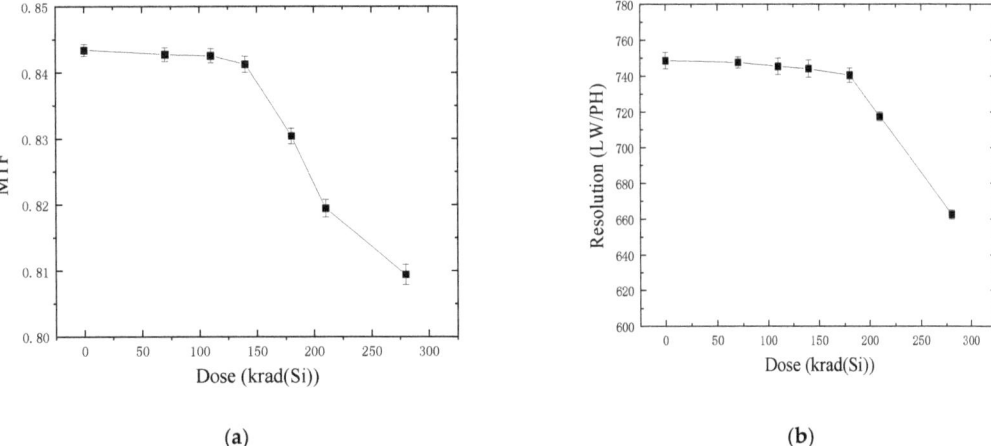

Figure 2. The camera resolution varies with the TID under γ-ray irradiation: (**a**) MTF calculation value; (**b**) resolution of camera varies with the TID under γ-ray irradiation.

For the camera system with only the irradiated CIS, the degradation of the MTF and the resolution is caused by the radiation damage of the CIS. Figure 3 illustrates the trend of camera resolution with the TID for a camera system with only the irradiated CIS. The CIS radiation damage has little effect on the resolution before reaching 210 krad(Si), but it sharply decreases after that.

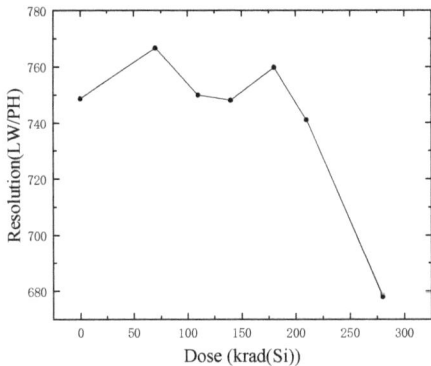

Figure 3. The camera resolution varies with the TID under γ-ray irradiation.

3.2. The Key Parameters of CIS Degradation

Dark current, Full Well Capacity, and spectral response are key parameters that evaluate the imaging performance of the CIS after irradiation, and their degradation has a significant impact on the overall performance of the camera system. Dark current is the current generated by the pixel cell of an image sensor when it absorbs spontaneously generated electrons due to the presence of defects (interface defects and body defects) under dark conditions, and it is usually measured in e^-/s [7]. When the CIS is exposed to ionizing radiation, the dark current signal mainly comes from the pixel cell and the peripheral circuit, with the peripheral circuit dark current being a fixed value independent of exposure time. Figure 4 presents the test results of the dark current of the CIS. The dark current of the image sensor increases with the increase of the TID, and it significantly increases at 75–100 and 175–210 krad(Si).

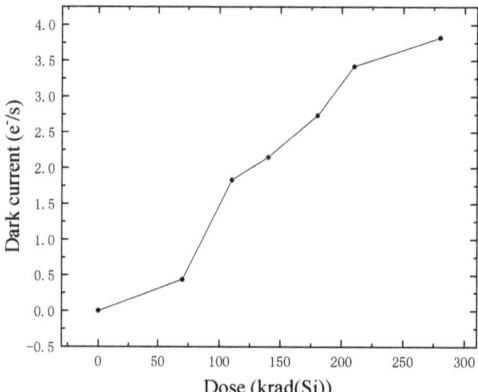

Figure 4. The changes of the dark current with the TID under irradiation at the dose rate of 28 rad(Si)/s.

The FWC is the maximum number of electrons that can be stored in the pinned photodiode (PPD) in a saturated state. Figure 5 illustrates the change in the FWC of the CIS with the increase of the TID during γ-ray irradiation at different dose rates. The variation of the FWC at different doses shows some differences: before 100 krad(Si), there is no significant change in the FWC of the CIS, while after 100 krad(Si), the FWC shows a significant decreasing trend. Moreover, the decreasing rate of the FWC increases significantly after 210 krad(Si).

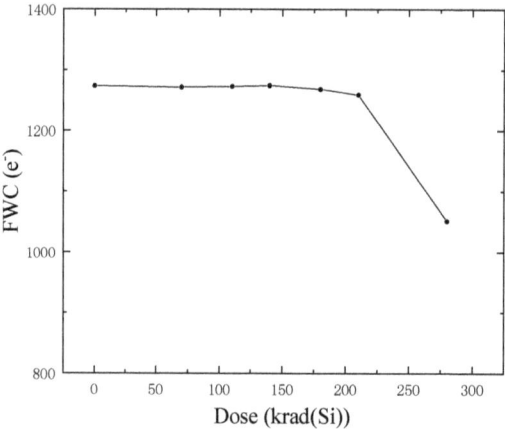

Figure 5. The FWC varies with the TID under γ-ray irradiation at the dose rate of 28 rad(Si)/s.

The spectral response of the CIS is an important parameter for evaluating its ability to convert incident photons of different wavelengths into electrical signals, and it is crucial for assessing color reproduction in color cameras. QE is used to characterize the responsiveness of the CIS to light signals of specific wavelengths. Figure 6 shows the degradation ratios of the spectral response curve of the CIS under γ-ray irradiation at 28 rad(Si)/s for incident light wavelengths of 420, 450, 516, 550, and 630 nm. The degradation of the CIS is greater in the short wavelength band, as indicated by the degradation ratios in Figure 6. The degradation ratio decreases gradually with increasing wavelength, and a larger dose is required to show significant degradation.

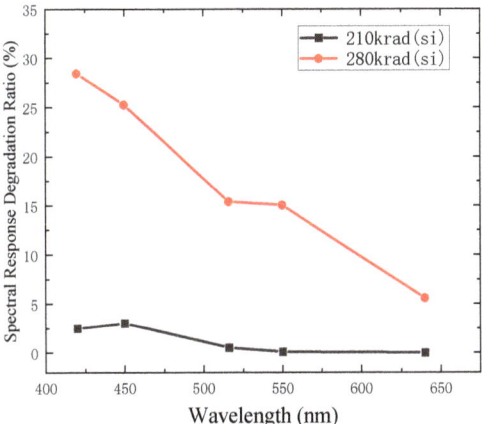

Figure 6. Degradation ratio of the CIS spectral response under γ-ray irradiation at the dose rate of 28 rad(Si)/s.

4. Discussion

For the CIS, the comprehensive MTF can characterize its detail resolution capability, which is composed of three types of MTF: geometric MTF, transfer MTF, and diffusion MTF. Usually, the comprehensive MTF function is obtained by multiplying these three types of MTF in the frequency domain. For the CIS in radiation cameras, because the internal pixel structure of the CIS remains unchanged, the geometric MTF remains unchanged as well. The transfer MTF refers to the charge loss generated during the charge transfer between pixels. The TID effect causes trap positive charges to be generated in the STI region near the Transfer Gate (TG), which induces the production of negative charges on the Si–SiO$_2$ surface of the STI due to the appearance of trap positive charges. The accumulation of these negative charges increases the regional electron density, reduces the TG channel potential barrier, and, ultimately, allows some photoelectrons in the PPD to transfer to the FD through the channel sidewalls without voltage applied to the TG [8], which leads the transfer MTF and the FWC to decrease with the increase of dose. The diffusion MTF refers to the difference in position of photogenerated carriers caused by the difference in the depth of incidence of incident light for different spectral bands after the incident light enters. The photogenerated carriers that are far away from the depletion region will diffuse freely before entering the depletion region. With increasing TID, the interface trap charge density formed at the SiO$_2$ layer surface due to the TID effect also increases. The energy level of interface trap charges is close to the center of the bandgap, and they can act as effective recombination centers, increasing the net recombination rate and reducing the lifetime of photogenerated carriers in this region. This directly reduces the diffusion length of carriers and, ultimately, lowers the efficiency of collecting photogenerated carriers in the depletion region. As such, the diffusion MTF also decreases with the increase of dose. Because incident lights of different wavelengths have different penetration depths, longer-wavelength light generates fewer photogenerated carriers near the interface and is less affected by the interface trap charge density [9,10]. Consequently, the degradation of the QE after irradiation is lower for longer-wavelength light.

According to Equation (2), the maximum and minimum gray values of the image before and after irradiation will have a certain impact on the MTF value, where the change in the minimum gray value is mainly affected by the CIS dark current noise, and the change in the maximum gray value is mainly affected by the FWC. The radiation-induced increase in interfacial trap charges at the Si–SiO$_2$ interface of the 4T pixel structured CIS, especially at the periphery of the shallow trench isolation (STI) region, the TG–PPD overlap region, and the PPD surface [11], which is the main mechanism behind the gradual increase in the CIS dark current with increasing radiation dose. During the γ-ray irradiation process, the

Si–SiO$_2$ interface precipitated in the STI region generates broken suspension bonds and forms interface defects [12]. Unstable gaseous substances, such as silicon monoxide, generated by incomplete reactions between silicon and oxygen at the interface can be emitted from the oxide layer at high temperatures, creating dangling bonds at the interface [13]. Therefore, during ionizing radiation, the density of dangling bonds and point defects will continue to increase with the increase of the TID, becoming one of the main sources of increased dark current after irradiation [14,15]. The dark current increases more significantly at 75–100 and 175–210 krad(Si), and Figure 4 reflects the introduction of different dark current sources with the increase of irradiation dose. By substituting the corresponding gray value parameters of the entire irradiation experimentally collected image measurement area into Equation (2) to calculate the post-irradiation MTF value, the effects of the CIS dark current noise and the FWC on camera resolution are compared and analyzed. Figure 7a shows the MTF value calculated by substituting the maximum gray value measured in the target area of the image under different doses and the minimum gray value measured in the target area of the image under unirradiated conditions, while Figure 7b shows the MTF value calculated by simultaneously substituting the maximum and minimum gray values measured in the target area of the image under different doses.

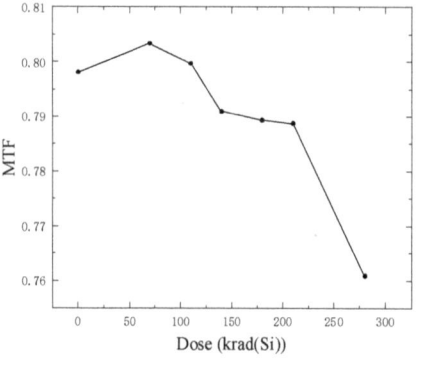
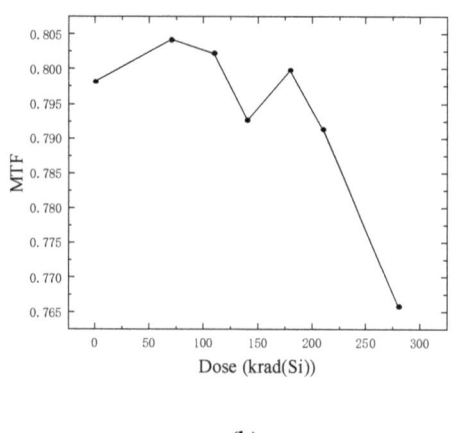

(a) (b)

Figure 7. The MTF calculation values vary with the TID under γ-ray irradiation (**a**) based on maximum gray value measurements; (**b**) based on maximum and minimum gray value measurements.

From Figure 7, it can be seen that the overall trend of the MTF calculation value decreases as the dose increases, and the degree of MTF reduction gradually increases as the dose reaches a certain level. At the same time, whether the minimum gray measurement value under different doses is substituted into Equation (2) has a certain impact on the MTF calculation value, but there is no significant difference in the overall trend. Therefore, the CIS dark current noise has a certain impact on the MTF, but the degradation of the CIS FWC after irradiation has a more significant impact on the MTF.

The test card image captured by the camera is a combination of effective signal and noise, where the noise can be mainly divided into image signal noise and background noise. The image signal noise is caused by scattered photons from external incident light, while the background noise includes the CIS noise and noise from the camera's peripheral circuit. Under light and dark fields, the RGB three-channel noise value of the CIS calculated using Imatest Master changes with the dose, as shown in Figure 8.

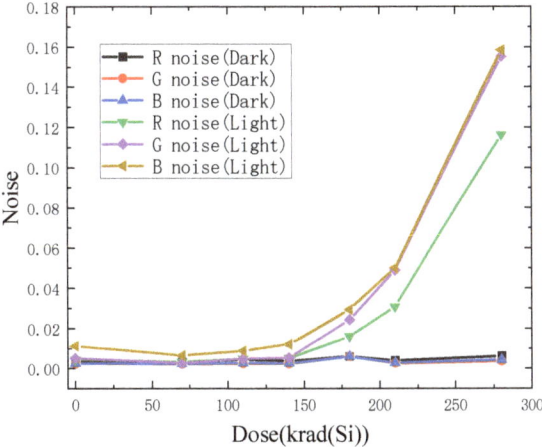

Figure 8. The noise of the CIS varies with the TID under γ-ray irradiation.

From Figure 8, the noise of the CIS increases with the increase of the TID, and the increase in the CIS noise under the light field condition is much larger than that under the dark field condition. This is because under sufficient light conditions, photon scatter noise is much greater than the dark current noise. In addition, the higher the dose rate, the more obvious the increase in noise. Finally, after 140 krad(Si), the growth rate of the CIS noise increases significantly.

Figure 9 shows the gray level values of the test card images captured by the test cameras with CIS combinations at different dose rates measured using Imatest Master software (imatest, 2020.2, Boulder, CO, USA). The gray level value with serial number 1 represents the maximum gray level value in the image, which is mainly affected by the degradation of the CIS FWC parameter after irradiation.

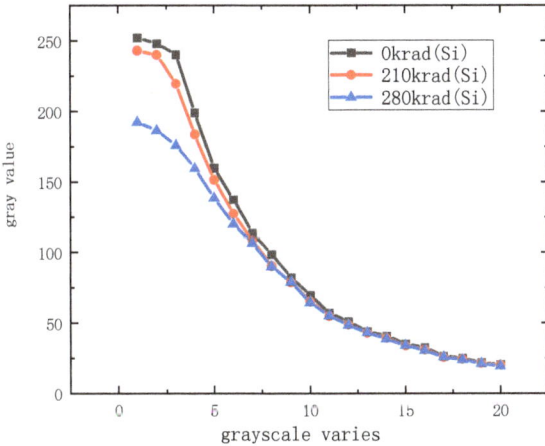

Figure 9. The gray values of grayscale vary with the TID under γ-ray irradiation.

Based on Figures 8 and 9, it can be seen that different gray level values show varying degrees of change after CIS irradiation. Meanwhile, the change in the CIS noise in Figure 8 results in a change value of less than 0.2 DN compared to the actual gray level value, indicating that the effect of the CIS noise on gray level values is not significant. The impact on the minimum gray level value of the image is greater than on the brightest gray level value. Therefore, combined with the previous analysis of the specific performance of the

MTF calculation values in Figure 9, it can be seen that the CIS noise has a certain impact on the MTF values, but it does not change the overall trend of MTF change.

The change in the diffusion MTF is affected by the change in the number of photo-generated carriers caused by QE degradation after irradiation. Therefore, the degradation of the image brightness Y after irradiation is calculated to estimate the degradation of the diffusion MTF. In addition, the camera resolution is determined by the software based on processing of the test card image using brightness Y, which is calculated from the RGB image using Equation (3).

$$Y_0 = 0.299 \times R_0 + 0.587 \times G_0 + 0.114 \times B_0 \tag{3}$$

In Equation (3), Y_0 is the gray level value of the transformed image captured by the camera before irradiation, R_0 is the red component of the image captured by the camera before irradiation, G_0 is the green component of the image captured by the camera before irradiation, and B_0 is the blue component of the image captured by the camera before irradiation. The degradation of the R, G, and B values of the image in the camera before and after irradiation is related to the QE degradation of the CIS in the corresponding red, green, and blue light bands after irradiation. Substituting the irradiation degradation rate of the CIS in the corresponding red, green, and blue incident light bands into Equation (3) yields Equation (4).

$$Y_1 = 0.299 \times R_0 \times m + 0.587 \times G_0 \times n + 0.114 \times B_0 \times l \tag{4}$$

In Equation (4), Y_1 is the gray level value of the transformed image captured by the camera after irradiation, m is the QE degradation rate of the CIS in the red light band, n is the QE degradation rate of the CIS in the green light band, and l is the QE degradation rate of the CIS in the blue light band. The coefficients for R, G, and B are the influence weights of the change rate of the QE in the corresponding band on the Y value.

Based on the above analysis, the change in the FWC directly affects the maximum gray level value of the image before and after irradiation; the minimum gray level value is affected by the CIS noise, but the effect of noise on the minimum gray level value does not change the overall trend of MTF change. The ratio of brightness Y before and after irradiation reflects the degree of degradation of the diffusion MTF. Considering these factors, Equation (2) is modified to obtain Equation (5), where K represents the conversion gain, which refers to the output image gray level value increase of the unit effective photo-generated electrons after system processing. According to relevant studies, the conversion gain is not a sensitive parameter for TID effects and can generally be regarded as a constant value in parameter calculations for the same device [16].

$$MTF_{camera-1} = MTF_{camera} \times \left[\left(\frac{FWC_1 \times K - I_0}{FWC_1 \times K + I_0} \right) / \left(\frac{FWC_0 \times K - I_0}{FWC_0 \times K + I_0} \right) \right] \times (Y_1/Y_0) \tag{5}$$

In Equation (5), MTF_{camera} is the camera resolution before irradiation, $MTF_{camera-1}$ is the calculated camera resolution after irradiation in LW/PH, FWC_0 is the Full Well Capacity when the camera is not irradiated, FWC_1 is the Full Well Capacity after irradiation in e^-, and K is the camera conversion gain in DN/e^-. I_0 is the minimum grayscale value of the captured image when the camera is not irradiated, and the unit is DN; Y_0 is the converted grayscale value of the captured image before irradiation, and the unit is DN; and Y_1 is the converted grayscale value of the captured image after irradiation, and the unit is DN. After substituting the experimental values into Equation (5), the theoretical calculation results of the camera resolution were consistent with the actual camera irradiation measurement values. This also indicates that the defects generated by radiation have a significant impact on the CIS photodetector signals, which is an important reason for the decrease in camera resolution. In the subsequent development of radiation-resistant cameras, STI and PPD reinforcement technologies can be used to strengthen the CIS against radiation, thereby reducing the generation of radiation-induced defects, inhibiting the impact of radiation-induced defects on the photodetector signals, and reducing the influence of radiation on camera resolution.

5. Conclusions

After γ-ray radiation, the QE and FWC of the CMOS image sensor degrade with the increase of the TID. Through CMOS image sensor irradiation experiments, camera system irradiation experiments, and result analysis, combined with theoretical deduction, it was found that the degradation of the QE and FWC of the CMOS image sensor is the main cause of the decrease in camera resolution. By revealing the impact mechanism of radiation damage to the CMOS image sensor on camera resolution, a quantitative evaluation formula for the influence of the FWC and QE on camera resolution was established, laying a theoretical foundation for the assessment of camera radiation resistance and the development of radiation-resistant cameras in strong nuclear radiation environments.

Author Contributions: Conceptualization, J.F. and H.-C.W.; methodology, H.-C.W., Y.-D.L. and J.F.; software, H.-C.W.; validation, H.-C.W. and Y.-D.L.; formal analysis, J.F.; investigation, Y.-D.L.; data curation, H.-C.W.; writing—original draft preparation, H.-C.W.; writing—review and editing, J.F. and Y.-D.L.; visualization, H.-C.W.; supervision, Y.-D.L.; project administration, L.W. and Q.G.; funding acquisition, J.F. and Y.-D.L. All authors have read and agreed to the published version of the manuscript.

Funding: This research was funded by the Science and Technology Innovation Leading Talent Project of Xinjiang Uygur Autonomous Region No. 2022TSYCLJ0042, the National Natural Science Foundation of China under grant No. 12175307, the West Light Talent Training Plan of the Chinese Academy of Sciences under grant No. 2022-XBQNXZ-010, and the Tianshan Innovation Team Program of Xinjiang Uygur Autonomous Region No. 2022D14003.

Data Availability Statement: Not applicable.

Conflicts of Interest: The authors declare no conflict of interest.

References

1. Kawabata, K. Toward technological contributions to remote operations in the decommissioning of the Fukushima Daiichi Nuclear Power Station. *Jpn. J. Appl. Phys.* **2020**, *59*, 050501. [CrossRef]
2. Bogue, R. Robots in the nuclear industry: A review of technologies and applications. *Ind. Robot. Int. J.* **2011**, *38*, 113–118. [CrossRef]
3. Kim, K.H.; Cho, G. Radiation effects on the resolution (MTF) of the scintillator coupled CMOS APS array imager for non-destructive test X-ray imaging. *Ann. Nucl. Energy* **2004**, *31*, 805–811.
4. Kim, K.H.; Kang, D.W.; Kim, D.K.; Kim, Y.K. Unified MTF for scintillator-coupled CMOS sensor. *Nucl. Instrum. Methods Phys. Res. Sect. A Accel. Spectrometers Detect. Assoc. Equip.* **2007**, *579*, 235–238. [CrossRef]
5. Jie, Y.; Size, C.; Lianxin, Z.; Zaodi, Z.; Xuning, C.; Rui, Z.; Zhanguo, Y.; Taosheng, L.; Jie, Y. Influence of CCD Camera Transient Noise on Imaging Resolution and Shielding Requirement in Neutron Radiography. *At. Energy Sci. Technol.* **2021**, *55*, 151–157.
6. Viallefont-Robinet, F.; Léger, D. Improvement of the edge method for on-orbit MTF measurement. *Opt. Express* **2010**, *18*, 3531–3545. [CrossRef] [PubMed]
7. Carrère, J.-P.; Place, S.; Oddou, J.-P.; Benoit, D.; Roy, F. CMOS image sensor: Process impact on dark current. In Proceedings of the 2014 IEEE International Reliability Physics Symposium, Waikoloa, HI, USA, 1–5 June 2014; pp. 3C.1.1–3C.1.6. [CrossRef]
8. Goiffon, V.; Estribeau, M.; Cervantes, P.; Molina, R.; Gaillardin, M.; Magnan, P. Influence of transfer gate design and bias on the radiation hardness of pinned photodiode CMOS image sensors. *IEEE Trans. Nucl. Sci.* **2014**, *61*, 3290–3301. [CrossRef]
9. Rao, P.R.; Wang, X.; Theuwissen, A.J.P. Degradation of spectral response and dark current of CMOS image sensors in deep-submicron technology due to γ-irradiation. In Proceedings of the ESSDERC 2007—37th European Solid State Device Research Conference, Munich, Germany, 11–13 September 2007; IEEE: New York, NY, USA, 2007; pp. 370–373.
10. Fu, J.; Wen, L.; Feng, J.; Wei, Y.; Zhou, D.; Li, Y.D.; Guo, Q. Quantum Efficiency Simulation and Analysis of Irradiated Complementary Metal-Oxide Semiconductor Image Sensors. *J. Nanoelectron. Optoelectron.* **2022**, *17*, 311–318. [CrossRef]
11. Tan, J.; Buttgen, B.; Theuwissen, A.J.P. Analyzing the radiation degradation of 4-transistor deep submicron technology CMOS image sensors. *IEEE Sens. J.* **2012**, *12*, 2278–2286. [CrossRef]
12. Goiffon, V.; Virmontois, C.; Magnan, P.; Girard, S.; Paillet, P. Analysis of Total Dose-Induced Dark Current in CMOS Image Sensors from Interface State and Trapped Charge Density Measurements. *IEEE Trans. Nucl. Sci.* **2010**, *57*, 3087–3094. [CrossRef]

13. Goiffon, V.; Estribeau, M.; Marcelot, O.; Cervantes, P.; Magnan, P.; Gaillardin, M.; Virmontois, C.; Martin-Gonthier, P.; Molina, R.; Corbiere, F.; et al. Radiation Effects in Pinned Photodiode CMOS Image Sensors: Pixel Performance Degradation Due to Total Ionizing Dose. *IEEE Trans. Nucl. Sci.* **2013**, *59*, 2878–2887. [CrossRef]
14. Goiffon, V.; Virmontois, C.; Magnan, P.; Cervantes, P.; Place, S.; Gaillardin, M.; Girard, S.; Paillet, P.; Estribeau, M.; Martin-Gonthier, P. Identification of Radiation Induced Dark Current Sources in Pinned Photodiode CMOS Image Sensors. *IEEE Trans. Nucl. Sci.* **2012**, *59*, 918–926. [CrossRef]
15. Wang, Z.; Liu, C.; Ma, Y.; Wu, Z.; Wang, Y.; Tang, B.; Liu, M.; Liu, Z. Degradation of CMOS APS Image Sensors Induced by Total Ionizing Dose Radiation at Different Dose Rates and Biased Conditions. *IEEE Trans. Nucl. Sci.* **2015**, *62*, 527–533.
16. Ma, L.; Li, Y.; Guo, Q.; Wen, L.; Zhou, D.; Feng, J.; Liu, Y.; Zeng, J.; Zhang, X.; Wang, T. Analysis of proton and γ-ray radiation effects on CMOS active pixel sensors. *Chin. Phys. B* **2017**, *26*, 114212. [CrossRef]

Disclaimer/Publisher's Note: The statements, opinions and data contained in all publications are solely those of the individual author(s) and contributor(s) and not of MDPI and/or the editor(s). MDPI and/or the editor(s) disclaim responsibility for any injury to people or property resulting from any ideas, methods, instructions or products referred to in the content.

Article

Total Ionizing Dose Effects of ⁶⁰Co γ-Ray Radiation on Split-Gate SiC MOSFETs

Haonan Feng [1,2], Xiaowen Liang [1,2], Xiaojuan Pu [1,2], Yutang Xiang [1,2], Teng Zhang [3], Ying Wei [1], Jie Feng [1,*], Jing Sun [1], Dan Zhang [1,2], Yudong Li [1], Xuefeng Yu [1,*] and Qi Guo [1]

[1] Key Laboratory of Functional Materials and Devices for Special Environments of CAS, Xinjiang Key Laboratory of Electric Information Materials and Devices, Xinjiang Technical Institute of Physics and Chemistry of CAS, Urumqi 830011, China; fenghaonan19@mails.ucas.edu.cn (H.F.); liangxiaowen17@mails.ucas.edu.cn (X.L.); puxiaojuan20@mails.ucas.edu.cn (X.P.); xiangyutang21@mails.ucas.edu.cn (Y.X.); weiying@ms.xjb.ac.cn (Y.W.); sunjing@ms.xjb.ac.cn (J.S.); zhangdan21@mails.ucas.edu.cn (D.Z.); lydong@ms.xjb.ac.cn (Y.L.); guoqi@ms.xjb.ac.cn (Q.G.)

[2] University of Chinese Academy of Sciences, Beijing 100049, China

[3] State Key Laboratory of Wide-Band Gap Semicond, Nanjing Electronic Devices Institute, Nanjing 210016, China; air.sola.zt@gmail.com

* Correspondence: fengjie@ms.xjb.ac.cn (J.F.); yuxf@ms.xjb.ac.cn (X.Y.)

Abstract: SiC power devices require resistance to both single-event effects (SEEs) and total ionizing dose effects (TIDs) in a space radiation environment. The split-gate-enhanced VDMOSFET (SGE-VDMOSFET) process can effectively enhance the radiation resistance of SiC VDMOS, but it has a certain impact on the gate oxide reliability of SiC VDMOS. This paper investigates the impact mechanism and regularity of using the SGE process to determine the radiation resistance and long-term reliability of SiC VDMOS under other identical processes and radiation conditions. Our experimental results show that after ⁶⁰Co γ-ray irradiation, the degradation degrees of the static parameters of SGE-VDMOSFET and planar gate VDMOSFET (PG-VDMOSFET) are similar. The use of the new process leads to more defects in the oxide layer, reducing the long-term reliability of the device, but its stability can recover after high-temperature (HT) accelerated annealing. This research indicates that enhancing the resistance of SEEs using an SGE-VDMOSFET structure requires simultaneously considering the demand for TIDs and long-term reliability.

Keywords: split-gate-enhanced VDMOSFET; planar gate VDMOSFET; total ionizing dose effect; long-term reliability

Citation: Feng, H.; Liang, X.; Pu, X.; Xiang, Y.; Zhang, T.; Wei, Y.; Feng, J.; Sun, J.; Zhang, D.; Li, Y.; et al. Total Ionizing Dose Effects of ⁶⁰Co γ-Ray Radiation on Split-Gate SiC MOSFETs. *Electronics* **2023**, *12*, 2398. https://doi.org/10.3390/electronics12112398

Academic Editor: Yahya M. Meziani

Received: 24 April 2023
Revised: 19 May 2023
Accepted: 19 May 2023
Published: 25 May 2023

Copyright: © 2023 by the authors. Licensee MDPI, Basel, Switzerland. This article is an open access article distributed under the terms and conditions of the Creative Commons Attribution (CC BY) license (https://creativecommons.org/licenses/by/4.0/).

1. Introduction

As the aerospace industry rapidly develops towards deep space exploration, electronic devices are facing increasingly diverse working environments, making it particularly important to ensure the stability of device operation in complex space environments. A space radiation environment is filled with a large number of high-energy particles such as electrons, protons, gamma-rays, and heavy ions, which pose a threat to semiconductor components in spacecrafts [1,2]. Silicon carbide power field effect transistors (SiC VDMOS) have significant advantages, including high temperature, high power, high efficiency, and reduced volume, meeting the requirements of the new generation of spacecraft power semiconductor devices. Therefore, SiC-based power semiconductor devices have great potential for application in space radiation environments [3,4].

Currently, the radiation effects of SiC metal-oxide semiconductor field-effect transistors (SiC MOSFETs) have gained attention [5–8]. In 2012, Akturk et al. conducted a TID experiment on 1200 V SiC MOSFET power devices using ⁶⁰Co γ-radiation. The results showed that the device still had good performance when the accumulated dose exceeded 100 krad (Si). When the accumulated dose exceeded 300 krad (Si), the capacitance between the gate and drain changed, which significantly affected the device's switching performance [9]. In 2014, Alexandru et al. examined the influence of proton and electron irradiation on the electrical parameters of 4H-SiC nMOSFETs [10]. The results showed that the threshold voltage decreased after proton irradiation and tended to stabilize over time. With an increase in proton dose, the threshold voltage showed a decreasing trend. Similar changes were observed before and after electron irradiation. Meanwhile, the gate leakage current remained almost unchanged under the highest flux proton irradiation and the maximum dose of electron irradiation [11]. In 2019, TID radiation experiments were conducted on different SiC power devices by Hazdra and Popelka. The results showed that SiC devices with an oxide layer were more susceptible to TIDs after radiation [12].

In recent years, the radiation damage effects and radiation hardening techniques of SiC VDMOS devices have gradually become a research hotspot due to their sensitivity to radiation-induced damage in the oxide layer. Among them, the function disablement characteristic of SEE radiation damage has received extensive attention both domestically and abroad [13–16]. Increasing the thickness of the VDMOS oxide layer is a commonly employed measure to strengthen resistance against SEEs. This method can effectively enhance the device's resistance to single-Event gate rupture (SEGR) and has already been extensively studied in silicon-based VDMOS. As early as 1986, A. E. Waskiewicz et al. first proposed the SEB effect of power VDMOS devices [17]. In 1987, T. Fischer proposed the SEGR effect of power VDMOS devices [18]. Cascio, A et al. proposed a method of thickening the gate dielectric, where they reinforced the device by separately depositing a thick oxide layer in the gate oxide region of the JFET region. The purpose of this structure is to increase the breakdown voltage of the gate dielectric layer by increasing the thickness of the oxide dielectric layer, thereby improving the device's resistance to SEEs [19]. The LOCOS (Local Oxidation of Silicon) structure adopted by Tang Zhaohuan and colleagues is also a reinforcement measure based on the principle of thickening the gate oxide layer. However, unlike the method of depositing a thick gate oxide layer alone, this structure consumes some of the silicon thickness in the JFET region [20]. Currently, conventional reinforcement technology improves the SEE resistance in power MOSFET, which can result in performance loss to factors such as conductivity and reliability. Therefore, determining how to improve SEE resistance while meeting the basic operating characteristics of the device has long been a focus of research for experts in radiation hardening.

Specific on-resistance (R_{onsp}) is an important indicator for evaluating the performance of unipolar power devices. Its physical meaning is the product of the on-resistance of the device and the active conducting area of the chip. A smaller value indicates a higher technical level, meaning that products with the same on-resistance value require smaller chip areas. In 1993, J.W. Palmour proposed a vertical UMOSFET structure based on an immature ion implantation process in a silicon carbide material. As a result, this structure eliminated lattice loss caused by ion implantation through epitaxy, enabling the device to withstand voltages up to 330 V with a R_{onsp} of 33 mΩcm^2. Due to various issues in the UMOS structure process, more researchers have focused on research on VDMOS devices in recent years. However, VDMOS devices have a large gate-drain capacitance, which greatly reduces their frequency response and performance when the device operates under a high-frequency state, leading to performance losses [21,22]. To optimize the working performance of VDMOS devices under high-frequency conditions, the split-gate structure (split gate) emerged as a solution, dividing the gate structure in two. This structure significantly reduced gate leakage capacitance and improved the performance of trench-gate VDMOS devices.

Modifying the oxide layer structure of VDMOS can effectively enhance the radiation tolerance of SiC VDMOS, but this complex process could potentially affect the reliability of the gate oxide. With the use of a new split-gate structure oxide process, defects of different types and spatial distributions will occur inside the gate oxide layer, which will affect the TID sensitivity of SiC VDMOS. Therefore, investigating the quality of the oxide layer of the split-gate structure and studying the changes in TID radiation damage are crucial.

This paper presents a comparative study on the TID sensitivity and long-term reliability of the traditional PG-VDMOSFET structure and the SG-VDMOSFET structure. The transfer characteristics curves (I_{DS}–V_{GS}), threshold voltage drift (ΔV_{TH}), on-resistance ($R_{DS(on)}$), leakage current (I_{GSS}), interface state density (N_{it}), and current density–electric field strength characteristic curves (J-E characteristic curves) were analyzed before and after radiation exposure. This research is based on the improvement of electrical parameters and reliability degradation of SiC VDMOS, providing experimental data to optimize the gate oxide process and mitigate TID damage for SiC VDMOS.

2. Samples and Experimental Setup

This experiment utilized two domestically produced 1200 V N-channel SiC VDMOS devices with a planar gate structure, both packaged in TO-247-3 packages and processed using the same process flow. P-type wells and N+ source regions were doped via ion implantation on a 10 um epitaxial layer to form the MOSFET channel region. The gate oxide layer was then grown via thermal oxidation and annealed in a NO atmosphere after a HT treatment to activate the carriers. The two devices featured the same cell structure and had a tox of 50 nm. Specifically, the split-gate structure SiC VDMOS has two symmetrical gate electrodes placed at an interval on top of the gate insulation layer, with a single gate length of 1.2–1.25 um and a distance of 1.3 um between the two gates. Figure 1 shows the schematic of two different structures of VDMOSFETs: (a) SGE-VDMOSFET; (b) PG-VDMOSFET.

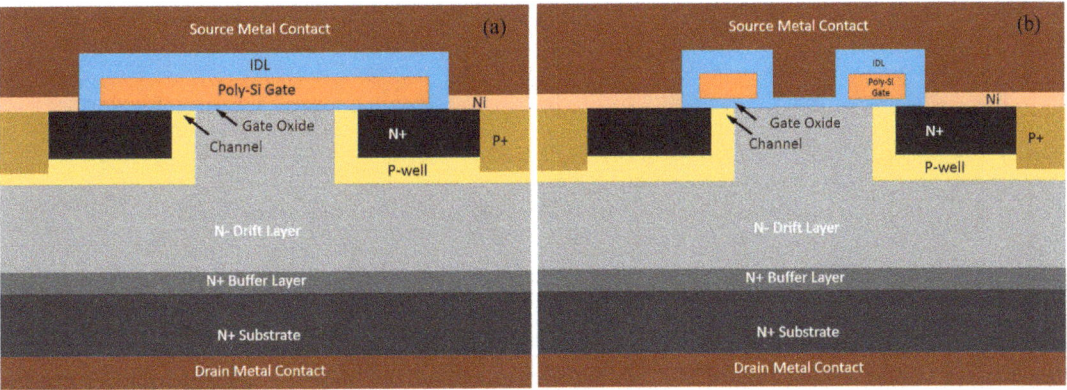

Figure 1. Typical-size VDMOSFET structure diagram: (**a**) PG-VDMOSFET; (**b**) SGE-VDMOSFET.

The experiment was carried out at the Xinjiang Institute of Physics and Chemistry, Chinese Academy of Sciences. The irradiation source was ^{60}Co, and the dose rate was 100 rad (Si)/s. The samples were irradiated to 300 krad (Si) with a positive gate bias (V_{GS} = 20 V, drain and source grounded) at room temperature (RT).

An Agilent B1500 A semiconductor device analyzer was used to measure the I_{DS}–V_{GS} curve before irradiation. Linear extrapolation was performed on the I_{DS}–V_{GS} curve to obtain the threshold voltage (V_{TH}) of the device. The $R_{DS\,(on)}$ was tested using a semiconductor isolation device testing system (BC3193) based on the device datasheet, and I_{GSS} was tested at V_{GS} =20 V and V_{DS} = 0 V. Some devices were subjected to TID irradiation followed by 168 h annealing at 100 °C, with the same bias as the radiation process maintained on all pins. The device characteristics were then retested after annealing.

3. Results and Analyses

3.1. The Influence of TID Radiation on the Static Characteristics of SGE-VDMOSFET and PG-VDMOSFET

Figure 2 shows the relationship between the I_{DS}–V_{GS} curves and cumulative dose for the two types of SiC VDMOS transistor. For the I_{DS}–V_{GS} curve, we mainly focus on its sub-threshold region, so the size of VD does not affect the area we need to observe. At the same time, when testing the I_{DS}–V_{GS} curve using the Agilent B1500 A semiconductor device analyzer, if VD is greater than 0.1 V and the device is conducting, the equipment will cause current limiting and cannot be tested. Therefore, to obtain the complete I_{DS}–V_{GS} curve of the device, VD needs to be set to less than or equal to 0.1 V. As the same process and cell size were used, the investigation of TIDs was mainly focused on the oxide thickness (tox) of the devices. In terms of radiation damage, the SGE-VDMOSFET exhibited less severe damage than the PG-VDMOSFET. Figure 3 shows the threshold voltage degradation curves of the two devices, with good consistency observed among multiple tested devices in terms of threshold voltage performance. As the cumulative dose increased, the threshold voltage decreased uniformly for both devices. The threshold voltage of the PG-VDMOSFET had a negative drift of 1.62 V at a cumulative dose of 300 krad (Si), while the SGE-VDMOSFET already exhibited a threshold voltage drift of 1.36 V at the same cumulative dose.

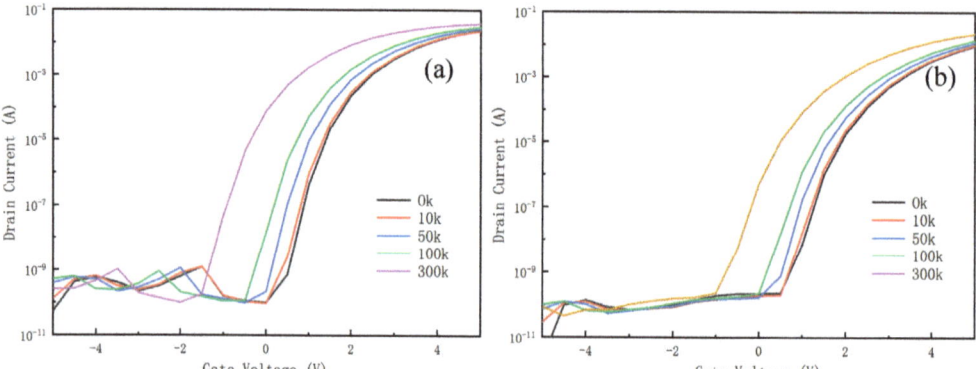

Figure 2. Transfer characteristic curve: (**a**) PG–VDMOSFET; (**b**) SGE–VDMOSFET.

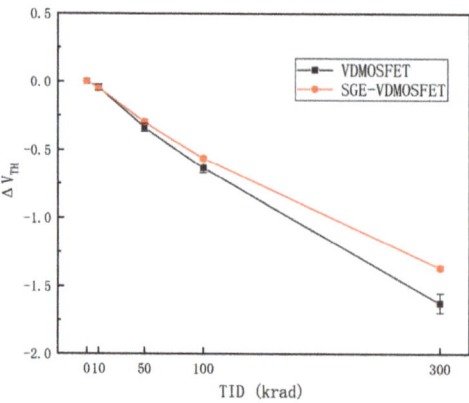

Figure 3. Threshold voltage shifts ΔV_{TH} as a function of TIDs for PG-VDMOSFET and SGE-VDMOSFET.

Radiation causes ionization in the gate oxide, resulting in the generation of a large number of electron–hole pairs. Although there are a small number of electron traps formed by carbon residues during the thermal oxidation growth process in the gate oxide, their density is extremely low, and the electrons with high mobility will be quickly swept out of the gate oxide layer under the influence of an electric field. The holes with lower mobility are more likely to be captured by hole traps in the oxide layer, becoming positively charged oxide traps. Oxide traps (mainly hole traps generated during the thermal oxidation process) will capture more positively charged holes after irradiation, leading to an increased inversion degree of the NMOS channel and a negative shift in the transfer characteristic curve towards the negative x-axis. The oxide trap charges that affect the threshold voltage of the device are mainly located above the channel, with a smaller influence on the gate oxide layer in the JFET region. Therefore, theoretically, under the same gate oxide thickness for both devices, the degradation of the threshold voltage should be almost the same. However, it can be clearly seen from the data of multiple devices that the threshold voltage degradation of the SGE-VDMOSFET is weaker. This phenomenon often occurs due to defects in the interface between the gate oxide layer and the silicon carbide substrate caused by changes in the gate oxide layer fabrication process. Therefore, in Section 3.2 of this article, this inference is validated.

Subthreshold Swing (SS) refers to the change in I_{DS} with a one-decade increase in V_{GS}. It represents the change in gate voltage required for the I_{DS} to vary by a factor of 10 and is also known as the S factor. A smaller S value indicates a faster ON/OFF switching speed.

$$S = \frac{dV_{GS}}{d(\lg I_{DS})} = \ln 10 \frac{dV_{GS}}{d(\ln I_{DS})} = \ln 10 \frac{d\psi_s}{dI_{DS}} I_{DS} \frac{dV_{GS}}{d\psi_s} \quad (1)$$

$$\frac{d\psi_s}{dI_{DS}} I_{DS} = \frac{kT}{q} \quad (2)$$

$$\eta = \frac{dV_{GS}}{d\psi_s} = \frac{C_{ox} + C_{dep} + C_{it}}{C_{ox}} \quad (3)$$

$$SS = \frac{kT}{q} \eta \ln 10 = \frac{kT}{q} \frac{C_{ox} + C_{dep} + C_{it}}{C_{ox}} \ln 10 \quad (4)$$

In this equation, $\frac{kT}{q}$ represents the thermal voltage, η is referred to as the body factor, and ψ_s is the surface potential. C_{ox} is the capacitance of the top-gate oxide layer, C_{dep} is the depletion capacitance, C_{it} is the interface trap capacitance, k is the Boltzmann constant, T is the temperature, and q is the electronic charge.

The factors affecting subthreshold swing are as follows: (1) An increase in temperature leads to an increase in subthreshold swing. (2) An increase in gate oxide capacitance leads to a decrease in subthreshold swing; the use of high-k dielectric materials or a reduction in the gate oxide thickness can result in a decrease in subthreshold swing. (3) A decrease in Si depletion layer capacitance results in a decrease in subthreshold swing; factors that may increase the depletion layer width, such as a decrease in substrate concentration Na or an increase in substrate bias voltage, lower the subthreshold swing. (4) Interface defects between the gate oxide layer and the substrate silicon can store charge and can effectively increase the capacitance, leading to an increase in the subthreshold swing. (5) A shorter channel length weakens the gate control capability, and hence, leads to an increase in subthreshold swing. (6) An increase in gate voltage results in stronger surface inversion, leading to weaker gate control of the channel and an increase in subthreshold swing.

Figure 4 shows the relationship between the S and the total irradiation dose for the two SiC VDMOS transistors. After TID radiation, the S value of the SGE-VDMOSFET changes slightly from 282 to 301, while the S of the PG-VDMOSFET remains almost unchanged.

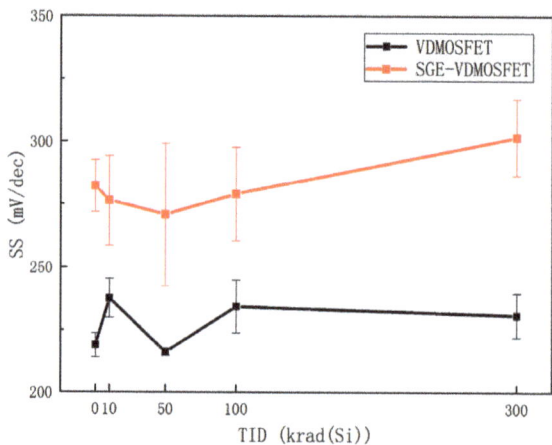

Figure 4. Subthreshold Swing as a function of TIDs for PG-VDMOSFET and SGE-VDMOSFET.

3.2. Differences in Threshold Voltage Degradation between SGE-VDMOSFET and PG-VDMOSFET

According to possible charge states, interface states are divided into acceptor-type interface states and donor-type interface states. Regarding acceptor-type interface traps, they are negatively charged when they are filled with electrons and electrically neutral when electrons are lost. For donor-type interface states, they are electrically neutral when they are filled with electrons and positively charged when electrons are lost. When the Fermi level at the interface coincides with the center of the bandgap (mid-gap), the interface states are approximately electrically neutral, and do not cause mid-gap voltage changes. Since the distribution of interface states in the bandgap is complex, it can only be assumed that the interface states are approximately electrically neutral at the mid-gap. Other factors affecting threshold voltage, such as work function difference and semiconductor space charge, do not change due to irradiation; therefore, the change in mid-gap voltage caused by irradiation is mainly due to the change in oxide trap charge [23–25].

Figure 5 shows the voltage changes caused by oxide trap charge-induced changes (ΔV_{ot}) and interface state-induced changes (ΔV_{it}) as separated using the mid-gap voltage method [26]. It can be seen that ΔV_{ot} is the main cause of the threshold voltage change. During the thermal oxidation process, the oxide layer hole traps generated by oxygen defects and the near-interface hole traps formed during nitrogen passivation can capture a large number of holes generated via ionizing radiation. The trapped holes mainly affect the threshold voltage drift of the device. After being exposed to total ionizing radiation, the ΔV_{TH} of PG-VDMOSFET decreased by 1.6 V, and the ΔV_{TH} of SGE-VDMOSFET decreased by 1.4 V, with weaker degradation in the ΔV_{TH} of SGE-VDMOSFET. The ΔV_{ot} of PG-VDMOSFET decreased by 1.9 V, and the ΔV_{ot} of SGE-VDMOSFET decreased by 1.8 V, with weaker degradation in the ΔV_{ot} of SGE-VDMOSFET. The ΔV_{it} of PG-VDMOSFET increased by 0.3 V, and the ΔV_{it} of SGE-VDMOSFET increased by 0.4 V, with stronger degradation in the ΔV_{it} of SGE-VDMOSFET. Our analysis indicates that there are two reasons for the weaker degradation in the ΔV_{TH} of SGE-VDMOSFET: (1) from the ΔV_{ot} perspective, the SGE-VDMOSFET utilizing the new process has fewer oxide defect potential hole traps in the oxide layer produced during the thermal oxidation process compared to PG-VDMOSFET, leading to weaker degradation in the ΔV_{ot} of SGE-VDMOSFET; (2) from the ΔV_{it} perspective, PG-VDMOSFET has fewer interface states at the gate oxide interface due to less etching, resulting in fewer captured electrons during the radiation process and weaker degradation in the ΔV_{it}. Overall, reason 2 is the main reason for the weaker degradation in the ΔV_{TH} of SGE-VDMOSFET.

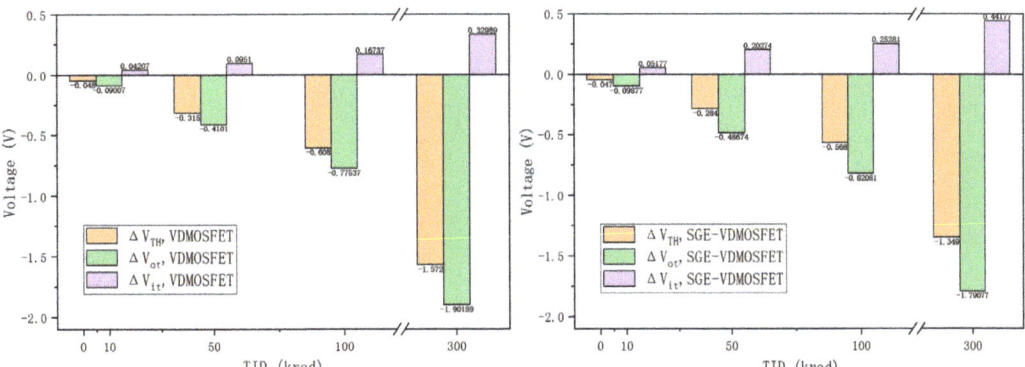

Figure 5. ΔV_{TH}, ΔV_{ot}, and ΔV_{it} as functions of TIDs for PG-VDMOSFET and SGE-VDMOSFET.

When designing a switching power supply or driving circuit using MOSFET, it is generally necessary to consider the on-resistance of the MOSFET. Because energy is consumed on this resistance when current flows through the drain and source, this energy consumption is called conduction loss. Choosing a MOSFET with a lower on-resistance can reduce conduction loss to a certain extent. If a higher breakdown voltage is required, the internal structure needs to be made thicker, so the on-resistance of a MOSFET with a higher breakdown voltage will be larger.

Meanwhile, gate leakage current is also a very important parameter in semiconductor devices. It describes the insulation effect of the transistor and has an important impact on the service life and stability of the device. The gate leakage current refers to the leakage current between the gate and drain of a transistor when it is in the off state. By measuring the leakage current value of the device, we can evaluate the insulation quality. As an important component of semiconductor devices, the gate leakage current of a transistor has a significant impact on its performance.

There are three factors that affect gate leakage current: 1. Temperature is an important factor that affects gate leakage current. As the temperature increases, the leakage current between the gate and drain of the transistor gradually increases. Therefore, when conducting temperature tests, it is necessary to ensure that the operating temperature of the transistor does not exceed its allowable maximum temperature. At the same time, it is also necessary to compare different devices under the same temperature conditions to determine the magnitude of their gate leakage current. 2. Voltage is also one of the factors that affect the magnitude of the transistor's gate leakage current. At a certain operating temperature, as the acceleration voltage increases, the gate leakage current of the transistor also shows an increasing trend. Therefore, when conducting gate leakage current tests, attention should be paid to the selection of the test voltage and the duration of the test. 3. The quality of insulation materials also affects the magnitude of the gate leakage current. Higher-quality insulation materials can effectively suppress the gate leakage current of the transistor. Therefore, when designing semiconductor devices, it is necessary to select high-quality insulation materials to ensure the stability and reliability of the device.

Figure 6 shows the changes in the electrical parameters of two SiC VDMOS transistors as a function of accumulated radiation dose. After irradiation, the gate oxide of the two devices accumulated positive charges in oxide traps to varying degrees. Defects that carry a positive charge attract electrons onto the semiconductor surface, increasing the surface electron concentration, and hence, reducing the device $R_{DS\,(on)}$. When the accumulated dose reaches 300 krad (Si), the I_{GSS} of SGE-VDMOSFET increases slightly but remains at the nA level. TIDs do not significantly affect the I_{GSS} of either device. The examination of gate leakage current is often carried out by observing whether it undergoes a sudden change from the nA level to the μA level. The variation in Figure 6 is very small, which is highly likely to be due to measurement errors caused by the equipment.

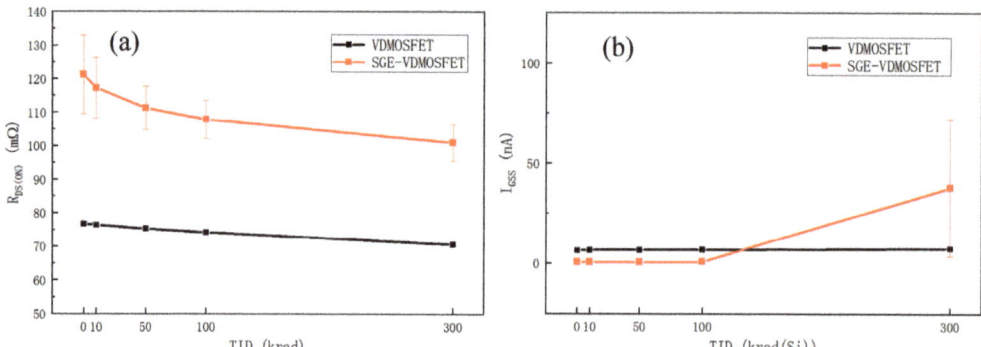

Figure 6. $R_{DS(on)}$ and I_{GSS} as a function of TIDs for samples (**a**,**b**).

3.3. The Influence of TID Radiation on Electron Tunneling in Gate Oxide

It is generally accepted that Fowler–Nordheim (F-N) tunneling becomes one of the main reasons for charge transport through the oxide layer when the oxide layer is thick or the gate voltage is high. F-N tunneling is an electron tunneling phenomenon caused by an electric field. As shown in Figure 7, when a high voltage is applied to the polycrystalline silicon/oxide/carbon structure, the barrier in the oxide layer becomes very steep, and electrons in the SiC conduction band face a triangular barrier that depends on the external electric field. When the voltage is high enough, the barrier becomes extremely narrow, and electrons can tunnel through the barrier, entering the oxide conduction band from the SiC conduction band, thus inducing F-N tunneling. The tunneling current density can be expressed as follows using a self-consistent electron model and WKB approximation [27]:

$$J_{FN} = \frac{q^3 E^2}{16\pi \hbar \phi_B} \exp\left(-\frac{4\sqrt{2m^* \phi_B^3}}{3q\hbar E}\right) \quad (5)$$

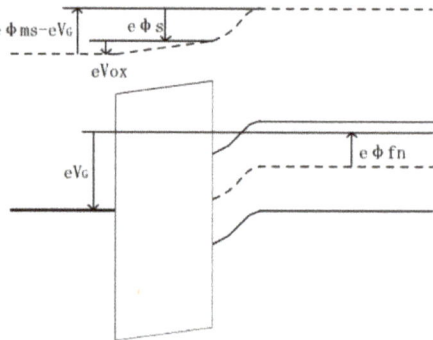

Figure 7. An energy band diagram of P-type substrate silicon carbide under positive gate bias.

The expression can be adjusted using a correction factor that includes $\hbar$ as the reduced Planck constant, ϕ_B as the barrier height, m^* as the effective mass of electrons in the oxide layer, and E as the electric field strength in the oxide layer. The correction factor can reflect the effect of the mirror potential as well as the influence of temperature on the tunneling process [28].

Figure 8 shows changes in the gate oxide J-E curves of two SiC VDMOS transistors as a function of accumulated total dose. For the PG-VDMOSFET device, the gate oxide current density increases uniformly when the accumulated dose reaches 300 krad (Si), but the characteristic gradually recovers after 168 h HT accelerated annealing. The J-E characteristic of non-irradiated SGE-VDMOSFET exhibits hysteresis at an electric field strength of 4 MV/cm. When the accumulated dose for SGE-VDMOSFET reaches 300 krad (Si), the tunneling characteristic of the device tends to be stable, and the device has better gate oxide characteristics after subsequent 168 h HT accelerated annealing. Further investigation is required to ensure the long-term reliability of the device.

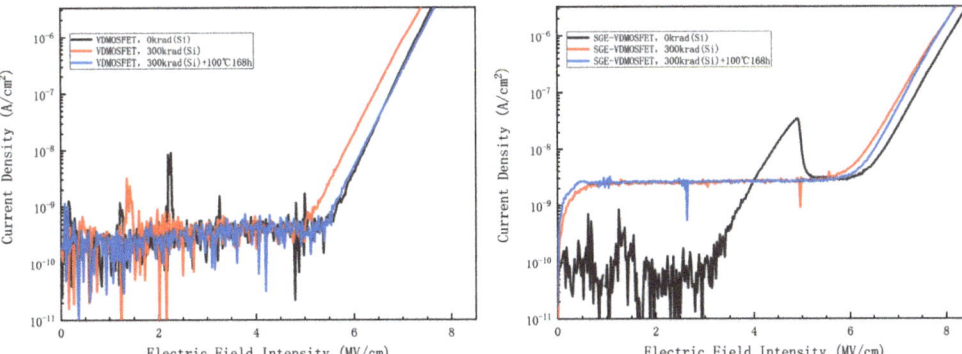

Figure 8. Current density–electric field strength (J-E) characteristic curves for devices irradiated with ^{60}Co γ-rays up to 300 krad (Si) at a dose rate of 100 rad(SiO$_2$)/s, and annealed for 168 h at 100 °C.

4. Conclusions

With an increase in accumulated dose during ^{60}Co γ-ray irradiation, both PG-VDMOSFET and SGE-VDMOSFET experience varying degrees of radiation damage. For V_{TH}, SGE-VDMOSFET experiences weaker degradation compared to PG-VDMOSFET. This is mainly due to the presence of more defects at the interface during the growth and etching of gate oxide in SGE-VDMOSFET, and an increase in interface states after irradiation. The electrical performance consistency of SGE-VDMOSFET devices is poor. At the same time, there are reliability problems with the gate oxide. The initial samples had more deep-level defects inside the gate oxide, which facilitated electron tunneling from the substrate through F-N tunneling under electric stress. The SGE process may lead to the generation of more deep-level defects in the gate oxide when solving the SEE problems of SiC VDMOS. J-E can be significantly improved after irradiation and 168 h high-temperature annealing. Improving the SEE resistance of SiC VDMOS via the SGE process requires consideration of the long-term reliability degradation of the devices.

Author Contributions: Conceptualization, H.F. and X.Y.; methodology, H.F., X.P., D.Z. and X.Y.; software, Y.X. and Y.W.; validation, H.F.; formal analysis, H.F.; investigation, H.F. and X.P.; resources, T.Z.; data curation, X.L.; writing—original draft preparation, H.F.; writing—review and editing, H.F., X.Y. and Q.G.; visualization, H.F.; supervision, X.L. and X.Y.; project administration, X.Y.; funding acquisition, J.F., J.S., Y.L., X.Y. and Q.G. All authors have read and agreed to the published version of the manuscript.

Funding: This work was jointly supported by the National Natural Science Foundation of China (grant nos. 11975305 and 11805268).

Data Availability Statement: Data are not available in a publicly accessible repository and they cannot be shared upon request.

Conflicts of Interest: The authors declare no conflict of interest.

References

1. Fleetwood, D.M. Total-Ionizing-Dose Effects, Border Traps, and 1/f Noise in Emerging MOS Technologies. *IEEE Trans. Nucl. Sci.* **2020**, *67*, 1216–1240. [CrossRef]
2. Adell, P.C.; Scheick, L.Z. Radiation Effects in Power Systems: A Review. *IEEE Trans. Nucl. Sci.* **2013**, *60*, 1929–1952. [CrossRef]

3. Jie, X.; Qing, K.; Xuan, Z.; Feng, L. Application Prospect of SiC Power Semiconductor Devices in Spacecraft Power Systems. In Proceedings of the 2017 13th IEEE International Conference on Electronic Measurement & Instruments (ICEMI), Yangzhou, China, 20–22 October 2017; pp. 185–190.
4. Roccaforte, F.; Fiorenza, P.; Greco, G.; Nigro, R.L.; Giannazzo, F.; Iucolano, F.; Saggio, M. Emerging trends in wide band gap semiconductors (SiC and GaN) technology for power devices. *Microelectron. Eng.* **2018**, *187–188*, 66–77. [CrossRef]
5. Sun, Y.; Wan, X.; Liu, Z.; Jin, H.; Yan, J.; Li, X.; Shi, Y. Investigation of total ionizing dose effects in 4H–SiC power MOSFET under gamma ray radiation. *Radiat. Phys. Chem.* **2022**, *197*, 110219. [CrossRef]
6. Murata, K.; Mitomo, S.; Matsuda, T.; Yokoseki, T.; Makino, T.; Onoda, S.; Takeyama, A.; Ohshima, T.; Okubo, S.; Tanaka, Y.; et al. Impacts of gate bias and its variation on gamma-ray irradiation resistance of SiC MOSFETs. *Phys. Status Solidi (A)* **2017**, *214*, 1600446. [CrossRef]
7. Zhang, T.; Allard, B.; Bi, J. The synergetic effects of high temperature gate bias and total ionization dose on 1.2 kV SiC devices. *Microelectron. Reliab.* **2018**, *88–90*, 631–635. [CrossRef]
8. Zhang, C.X.; Shen, X.; Zhang, E.X.; Fleetwood, D.M.; Schrimpf, R.D.; Francis, S.A.; Roy, T.; Dhar, S.; Ryu, S.-H.; Pantelides, S.T. Temperature Dependence and Postirradiation Annealing Response of the 1/f Noise of 4H-SiC MOSFETs. *IEEE Trans. Electron Devices* **2013**, *60*, 2361–2367. [CrossRef]
9. Akturk, A.; McGarrity, J.M.; Potbhare, S.; Goldsman, N. Radiation Effects in Commercial 1200 V 24 A Silicon Carbide Power MOSFETs. *IEEE Trans. Nucl. Sci.* **2012**, *59*, 3258–3264. [CrossRef]
10. Alexandru, M.; Florentin, M.; Constant, A.; Schmidt, B.; Michel, P.; Godignon, P. 5 MeV proton and 15 MeV electron radiation effects study on 4H-SiC nMOSFET electrical parameters. *IEEE Trans. Nucl. Sci.* **2014**, *61*, 1732–1738. [CrossRef]
11. Lebedev, A.A.; Kozlovski, V.V.; Levinshtein, M.E.; Ivanov, A.E.; Strel, A.M.; Zubov, A.V.; Fursin, L. Impact of 0.9 MeV electron irradiation on main properties of high voltage vertical power 4H-SiC MOSFETs. *Radiat. Phys. Chem.* **2020**, *177*, 109200. [CrossRef]
12. Hazdra, P.; Popelka, S. Displacement damage and total ionisation dose effects on 4H-SiC power devices. *IET Power Electron.* **2019**, *12*, 3910–3918. [CrossRef]
13. Peng, C.; Lei, Z.; Zhang, Z.; Chen, Y.; He, Y.; Yao, B.; En, Y. Influence of Drain Bias and Flux on Heavy Ion-Induced Leakage Currents in SiC Power MOSFETs. *IEEE Trans. Nucl. Sci.* **2022**, *69*, 1037–1043. [CrossRef]
14. Zhang, H.; Guo, H.-X.; Zhang, F.-Q.; Lei, Z.-F.; Pan, X.-Y.; Liu, Y.-T.; Gu, Z.-Q.; Ju, A.-A.; Zhong, X.-L.; Ouyang, X.-P. Study on proton-induced single event effect of SiC diode and MOSFET. *Microelectron. Reliab.* **2021**, *124*, 114329. [CrossRef]
15. Martinella, C.; Alia, R.G.; Stark, R.; Coronetti, A.; Cazzaniga, C.; Kastriotou, M.; Kadi, Y.; Gaillard, R.; Grossner, U.; Javanainen, A. Impact of Terrestrial Neutrons on the Reliability of SiC VD-MOSFET Technologies. *IEEE Trans. Nucl. Sci.* **2021**, *68*, 634–641. [CrossRef]
16. Zhou, X.; Tang, Y.; Jia, Y.; Hu, D.; Wu, Y.; Xia, T.; Gong, X.; Pang, H.Y. Single-Event Effects in SiC Double-Trench MOSFETs. *IEEE Trans. Nucl. Sci.* **2019**, *66*, 2312–2318. [CrossRef]
17. Waskiewicz, A.E.; Groninger, J.W.; Strahan, V.H.; Long, D.M. Burnout of Power MOS Transistors with Heavy Ions of Californium-252. *IEEE Trans. Nucl. Sci.* **1986**, *33*, 1710–1713. [CrossRef]
18. Fischer, T.A. Heavy-Ion-Induced, Gate-Rupture in Power MOSFETs. *IEEE Trans. Nucl. Sci.* **1987**, *34*, 1786–1791. [CrossRef]
19. Cascio, A.; Curro, G. MOS Device Resistant to Ionizing Radiation. U.S. Patent EP1918986A2, 22 February 2010.
20. Zhaohuan, T.; Gangyi, H.; Guangbing, C.; Kaizhou, T.; Yong, L.; Jun, L.; Xueliang, X. A novel structure for improving the SEGR of a VDMOS. *J. Semicond.* **2012**, *33*, 044002.
21. Wei, J.X.; Liu, S.Y.; Zhao, H.B.; Fu, H.; Zhang, X.B.; Li, S.Y.; Sun, W.F. Verification of Single-Pulse Avalanche Failure Mechanism for Double-Trench SiC Power MOSFETs. *IEEE J. Emerg. Sel. Top. Power Electron.* **2021**, *9*, 2190–2200. [CrossRef]
22. Mantia, S.L.; Pulvirenti, M.; Sciacca, A.G.; Nania, M. SiC MOSFETs Applications and Technology Robustness Evaluation under Avalanche Conditions. In Proceedings of the PCIM Europe Digital Days 2020; International Exhibition and Conference for Power Electronics, Intelligent Motion, Renewable Energy and Energy Management, Nuremburg, Germany, 7–8 July 2020.
23. Sharov, F.V.; Moxim, S.J.; Haase, G.S.; Hughart, D.R.; Lenahan, P.M. A Comparison of Radiation-Induced and High-Field Electrically Stress-Induced Interface Defects in Si/SiO$_2$ MOSFETs via Electrically Detected Magnetic Resonance. *IEEE Trans. Nucl. Sci.* **2022**, *69*, 208–215. [CrossRef]
24. Lelis, A.J.; Green, R.; Habersat, D.B. SiC MOSFET Reliability and Implications for Qualification Testing. In Proceedings of the 2017 IEEE International Reliability Physics Symposium (IRPS), Monterey, CA, USA, 2–6 April 2017; pp. 2A-4.1–2A-4.4.
25. Chaturvedi, M.; Dimitrijev, S.; Haasmann, D.; Moghadam, H.A.; Pande, P.; Jadli, U. Quantified density of performance-degrading near-interface traps in SiC MOSFETs. *Sci. Rep.* **2022**, *12*, 4076. [CrossRef]
26. Terman, L.M. An investigation of surface states at a silicon/silicon oxide interface employing metal-oxide-silicon diodes. *Solid-State Electron.* **1962**, *5*, 285–299. [CrossRef]
27. Moll, J.L. *Physics of Semiconductors*, 1st ed.; McGraw-Hill: New York, NY, USA, 1964; pp. 145–166.
28. Lenzlinger, M.; Snow, E.H. Fowler-Nordheim Tunneling into Thermally Grown SiO$_2$. *J. Appl. Phys.* **1969**, *40*, 278–283. [CrossRef]

Disclaimer/Publisher's Note: The statements, opinions and data contained in all publications are solely those of the individual author(s) and contributor(s) and not of MDPI and/or the editor(s). MDPI and/or the editor(s) disclaim responsibility for any injury to people or property resulting from any ideas, methods, instructions or products referred to in the content.

Article

Neutron Irradiation Testing and Monte Carlo Simulation of a Xilinx Zynq-7000 System on Chip

Weitao Yang [1,2,3], Yonghong Li [2,*], Yang Li [2], Zhiliang Hu [2,4,5], Jiale Cai [6], Chaohui He [2], Bin Wang [1,*] and Longsheng Wu [1]

1 School of Microeletronics, Xidian University, Xi'an 710071, China
2 School of Nuclear Science & Technology, Xi'an Jiaotong University, Xi'an 710049, China
3 Dipartimento di Automatica e Informatica, Politecnico di Torino, 10129 Torino, Italy
4 Spallation Neutron Source Science Center, Dongguan 523803, China
5 Institute of High Energy Physics, Chinese Academy of Sciences (CAS), Beijing 100049, China
6 Dipartimento di Elettronica e Telecomunicazioni, Politecnico di Torino, 10129 Torino, Italy
* Correspondence: yonghongli@mail.xjtu.edu.cn (Y.L.); wbin@xidian.edu.cn (B.W.)

Abstract: The reliability of nanoscale electronic systems is important in various applications. However, they are becoming increasingly vulnerable to atmospheric neutrons. This research conducted spallation neutron irradiations on a Xilinx Zynq-7000 system on a chip using the China Spallation Neutron Source. The results were analyzed in combination with a Monte Carlo simulation to explore the impact of atmospheric neutrons on the single event effects of the target system on chip. Meanwhile, the contribution of thermal neutrons to the chip's single event effect susceptibility was also assessed. It was found that absorbing thermal neutrons with a 2 mm Cd sheet can protect against the single event effect on the system on the chip by about 44.4%. The effects of B and Hf elements, inside the device, on a single event effect of the Xilinx Zynq-7000 system on chip were evaluated too. Additionally, it was discovered that ^{10}B interacting with thermal neutrons was the primary cause of the thermal neutron-induced single event effect in the system on chip. Although Hf has a high neutron capture cross section, its presence does not significantly affect the sensitivity to single event effects. However, during atmospheric neutron irradiation, the presence of Hf increases the possibility of depositing the total dose in the tested chip.

Keywords: spallation neutron; thermal neutron; Monte Carlo; system on chip; single event effect

1. Introduction

The atmospheric neutron comes from the interaction of cosmic rays with the atmospheric nuclei. These neutrons have a wide range of energies, ranging from thermal neutron to GeV [1]. In the past decades, this has been serious concern in the field of avionics, regarding atmospheric neutrons resulting in single event effects (SEE) [2,3]. As semiconductor manufacturing technology rapidly develops, concerns have also shifted to the potential of atmospheric neutrons to cause single event effects (SEE) in terrestrial electronic systems [4–6].

The sensitivity of electronic systems to atmospheric neutrons can be explored in two ways [7]. One involves conducting high altitude tests in real atmospheric environments. For instance, Xilinx's Rosetta experiment assessed the soft errors of various technology field programmable gate arrays in different locations globally in the past few years [8–10]. Similarly, in China, Chen et al. investigated the real-time atmospheric neutron induced soft errors on different static random access memories at the Yangbajing international cosmic ray observatory of the Chinese Academy of Science in Tibet, China [11–13]. However, the major drawback of this method is that it can be quite time consuming, even though the results obtained are the most authentic.

The other method involves performing irradiation tests using various irradiation equipment to evaluate the effects of years of exposure in real surroundings in just a few hours with intense flux [7]. This can be achieved by using sources such as a reactor, monoenergetic neutrons, and spallation neutrons [14]. Among these, spallation neutrons are considered to be closer to the real situation and are the ideal surrogate for accelerating atmospheric neutron SEE tests when compared to the former two [15]. In addition, by spallation neutron irradiation, it is possible to conduct a more comprehensive analysis on the total single-event effects (SEE) caused by atmospheric neutrons, including the assessment of the contribution of different energy ranges of neutrons [16]. For instance, the assessment of SEE soft errors comes from the thermal neutron, the neutron above 1 or 10 MeV. In particular, it is more convenient to further investigate the impact of thermal neutrons on an atmospheric environment.

In 2001, R. C. Baumann was the first to report that ^{10}B interacting with the thermal neutron is a dominant factor in soft errors for deep-submicron static random access memory with borophosphosilicate glass (BPSG) packages [17]. Since then, advanced integrated circuit development has led to chip packages no longer requiring the BPSG package [18–20]. However, for the nanoscale electronic systems, even though the BPSG package is no longer used, they still face the threats of ^{10}B interacting with thermal neutrons [21–24]. This is because ^{10}B contamination might occur in the semiconductor contact and doping processes, and the rapidly developed semiconductor manufacturing technology pushes the supply voltage and SEE critical charges lower and lower.

In [21], M. Cecchetto et al. pointed out that the thermal neutron can induce almost 90% of upset events in some cases. In [22], C. Weulersse et al. analyzed the SEE soft errors for 90, 65, and 28 nm technology memories under thermal and high energy neutron conditions, and confirmed that 28 nm technology devices are strongly impacted by the thermal neutron. Recent research has investigated the influence of ^{10}B contamination on SEE susceptibility by exposing a 65 nm technology microcontroller unit to thermal and high-energy neutrons at the China Spallation Neutron Source [25]. The results showed that ^{10}B interacting with thermal neutrons dominated the atmospheric neutron SEE in the device, with a SEE ratio of 1.89:1 induced by thermal and higher energy neutrons on the 65 nm technology microcontroller unit [25].

The 65 nm technology microcontroller unit test results in our previous work also indicate that the interaction of ^{10}B with thermal neutrons is still a serious concern for advanced integrated chips. Additionally, the obtained results have also further motivated us to explore the thermal neutron impact on the smaller technology system on chip using China spallation neutron source, for instance, the Xilinx Zynq-7000 system on chip (SoC) which is manufactured with the 28 nm complementary metal oxide semiconductor (CMOS) technology.

For the Xilinx Zynq-7000 SoC thermal neutron SEE analysis, in addition to the possible boron contamination, another element should also be considered. This element is the hafnium (Hf) element. Compared to boron (B), the neutron capture cross section with Hf is higher at several eV intervals. Figure 1 displays the neutron cross section spectra of ^{10}B (19.9% abundance), ^{178}Hf (27.1% abundance), and ^{28}Si (92.2% abundance) [26]. It can be seen that the peak cross section of ^{178}Hf even reaches 10^5 barns. The cross sections of ^{178}Hf with thermal neutrons are also higher than those of ^{28}Si by two orders of magnitude. Another significant fact is that the element boron exists in the Xilinx Zynq-7000 SoC as a result of contamination from manufacturing processes; however, its region cannot be confirmed by measurements such as secondary ion mass spectrometry. Hafnium is different from the element boron, and does indeed exist in the high-K dielectric materials used in the manufacturing of the Xilinx Zynq-7000 SoC. This makes the atmospheric neutron-induced SEE assessment on the Xilinx Zynq-7000 SoC even more complicated.

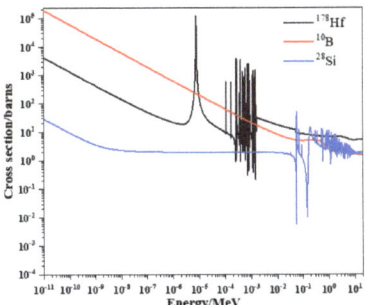

Figure 1. The cross sections of the neutron with ^{10}B (19.9% abundance), ^{178}Hf (27.1% abundance), and ^{28}Si (92.2% abundance) [26].

The atmospheric neutron-induced SEE on the Xilinx Zynq-7000 SoC was examined via an irradiation test conducted at the China Spallation Neutron Source (CSNS)-BL09 [27]. In the irradiation test, the Xilinx Zynq-7000 SoC chip was directly exposed to the ejected neutron beam without any shielding, which covered both thermal and high-energy components. To achieve a deeper understanding of the atmospheric neutron SEE on the Xilinx Zynq-7000 SoC, a second irradiation was conducted with the inclusion of a 2 mm cadmium (Cd) sheet to absorb the thermal neutrons in front of the chip. By comparing the results of the two irradiation tests, it is possible to investigate the contribution of thermal neutrons to the tested SoC. Furthermore, the impact of elements such as B and Hf can be analyzed through both irradiation and Monte Carlo simulations.

2. Irradiation Tests

As mentioned above, performing an actual atmospheric neutron SEE test would be time consuming, and the spallation neutron source is the closest to the real atmospheric neutron spectrum. The implementation of the China Spallation Neutron Source in 2018 has made it convenient to conduct atmospheric neutron SEE tests in China using the spallation neutron source [28]. We have obtained some SEE test results from [25,27]. Figure 2 illustrates the calculated differential flux of the neutron beam from the China Spallation Neutron Source (CSNS) (10^9 for the Beijing terrestrial system). The CSNS spectrum is very similar to that of the atmospheric neutron spectrum at sea level, as observed for the Beijing terrestrial system, though larger by a factor of 10^9. In the actual environment, even though the neutron spectrum impinging on a chip will be different from the spallation neutron beam due to the surrounding factors, the detected results are close to the actual situation.

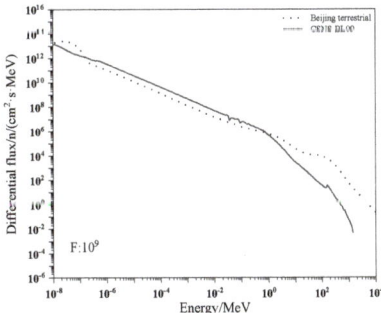

Figure 2. The differential flux of the neutron beam at CSNS [25,27,29].

Two separate irradiation tests were performed on the same series Xilinx Zynq-7000 SoC using the CSNS-BL09 facility. In the first test (described in reference [27]), the tested SoC was exposed to the neutron beam without any shielding. In the second test, which is the main effort of this work, a 2 mm thick Cd sheet was inserted between the beam ejection stop and the tested chip to absorb thermal neutrons. The effectiveness of the sheet in absorbing neutrons with energies below 0.5 eV is demonstrated in Figure 3, which shows a comparison of the neutron spectrum at the terminal with and without the Cd sheet. In Figure 3, the spectrum of CSNS-BL09 + 2 mm Cd is measured behind the 2 mm Cd sheet before the irradiation test. Even though the thermal neutron that interacts with the Cd might produce some new neutrons, the figure indicates that the neutron fluence was reduced by two to three orders of magnitude with the 2 mm Cd sheet shielding in place at energies lower than 0.5 eV.

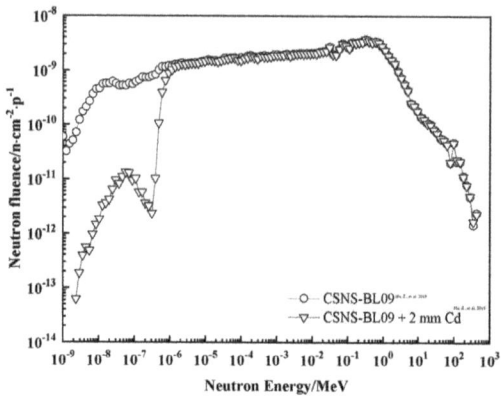

Figure 3. Neutron fluence with and without 2 mm Cd sheet [25].

The on-chip memory (OCM) block of the Xilinx Zynq-7000 SoC (Xilinx-Zynq 7020 CLG484) was examined in the first irradiation. For comparison, it was tested in this work again. The 64 kB data stored in the OCM were dynamically tested; the checked pattern data were written into the OCM addresses and subsequently read back by the SoC. The read back data were compared with the check pattern data to identify any SEE. The comparison results were transferred to a PC and displayed in a terminal. To allow for comparison with the first irradiation, which examined the normal condition without any SEE mitigation techniques, the same conditions were replicated in this effort.

The test setups for both irradiation tests were almost identical, except for the addition of a 2 mm Cd sheet in the second test to absorb thermal neutrons. The test board was powered by a 2260B programming DC power supply, and the real-time current was monitored and recorded by a remote host computer. Additionally, potential single event latch-up was monitored during the tests. Communication between the host computer and test board was established through a universal serial bus cable, and running messages were recorded in real-time.

3. Results and Discussions

During both irradiations, four types of soft errors were detected: single bit upset (SBU), dual cell upset (DCU), multi-cell upset (MCU), and single event functional interruption (SEFI). No abnormal current was detected, indicating that no latch-up event occurred during the atmospheric neutron SEE irradiation tests of the chip. However, there were discrepancies between the two irradiations in terms of SEE cross section, suggesting that thermal neutrons had an impact on the Xilinx Zynq-7000 SoC during the atmospheric neutron irradiation tests.

3.1. Detected Events

In the current irradiation test, 19 events were detected. Table 1 presents the number of each type of error, with SBU events being the most frequent, which is similar to the first irradiation. During the current irradiation, the neutron flux above 1 MeV was approximately 6.85×10^5 n·cm^{-2}·s^{-1}, and the corresponding fluence was 2.47×10^{10} n·cm^{-2}. As a result, the SBU cross section was calculated to be $(5.26 \pm 0.26) \times 10^{-10}$ cm^2 and $(1.00 \pm 0.05) \times 10^{-15}$ cm^2·bit^{-1} for the irradiation with few thermal neutrons.

Table 1. The detected SEE in irradiation with few thermal neutrons.

SBU	DCU	MCU	SEFI
13	2	2	2

Figure 4 displays the detected SEE during the first irradiation [27]. The figure shows the number of SBU 21, which is the highest among all types of events. Table 1 and Figure 4 show that SBU events dominate the detected soft errors in both irradiations.

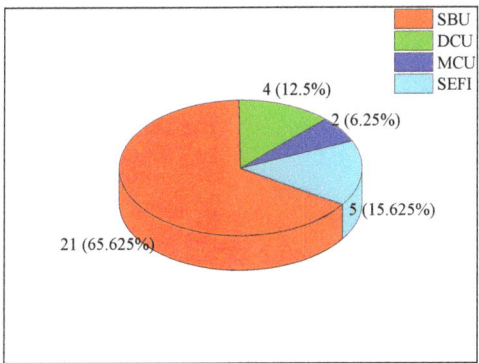

Figure 4. The detected numbers of different SEE in the first irradiation test at CSNS-BL09 [27].

Table 2 presents the SBU cross sections for two different irradiations.

Table 2. The SBU cross sections in two irradiations [27].

Neutron Beam	Fluence 10^{10} cm^{-2}	SBU	Cross Section 10^{-10} cm^2	Bit Cross Section 10^{-15} cm^2·bit^{-1}
CSNS-BL09 [27]	2.22	21	9.46 ± 0.47	1.80 ± 0.09
CSNS-BL09 + 2 mm Cd	2.47	13	5.26 ± 0.26	1.00 ± 0.05

In the current irradiation, the neutron fluence is 2.47×10^{10} n·cm^{-2}, which is 11.26% higher than in the first irradiation. Surprisingly, the number of SBU events is only 13 in the current irradiation, whereas it was more in the first irradiation, achieving 21. In general, a high fluence should correspond to a high number of SEE, this anomaly indicates that thermal neutrons may be contributing to the atmospheric neutron SEE on the tested SoC. The difference in bit cross sections between the two irradiations is 0.8×10^{-15} cm^2·bit^{-1}, which could be attributed to the shielding of thermal neutrons in the second irradiation. According to Formula (1), this indicates that shielding thermal neutrons with a 2 mm Cd sheet could make the SEE cross section smaller by about 44.4%. These findings demonstrate that the impact of thermal neutrons cannot be disregarded when it comes to SEE caused by atmospheric neutrons in the Xilinx Zynq-7000 SoC, even though the tested chips no longer employ BPSG in their packaging.

$$d = \frac{\sigma - \xi}{\sigma} \qquad (1)$$

where d is the discrepancy, σ and ξ are the bit cross section of the first and the current irradiation test with cm^2·bit^{-1}.

3.2. B Influence

The atmospheric neutron irradiation test results of the 65 nm technology microcontroller unit showed that secondary particles from thermal neutrons interacting with ^{10}B could cause SEU in advanced electronic systems and make a significant contribution to the SEE cross section. When compared to the 65 nm technology memory cell, the SEU critical charge of the 28 nm complementary metal oxide semiconductor technology memory cells is lower, making the Xilinx Zynq-7000 SoC memory cells more susceptible to soft errors induced by thermal neutrons.

$$^{10}\text{B} + n_{th} \rightarrow {}^{7}\text{Li}(1.01 \text{ MeV}) + \alpha(1.78 \text{ MeV}) \qquad (2)$$

$$^{10}\text{B} + n_{th} \rightarrow {}^{7}\text{Li}(0.84 \text{ MeV}) + \alpha(1.47 \text{ MeV}) + \gamma(0.48 \text{ MeV}) \qquad (3)$$

The sum of the energies of generated secondary α and ^{7}Li is constant. They are located within a determined region in the pulse amplitude distribution spectrum. Formulas (2) and (3) describe the mechanisms of the thermal neutron (nth) interacting with ^{10}B. The probability of (2) is 6%, while that of (3) is 94% [30,31]. In (3), although 0.48 MeV γ is also produced, unlike the generated ^{7}Li and α ions, it is uncharged. It needs to generate secondary electrons or other ionization particles to trigger SEE; this case's probability is rather low. In addition, the γ ray has a high penetration depth, making it deposit energy over a long trajectory, while the size of the sensitive volume of 28nm technology memory cells is extremely small. This means the sensitive volume cannot collect as many charges to induce SEE when a gamma ray passes. Thus, SEE induced by the produced gamma ray can be disregarded here. Given the above analysis, it could be speculated that the key factors of SEE cross section discrepancy on the tested SoC are generated by the secondary α and ^{7}Li. Table 3 illustrates the ranges and linear energy transfers (LETs) of the generated ionized secondary particles in silicon [32].

Table 3. The ranges and LETs of secondary particles of ^{10}B with the thermal neutron.

Range in Silicon/μm				LET/MeV·cm^2·mg^{-1}			
^{7}Li		α		^{7}Li		α	
0.84 MeV	1.01 MeV	1.47 MeV	1.78 MeV	0.84 MeV	1.01 MeV	1.47 MeV	1.78 MeV
2.50	2.80	5	6.36	2.10	2.16	1.15	1.06

Due to the significant discrepancy between Formulas (3) and (2), the following analysis will primarily focus on the secondary particles of (3), which are similar to (2). The α (1.47 MeV) and ^{7}Li (0.84 MeV) ions have ranges of only 5 μm and 2.5 μm in silicon, respectively. These are much lower than the thickness of the Xilinx Zynq-7000 SoC from its top passive layers to the substrate's surface [33]. This phenomenon suggests that ^{10}B contamination indeed exists within the chip, which is approaching the sensitive volumes of the memory cell of the SoC. The SEE LET threshold of the 28 nm CMOS memory cell is approximately 0.50 MeV·cm^2·mg^{-1} [34]. This is because the LETs are about 2.10 and 1.15 MeV·cm^2·mg^{-1} for the ^{7}Li (0.84 MeV) and α (1.47 MeV), respectively, meaning they are higher than the threshold. This means that both secondary particles can induce SEE in the tested SoC.

The current tested Xilinx Zynq-7000 SoC also includes the Hf element in the high-K dielectric materials. Furthermore, the cross section of the Hf element with the thermal neutron is greater than that of silicon. As a result, it is not yet possible to conclude that the disparity between the two can be attributed solely to the presence or absence of ^{10}B.

3.3. Hf Influence

^{10}B interacts with thermal neutrons, leading to SEE primarily caused by nuclear reactions. Meanwhile, the main interaction between the thermal neutron and the Hf element is the (n, γ) reaction, as depicted in Figure 5, which produces γ-rays that typically result in the total ionizing dose rather than SEE in the device [35]. As mentioned above, the possibility of SEE occurring from the interaction between generated γ-rays is relatively low.

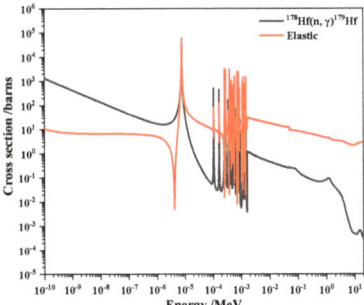

Figure 5. Neutron cross section spectrums of (n, γ) and elastic reactions with ^{178}Hf [26].

The cross sections of ^{178}Hf interacting with eV-level neutrons are remarkably high, reaching 10^5 barns when compared to high-energy neutrons. Therefore, it is crucial to thoroughly evaluate whether the presence of hafnium in the tested SoC contributes to atmospheric neutron-induced SEE.

While the contribution of the (n, γ) reaction to the induction of SEE in the SoC is relatively low, there is a potential for the transfer of energy from neutrons to hafnium nuclei in elastic interactions, which may increase the likelihood of causing SEE. The maximum transfer energy to the Hf nuclei from a neutron can be calculated using Formula (4) [36].

$$Et = \frac{4MnMt}{(Mn + Mt)^2} En \qquad (4)$$

Et is the max energy transfer to Hf nuclei with keV; Mn is the mass of the neutron, which is 1.67×10^{-27} kg; Mt is the Mass of Hf and it is 2.96×10^{-25} kg; and En is the energy of neutron with keV.

The focus of this article is on SEE soft errors for the Xilinx Zynq-7000 SoC induced by thermal neutrons reacting with boron and hafnium elements. As stated in Section 2, neutrons with an energy lower than 0.5 eV are absorbed by the inserted 2 mm Cd sheet. For the 0.5 eV neutron, the maximum energy transferred to the Hf atom is approximately 0.01 eV. The resonance in the Hf cross section has a high peak, while being intensely narrow. Additionally, even with an extension of the neutron energy to the rightmost peak, the corresponding maximum energy transferred is lower than 0.03 keV. Based on their corresponding LET values (which are lower than 0.50 MeV·cm^2·mg^{-1}), it is unlikely that they would result in SEE on the tested SoC. Therefore, the high cross section elastic interaction between Hf and thermal and eV neutrons does not have an impact on atmospheric neutron-induced SEE in the Xilinx Zynq-7000 SoC. It can be concluded that the difference in SEE cross section between the two irradiations is mainly caused by the presence of ^{10}B contamination, which indeed exists within the chip. It verifies the thermal neutron's influence on the nanoscale device's SEE susceptibility again.

3.4. Monte Carlo Simulation

The thickness and materials of the passive layers on the cut cross section of the chip were obtained [27,33]. The 28 nm high-K metal gate technology consists of TiN (8 nm), HfO$_2$ (10 nm), and SiON (1.2 nm) [37]. Additionally, the ultra-thin SiON layer in the high-k metal gate technology can also be an ultra-thin SiO$_2$ layer [38]. Based on this information, two Geant4 Monte Carlo simulation models were developed to examine the influence of the Hf element [39,40]. In Figure 6a, only the TiN and ultra-thin SiO$_2$ layers were considered in the first model, while in Figure 6b, the TiN, HfO$_2$, and ultra-thin SiO$_2$ layers were considered simultaneously in the second model. The remaining parameters for the two simulation models are the same. As trace amounts of boron impurities may be introduced by manufacturing processes or other means, and their abundance and region cannot be determined precisely, boron was not considered in the simulation.

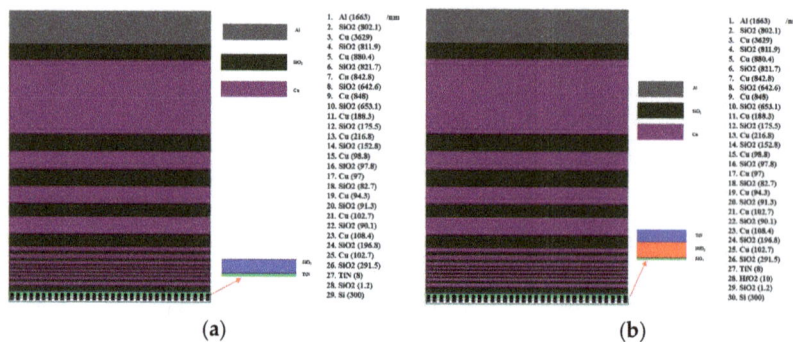

Figure 6. The two built Geant4 simulation models. (**a**) TiN and ultra-thin SiO$_2$ layers above the substrate, (**b**) TiN, HfO$_2$, and ultra-thin SiO$_2$ layers above the substrate.

In the simulation, the spectrum of neutron sources accurately reflects that of the first irradiation test, comprising both thermal and high energy neutrons. The model's surface area measures 10 µm × 10 µm, and contains a total of 10^7 neutrons. To detect single event upsets (SEUs), 32 × 32 sensitive volumes (SVs) have been strategically placed throughout the geometry, with each SV measuring 130 nm × 130 nm × 130 nm. An SEU is detected when the deposited energy in an SV exceeds the critical charge of 0.18 fC.

Table 4 presents the recorded number of the detected SEU events in the cells and the deposited doses in the ultra-thin SiO$_2$ layer in both simulations. The results show that the number of upset events and the cross section remain consistent between the two simulations. However, the deposited total dose in the ultra-thin SiO$_2$ layers differs by almost five times. This confirms that the presence of the Hf element does not impact the Xilinx Zynq-7000 SoC atmospheric neutron's SEE vulnerability. Nevertheless, the existence of hafnium may increase the total deposited dose during atmospheric neutron SEE irradiation, as highlighted by the high (n, γ) cross section shown in Figure 5. The larger number of γ rays implies a greater potential for total ionization dose on the examined SoC. These simulation results support the need for total dose threat monitoring in future high fluence atmospheric neutron SEE irradiation tests for similar series SoC.

Table 4. The upset number and deposited doses in two simulations.

Upset Number		Bit Cross Section/cm^2·bit^{-1}		Deposited Dose/rad	
First Model	Second Model	First Model	Second Model	First Model	Second Model
5	5	5×10^{-16}	5×10^{-16}	12.6	63.3

Based on the first model, the gamma ray striking simulation has also been executed, but no SEE events were detected. This confirms that the SEE cross section discrepancy between the two irradiations is produced by B interacting with thermal neutron.

The current two built models aimed to examine the influence of Hf on the SEE sensitivity of the Xilinx Zynq-7000, and this was achieved (five times total dose discrepancy was observed). In the future, if the information about the boron contamination could also be further confirmed in the SoC, the update models could be constructed and more detailed simulations could be executed. In addition, if possible, different energies of gamma ray striking experiments could be performed to investigate gamma influence on the SEE and confirm the findings of the current effort further.

4. Conclusions

The Xilinx Zynq-7000 SoC manufactured with 28 nm CMOS technology was exposed to two rounds of spallation neutron irradiation at CSNS-BL09. In the first irradiation, the spectrum covered both thermal and high-energy neutron components, while the second irradiation shielded thermal neutrons. The analysis of both irradiation tests revealed discrepancies. These discrepancies should be attributed to the interaction of ^{10}B with thermal neutrons. To mitigate the sensitivity of the Xilinx Zynq-7000 SoC to atmospheric neutron single event effects, a 2 mm Cd sheet can be employed to shield thermal neutrons against SEE sensitivity by approximately 44.4%. Although hafnium indeed exists in the Xilinx Zynq-7000 SoC and has a high interaction cross section with thermal neutrons, it does not affect atmospheric neutron SEE. However, it is important to pay attention to the total dose hazard during simulated atmospheric neutron SEE irradiation, as hafnium can increase the probability of a deposited total dose.

Author Contributions: Conceptualization, W.Y. and C.H.; methodology, Y.L.(Yonghong Li); software, Y.L.(Yang Li), J.C.; validation, Z.H.; formal analysis, L.W.; resources, Z.H.; writing—original draft preparation, W.Y and B.W. All authors have read and agreed to the published version of the manuscript.

Funding: This project was supported by the National Natural Science Foundation of China (Grant Nos. 11835006, 11690040, and 11690043), Natural Science Basic Research Plan in the Shaanxi Province of China (Grant No. 2023-JC-QN-0015), and the Fundamental Research Funds for the Central Universities (Grant No. XJSJ23049).

Data Availability Statement: The data used to support the findings of this study are available from the corresponding author upon request.

Conflicts of Interest: The authors declare no conflict of interest.

References

1. Dyer, C.; Hands, A.; Ford, K.; Frydland, A.; Truscott, P. Neutron-induced single event effect testing across a wide range of energies and facilities and implications for standards. *IEEE Trans. Nucl. Sci.* **2006**, *53*, 3596–3601. [CrossRef]
2. Normand, E. Single-Event Effects in Avionics. *IEEE Trans. Nucl. Sci.* **1996**, *43*, 461–474. [CrossRef]
3. Leray, J.L. Effects of atmospheric neutrons on devices, at sea level and in avionics embedded systems. *Microelectron. Reliab.* **2007**, *47*, 1827–1835. [CrossRef]
4. Song, Y.; Tu, X.; Li, Z. A Detection Method of Atmospheric Neutron Profile for Single Event Effects Analysis of Civil Aircraft Design. *Atmosphere* **2022**, *13*, 1441. [CrossRef]
5. Baumann, R.C. Landmarks in Terrestrial Single-Event Effects. In Proceedings of the Nuclear and Space Radiation Effects Conference (NSREC), San Francisco, CA, USA, 8–12 July 2013.
6. Autran, J.L.; Munteanu, D.; Roche, P.; Gasiot, G.; Martinie, S.; Uznanski, S.; Sauze, S.; Semikh, S.; Yakushev, E.; Rozov, S.; et al. Soft-errors induced by terrestrial neutrons and natural alpha-particle emitters in advanced memory circuits at ground level. *Microelectron. Reliab.* **2010**, *50*, 1822–1831. [CrossRef]
7. Bisello, D.; Candelori, A.; Dzysiuk, N.; Mastinu, P.; Mattiazzo, S.; Prete, G.; Silvestrin, L.; Wyss, J. Neutron production targets for a new Single Event Effects facility at the 70 MeV Cyclotron of LNL-INFN. *Phys. Procedia* **2012**, *26*, 284–293. [CrossRef]
8. Lesea, A.; Drimer, S.; Fabula, J.J.; Carmichael, C.; Alfke, P. The Rosetta experiment: Atmospheric soft error rate testing in differing technology FPGAs. *IEEE Trans. Dev. Mat. Reliab.* **2005**, *5*, 317–328. [CrossRef]
9. Lesea, A.; Castellani-Coulié, K.; Waysand, G.; Le Mauff, J.; Sudre, C. Qualification Methodology for Sub-Micron ICs at the Low Noise Underground Laboratory of Rustrel. *IEEE Trans. Nucl. Sci.* **2008**, *55*, 2148–2153. [CrossRef]
10. Xilinx. Continuing Experiments of Atmospheric Neutron Effects on Deep Submicron Integrated Circuits WP286 (v1.1). 13 October 2011. Available online: https://citeseerx.ist.psu.edu/document?repid=rep1&type=pdf&doi=51222cecfc39e38d06cf94d6f95c462ecf910a2d (accessed on 3 March 2023).
11. Chen, W.; Guo, X.; Wang, C.; Zhang, F.; Qi, C.; Wang, X.; Jin, X.; Wei, Y.; Yang, S.; Song, Z. Single-event upsets in SRAMs with scaling technology nodes induced by terrestrial, nuclear reactor, and monoenergetic neutrons. *IEEE Trans. Nucl. Sci.* **2019**, *66*, 856–865. [CrossRef]
12. Chen, W. Irradiation testing and simulation of neutron-induced single event effects. In Proceedings of the 26th International Seminar on Interaction of Neutrons with Nuclei, Xi'an, China, 28 May–1 June 2018.
13. Zhang, J.L.; Tan, Y.H.; Wang, H.; Lu, H.; Meng, X.C.; Muraki, Y. The Yangbajing muon–neutron telescope. *Nucl. Inst. Meth. Phys. Res. A* **2010**, *623*, 1030–1034. [CrossRef]

14. Jin, X.M.; Chen, W.; Li, J.L.; Qi, C.; Guo, X.Q.; Li, R.B.; Liu, Y. Single event upset on static random access memory devices due to spallation, reactor, and monoenergetic neutrons. *Chin. Phys. B* **2019**, *28*, 104212. [CrossRef]
15. Andreani, C.; Senesi, R.; Paccagnella, A.; Bagatin, M.; Gerardin, S.; Cazzaniga, C.; Frost, C.D.; Picozza, P.; Gorini, G.; Mancini, R.; et al. Fast neutron irradiation tests of flash memories used in space environment at the ISIS spallation neutron source. *AIP Adv.* **2018**, *8*, 025013. [CrossRef]
16. Hu, Z.L.; Yang, W.T.; Zhou, B.; Liu, Y.N.; He, C.; Wang, S.L.; Yu, Q.Z.; Liang, T.J. Neutron-induced single event effect in Xilinx 16nm MPSoC configuration RAM (CRAM) using white neutron and 2.72~81.8meV neutron in CSNS-BL20. *J. Nucl. Sci. Technol.* **2023**, *60*, 473–478. [CrossRef]
17. Baumann, R.C.; Smith, E.B. Neutron-induced ^{10}B fission as a major source of soft errors in high density SRAMs. *Microelectron. Reliab.* **2001**, *41*, 211. [CrossRef]
18. Lucas, M.L.; Daniel, S.; Helmut, P.; Rubén, G.A.; Manon, L.; Carlo, C.; Alberto, B.; Luigi, D. Neutron-induced effects on a self-refresh DRAM. *Microelectron. Reliab.* **2022**, *128*, 114406.
19. Kumar, S.; Agarwal, S.; Jung, J.P. Soft error issue and importance of low alpha solders for microelectronics packaging. *Rev. Adv. Mater. Sci.* **2013**, *34*, 185–202.
20. Wen, S.J.; Pai, S.Y.; Wong, R.; Romain, M.; Tam, N. B10 finding and correlation to thermal neutron soft error rate sensitivity for SRAMs in the sub-micron technology. In Proceedings of the IEEE International Integrated Reliability Workshop Final Report, South Lake Tahoe, CA, USA, 15–18 October 2010.
21. Cecchetto, M.; Alía, R.G.; Wrobel, F.; Tali, M.; Stein, O.; Lerner, G.; Bilko, K.; Esposito, L.; Castro, C.B.; Kadi, Y.; et al. Thermal neutron-induced SEUs in the LHC accelerator environment. *IEEE Trans. Nucl. Sci.* **2020**, *67*, 1412–1420. [CrossRef]
22. Weulersse, C.; Houssany, S.; Guibbaud, N.; Segura-Ruiz, J.; Beaucour, J.; Miller, F.; Mazurek, M. Contribution of Thermal Neutrons to Soft Error Rate. *IEEE Trans. Nucl. Sci.* **2018**, *65*, 1851–1857. [CrossRef]
23. Fang, Y.P.; Oates, A.S. Thermal neutron-induced soft errors in advanced memory and logic devices. *IEEE Trans. Device Mater. Rel.* **2014**, *14*, 583–586. [CrossRef]
24. Yamazaki, T.; Kato, T.; Uemura, T.; Matsuyama, H.; Tada, Y.; Yamazaki, K.; Soeda, T.; Miyajima, T.; Kataoka, Y. Origin analysis of thermal neutron soft error rate at nanometer scale. *J. Vac. Sci. Technol. B* **2015**, *33*, 020604. [CrossRef]
25. Hu, Z.; Yang, W.; Li, Y.; Li, Y.; He, C.; Wang, S.; Zhou, B.; Yu, Q.; He, H.; Xie, F.; et al. Atmospheric neutron single event effect in 65 nm microcontroller units by using CSNS-BL09. *Acta Phys. Sin.* **2019**, *68*, 238502. [CrossRef]
26. Brookhaven National Laboratory; National Nuclear Data Center (NNDC); Evaluated Nuclear Data File (ENDF). Available online: https://www.nndc.bnl.gov/endf/ (accessed on 2 February 2023).
27. Yang, W.; Li, Y.; Li, Y.; Hu, Z.; Xie, F.; He, C.; Wang, S.; Zhou, B.; He, H.; Khan, W.; et al. Atmospheric Neutron Single Event Effect Test on Xilinx 28nm System on Chip at CSNS-BL09. *Microelectron. Reliab.* **2019**, *99*, 119–124. [CrossRef]
28. Chen, Y. China Spallation Neutron Source. *Bull. Chin. Acad. Sci.* **2011**, *26*, 726–729.
29. Yu, Q.; Shen, F.; Yuan, L.; Lin, L.; Hu, Z.; Zhou, B.; Liang, T. Physical design of an Atmospheric Neutron Irradiation Spectrometer at China Spallation Neutron Source. *Nucl. Eng. Des.* **2022**, *386*, 111579. [CrossRef]
30. Orban, J.; Fuzi, J.; Rosta, L. Development of area detectors for neutron beam instrumentation at the Budapest neutron centor, 2020, IAEA-TECDOC-1935, Modern Neutron Detection Proceedings of a Technical Meeting, Vienna, Austria. Available online: https://www-pub.iaea.org/MTCD/Publications/PDF/TE-1935_web.pdf (accessed on 1 February 2023).
31. Hunt, S.; Iliadis, C.; Longland, R. Characterization of a 10B-doped liquid scintillator as a capture-gated neutron spectrometer. *Nucl. Instrum. Methods Phys. Res. Sect. A Accelerators. Spectrometers Detect. Assoc. Equip.* **2016**, *811*, 108–114. [CrossRef]
32. SRIM. Particle Interactions with Matter. 2013. Available online: http://www.srim.org/ (accessed on 4 March 2023).
33. Yang, W.; Yin, Q.; Li, Y.; Guo, G.; Li, Y.H.; He, C.H.; Zhang, Y.W.; Zhang, F.Q.; Han, J.H. Single-event effects induced by medium-energy protons in 28 nm system-on-chip. *Nucl. Sci. Technol.* **2019**, *30*, 151. [CrossRef]
34. Amrbar, M.; Irom, F.; Guertin, S.M.; Allen, G. Heavy ion single event effects measurements of Xilinx Zynq-7000 FPGA. In Proceedings of the IEEE Radiation Effects Data Workshop (REDW), Boston, MA, USA, 13–17 July 2015.
35. Di Mascioa, S.; Menicuccia, A.; Furano, G.; Szewczyk, T.; Campajola, L.; Di Capua, F.; Lucaroni, A.; Ottavi, M. Towards defining a simplified procedure for COTS system-on-chip TID testing. *Nucl. Eng. Technol.* **2018**, *50*, 1298–1305. [CrossRef]
36. Leroy, C.; Rancoita, P.G. *Principles of Radiation Interaction in Matter and Detection*, 2nd ed.; Word Scientific Publishing: Singapore, 2009.
37. Uwe, S.; Hwang, C.S.; Funakubo, H. *Ferroelectricity in Doped Hafnium Oxide: Materials, Properties and Devices*; Woodhead Publishing: Sawston, UK, 2019.
38. Czernohorsky, M.; Seidel, K.; Kühnel, K.; Niess, J.; Sacher, N.; Kegel, W.; Lerch, W. High-K metal gate stacks with ultra-thin interfacial layers formed by low temperature microwave-based plasma oxidation. *Microelectron. Eng.* **2017**, *178*, 262–265. [CrossRef]
39. Agostinelli, S.; Allison, J.; Amako, K.; Apostolakis, J.; Araujo, H.; Arce, P.; Asai, M.; Axen, D.; Banerjee, S.; Barrand, G.; et al. GEANT4-a simulation toolkit. *Nucl. Instrum. Methods Phys. Res. A* **2003**, *506*, 250–303. [CrossRef]
40. Yang, W.; Li, Y.; Zhang, W.; Guo, Y.; Zhao, H.; Wei, J.; Li, Y.; He, C.; Chen, K.; Guo, G. Electron inducing soft errors in 28 nm system-on-Chip. *Radiat. Eff. Defects Solids* **2020**, *175*, 745–754. [CrossRef]

Disclaimer/Publisher's Note: The statements, opinions and data contained in all publications are solely those of the individual author(s) and contributor(s) and not of MDPI and/or the editor(s). MDPI and/or the editor(s) disclaim responsibility for any injury to people or property resulting from any ideas, methods, instructions or products referred to in the content.

Article

Proton Radiation Effects of CMOS Image Sensors on Different Star Map Recognition Algorithms for Star Sensors

Yihao Cui [1,2,3], Jie Feng [1,2,*], Yudong Li [1,2,*], Lin Wen [1,2] and Qi Guo [1,2]

[1] Xinjiang Technical Institute of Physics and Chemistry, Chinese Academy of Sciences, Urumqi 830011, China
[2] Xinjiang Key Laboratory of Electronic Information Material and Device, Urumqi 830011, China
[3] University of Chinese Academy of Sciences, Beijing 100049, China
* Correspondence: fengjie@ms.xjb.ac.cn (J.F.); lydong@ms.xjb.ac.cn (Y.L.)

Abstract: Star sensors are widely used by satellites for their precise pointing accuracy. However, protons in space will cause cumulative effects and single-event transients in the imaging systems of star sensors. These effects will affect the success rate of star map recognition of star sensors. In this paper, proton irradiation experiments and field tests were carried out in turn, and three typical star recognition algorithms were used to recognize the star maps. The results showed that cumulative effects led to a decrease in the number of identifiable stars, which greatly affected the recognition success rate of the grid algorithm. Hot pixels caused by displacement damage effects increased the star centroid positioning error, leading to a decrease in the recognition success rate of the triangle algorithm and pyramid algorithm. Single-event transients produced by protons hitting the image sensor are similar to the grayscale value and shape of a star, and were recognized as "false stars", which had a significant impact on the success rate of the three recognition algorithms. In general, the pyramid algorithm was more effective than the other two algorithms in identifying the affected star map, and the recognition success rate of the grid algorithm was significantly reduced.

Keywords: CMOS image sensor; star sensor; hot pixel; single-event transient; star map recognition algorithm

1. Introduction

A star sensor is an optical measurement device that takes stars as observation targets and outputs the attitude information of its measurement coordinate system in the inertial coordinate system, and it is the most accurate attitude sensor at present [1,2]. The imaging system is an important component of a star sensor, and it determines the stellar detection sensitivity and attitude measurement accuracy of the star sensor. Since complementary metal oxide semiconductor image sensors (CMOS image sensors) have the advantages of high integration and low power consumption, most star sensors use CMOS image sensors as imaging devices [3–5]. However, CMOS image sensors in star sensors are subject to cumulative effects and single-event transients by the widespread presence of protons in space. The cumulative effects cause the degradation of image sensor parameters such as dark current, uniformity, and full well capacity [6–8]. Moreover, protons can cause bulk defects inside the image sensor, resulting in hot pixels in the image. As the radiation fluence increases, the detection performance and accuracy of the star sensor decrease [9]. Single-event transients produced by protons hitting CMOS image sensors appear as clusters in an image. Some of the clusters are similar to the grayscale value and shapes of stars, resulting in "false stars" in the field of view of the star sensor. The star sensor in JASON-1 can only be temporarily turned off when it is disturbed by transient clusters, and the satellite uses a gyroscope for attitude positioning [10,11].

Star map recognition is the matching of stars in the current field of view of the star sensor with reference stars in the existing star catalogue, being an important prerequisite for the accurate determination of the space attitude of the spacecraft. At present, the most widely used star map recognition algorithms can be divided into sub-graph isomorphism recognition algorithms and pattern recognition algorithms [12]. After years of development, star map recognition algorithms have greatly improved in terms of recognition accuracy and speed. However, the improvement in and research on these star map recognition algorithms are based on ideal conditions. Proton irradiation in space will degrade the performance of the image sensor and affect the recognition success rate. This will affect the output of high-precision attitude information from the star sensor. The specific impact mechanism of proton irradiation on different star map recognition algorithms has not yet been studied.

In this paper, we researched the mechanism of the cumulative effects and single-event transients caused by proton irradiation of CMOS image sensors on the star map recognition algorithm. The paper first introduces the test details and star map recognition algorithms used in this paper, then analyzes the effects of proton irradiation on the star map recognition algorithm of a star sensor such as reducing the identifiable number of stars due to the degradation of device performance caused by cumulative effects, star centroid positioning error increase by hot pixels, and "false stars" caused by single-event transients. This paper provides theoretical support for the improvement in star map recognition algorithms for long-term on-orbit star sensors.

2. Test and Algorithms

2.1. Test Details

The CMOS image sensor used in the experiment is a global exposure image sensor produced by CMOSIS; the number of pixels is 2048 × 2048, and the pixel size is 5.5 µm × 5.5 µm. Proton irradiation experiments were divided into the cumulative radiation test and single-event effects test. Cumulative irradiation test was carried out at Peking University. This test was an offline test, and the energy of proton irradiation was 3 MeV. Three image sensors were subjected to irradiation fluences of 3.68×10^9 p/cm^2, 1.47×10^{10} p/cm^2, and 3.6×10^{10} p/cm^2, separately. The single-event effects test was carried out at the Northwest Institute of Nuclear Technology, and the energy of proton irradiation was 60 MeV. This test was an online test, the image sensor was kept working during the irradiation, and 10-bit RAW format images containing transient bright clusters were obtained from the image sensor driver board to the computer through the cameralink cable. The characteristics of single-event transient bright clusters were extracted and analyzed after the test. After all the irradiation experiments were completed, the field test was carried out at the Lijiang Observatory. Three CMOS image sensors with different irradiation fluences and one unirradiated CMOS image sensor were installed into the star sensor test system in turn. The integration time of the CMOS image sensor was adjusted to 95.6 ms after the calibration of the principal point and focal length of the test system in the laboratory, and the platform was kept stable while pictures were taken of the Orion sky area.

2.2. Star Map Recognition Algorithms

The star map recognition algorithm relies on accurate locations of stars in a star map, but the star map captured by CMOS image sensor also contain noise besides stars, and thus the star map needs to be filtered to obtain identifiable stars through threshold segmentation. In this paper, a low pass filter was used, and the threshold segmentation value was the average grayscale value of the star map plus three times the standard deviation of the grayscale value. The classic star map recognition algorithms, including the triangle algorithm, pyramid algorithm, and grid algorithm, were used to complete the star map recognition. Although many new recognition algorithms have been proposed since these algorithms were first introduced, most of the new recognition algorithms were improved on the basis of these algorithms. Therefore, it is very meaningful to use these three algorithms as research objects. Among them, the triangle algorithm and the pyramid algorithm belong to the sub-graph isomorphism algorithms, and the grid algorithm is a pattern recognition algorithm.

The triangle recognition algorithm works by creating triangles out of three stars from the star map, and utilizing the diagonal distances between these stars as matching features [13]. The identification process involves filtering the star catalogue based on the star sensor's limit magnitude, then calculating the cosine value of the diagonal distance between each pair of filtered stars; the calculation formula is shown in Equation (1). Finally, a star pair information catalogue is constructed, which includes the numbers of the two stars and the cosine value of the angle between them.

$$\begin{aligned} \theta_{The} &= V_A \cdot V_B \\ &= \begin{pmatrix} \cos\delta_A \cdot \cos\alpha_A \\ \cos\delta_A \sin\alpha_A \\ \sin\delta_A \end{pmatrix} \cdot \begin{pmatrix} \cos\delta_B \cdot \cos\alpha_B \\ \cos\delta_B \sin\alpha_B \\ \sin\delta_B \end{pmatrix} \end{aligned} \quad (1)$$

In Equation (1), θ_{The} indicates the cosine of the theoretical diagonal distances between star A and B in the star catalog; V_A and V_B represent the direction vectors of stars A and B in the celestial coordinate system; and (α_A, δ_A) and (α_B, δ_B) signify the right ascension and declination of stars A and B, respectively. Equation (2) determines the diagonal distance of the triangle in the star map structure.

$$\begin{aligned} \theta_{Mea} &= V_\alpha \cdot V_\beta \\ &= \begin{bmatrix} -x_\alpha - x_0 \\ -y_\alpha - y_0 \\ f \end{bmatrix} / r_\alpha \cdot \begin{bmatrix} -x_\beta - x_0 \\ -y_\beta - y_0 \\ f \end{bmatrix} / r_\beta \\ r_i &= \sqrt{(x_i - x_0)^2 + (y_i - y_0)^2 + f^2}, i = \alpha \text{ or } \beta \end{aligned} \quad (2)$$

In Equation (2), θ_{Mea} represents the cosine of the measured diagonal distance between stars α and β in the star map. The direction vectors of stars α and β in the star sensor coordinate system are denoted as V_α and V_β, respectively. (x_α, y_α) and (x_β, y_β) represent the coordinate positions of stars α and β in the star map, respectively. The principal point of the star sensor is represented by (x_0, y_0), and the focal length of the star sensor is represented by f. The matching process is depicted in Figure 1a. If stars i, j, and k can be found from a star pair information catalogue such that their relationships with stars α, β, and γ in the star map satisfy Equation (3), and the result is unique, then the matching process is considered successful.

$$\begin{cases} \left| \theta_{ij}^t - \theta_{\alpha\beta}^m \right| \leq \varepsilon \\ \left| \theta_{ik}^t - \theta_{\alpha\gamma}^m \right| \leq \varepsilon \\ \left| \theta_{jk}^t - \theta_{\beta\gamma}^m \right| \leq \varepsilon \end{cases} \quad (3)$$

In Equation (3), θ_{ij}^t, θ_{ik}^t, and θ_{jk}^t represent the cosine of the theoretical diagonal distance between stars i, j, and k in the star catalog, and $\theta_{\alpha\beta}^m$, $\theta_{\alpha\gamma}^m$, and $\theta_{\beta\gamma}^m$ represent the cosine of the measured diagonal distance between stars α, β, and γ in the star map. The value of ε represents the threshold for the allowable error. Based on the previous algorithm, the pyramid algorithm has been enhanced by employing a pyramid structure for matching that involves using four stars. Compared with the triangle algorithm, this algorithm not only determines three more sets of diagonal distances during recognition but also verifies the previous recognition results when identifying new stars. The pyramid algorithm schematic is shown in Figure 1b [14].

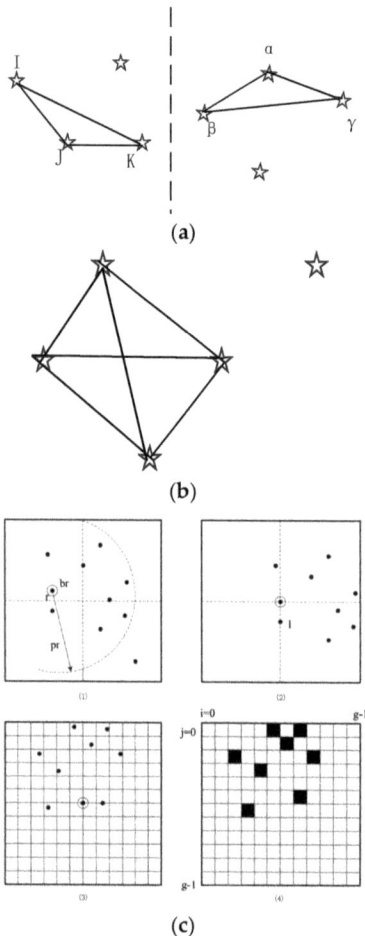

Figure 1. Schematic diagram of several different star map recognition algorithms. (**a**) Diagram of the triangulation algorithm. (**b**) Diagram of the pyramid algorithm. (**c**) Diagram of the grid algorithm.

The grid algorithm is different from the above-mentioned types of algorithms. The main idea of this algorithm is to select a star in the field of view as the reference star S1, and use "R" as the radius to determine the influence range, finding the nearest star as the closest-neighbor star S2. Then, a coordinate system with the connection line between star S1 and S2 serving as the positive x-axis is established and a grid network is constructed to project all the stars within the influence range into this grid, with the grid with stars recorded as 1 and the grid without stars recorded as 0. Following this, the pattern vector of the reference star is constructed in order. It is matched with the navigation star pattern vector constructed by stars in the star catalog, and the navigation star with the greatest consistency is the star. A schematic diagram of the grid algorithm is shown in Figure 1c [12,15].

Since these recognition algorithms usually only output the final recognition results, in order to explore the reasons for the decrease in the success rate of star map recognition caused by proton irradiation, these three algorithms were modified in this paper. In addition to outputting the final recognition results, the matching results between all the star pairs also outputted. For example, after identifying three stars, the original triangle algorithm will remove these three stars from the identification queue and replace other unidentified stars for recognition. The algorithm used in this paper keeps these three stars in the identification queue, allowing other unrecognized stars to be matched with them until the matching results of all triangular structures are outputted.

Three kinds of recognition algorithms were used to recognize the star map taken by the unirradiated CMOS image sensor, and the parameters of the star map recognition algorithm were optimized under the condition of considering the recognition success rate and recognition speed. The allowable error ε for the triangle algorithm and the pyramid algorithm was set to 0.000025, of the grid algorithm was set to 50×50, and the influence radius set to 1200 pixels. These three identification algorithms can correctly identify the eight stars in the star map. The star map recognition results are shown in Figure 2. The numbers in the figure correspond to the numbers of the stars in SAO Star Catalog (Smithsonian Astrophysical Observatory Star Catalog). The same parameters were used to identify and analyze star maps taken by the CMOS image sensors with different irradiation fluences and star maps taken by unirradiated CMOS image sensors with the addition of single-event transient clusters.

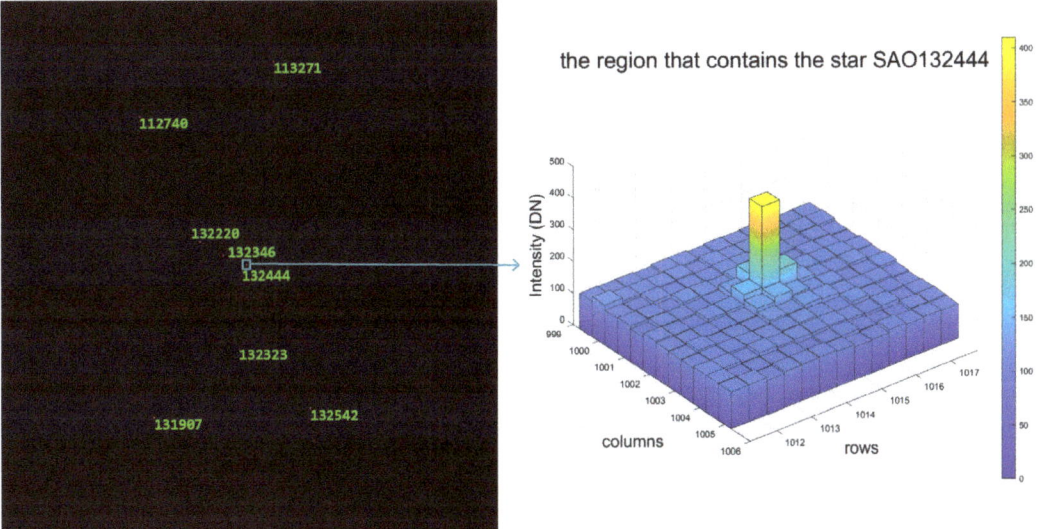

Figure 2. Recognition result of the star map taken by the non-irradiated CMOS image sensor. The numbers in the picture are the serial numbers of these stars in the SAO Star Catalog. In the zoomed version, DN (Digital Number) is the unit of gray value.

3. Results

Figure 3a,b show respectively images of the Orion's Belt captured by star sensor test systems with unirradiated CMOS image sensor and irradiated with a fluence of 1.47×10^{10} p/cm². The nonuniformity of the image taken by the irradiated CMOS image sensor was significantly greater compared to that of the unirradiated CMOS image sensor. Figure 3c,d show 3D point maps constructed from star maps that have been processed with low-pass filtering. The star map acquired by the unirradiated CMOS image sensor was smooth, except for the stars' region, while the star map taken by the CMOS image sensor with an irradiation fluence of 1.47×10^{10} p/cm² still had many spikes, even after the filtering process.

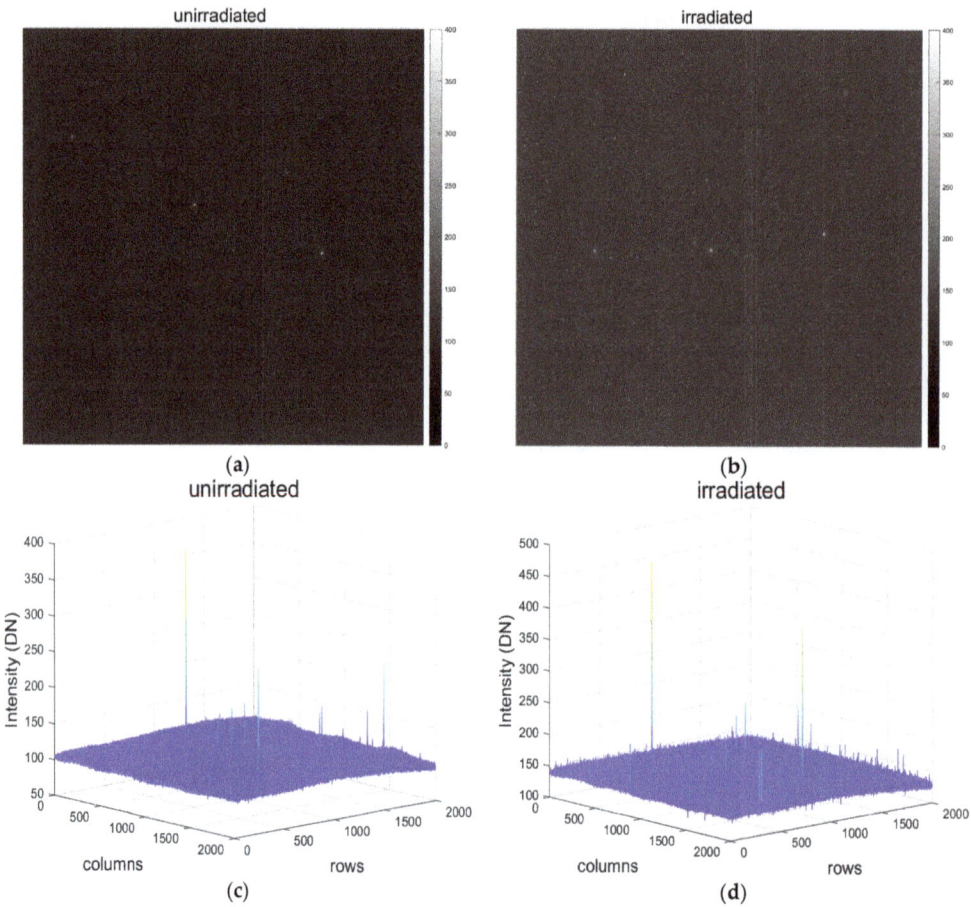

Figure 3. Star maps containing Orion's Belt and 3D point maps. (**a**) Star map captured with unirradiated CMOS image sensor; (**b**) star map captured with CMOS image sensor irradiated with a fluence of 1.47×10^{10} p/cm^2. (**c**) 3D point map constructed from low-pass filtered unirradiated image; (**d**) 3D point map constructed from low-pass filtered irradiated image.

The above three algorithms were used to identify star maps captured by the CMOS image sensors with different irradiation fluences. It was observed that as the irradiation fluence increased, the threshold segmentation value used in the pre-processing of star maps gradually increased and the number of stars entering the identification procedure decreased. This means that the final number of stars that could eventually be correctly identified decreased. When proton irradiation fluences were 3.68×10^9 p/cm^2, 1.47×10^{10} p/cm^2, and 3.6×10^{10} p/cm^2, the numbers of stars identified were six, four, and three, respectively. Moreover, the three recognition algorithms had different success rates. When the irradiation fluence was 1.47×10^{10} p/cm^2, the grid algorithm was no longer able to output the correct recognition result. When the proton irradiance was 3.6×10^{10} p/cm^2, the pyramid algorithm was unable to output a result because it requires a number of identifiable star points greater than four. The outputs of each triangular and pyramidal structure were compared with the correct SAO catalog number for each star; when the proton irradiation fluence was 1.47×10^{10} p/cm^2, some of these two structures outputted the wrong results.

An image containing single-event transient bright clusters captured by the CMOS image sensor is shown in Figure 4a. The shapes and gray values of these transient bright clusters appeared to be similar to stars, as can be seen in the figure. The average value of the gray value of each position of the 3 × 3 window centered on the maximum value was calculated. They were taken as the gray values of the single-event transient clusters in the follow-up experiment. These clusters were randomly added to the star map detected by the unirradiated image sensor. The star map after the addition of the transient star clusters is shown in Figure 4b.

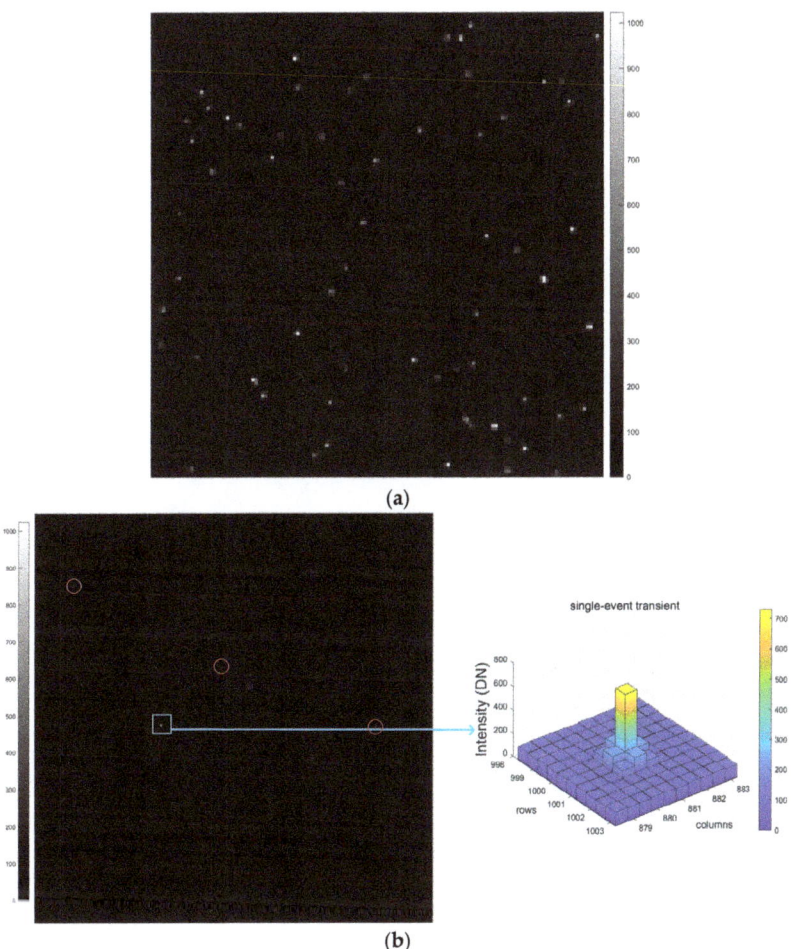

Figure 4. Image containing single-event transient clusters. (**a**) Part of an image collected online during a proton irradiation experiment. (**b**) Part of a star map with single-event transient clusters. The real stars are shown in the red circle and the transient cluster is shown in the blue box.

A single transient bright cluster is similar in size to a star in an image, and thus it is difficult for a small number of bright clusters to have a great impact on the threshold segmentation value of a star map. These transient clusters are recognized as identifiable stars along with true stars. As transient bright clusters are added at random locations in the map they may also overlap with real stars and affect the centroid positioning of stars. However, the probability of this being the case is low, and the corresponding treatment is now available [16]; therefore, this paper focused on the effect of transient bright clusters as identifiable stars in terms of the star map recognition algorithm.

Three identification algorithms were used to identify the star maps with transient clusters. For the triangle and pyramid algorithms, false results occurred as the number of transient clusters increased, but the pyramid algorithm had a higher recognition success rate than the triangle algorithm. After multiple recognitions, it was found that the triangle algorithm was very sensitive to the order of stars and transient clusters entering the identification queue. When the transient bright clusters were at the end of the recognition queue, the recognition success rate was higher than at the head. For the grid algorithm, both the number of transient bright clusters and their position in the star map caused the program to output different error results. The effects of cumulative radiation damage and single-event transients on different star map identification algorithms are summarized in Table 1.

Table 1. The effect of proton irradiation on the success rate of different star map recognition algorithms.

Star Map Recognition Algorithm	Cumulative Radiation Damage	Single-Event Transient
Triangle algorithm	The algorithm's recognition success rate is almost unaffected but the number of stars outputted by the algorithm decreased as the irradiation fluence increased. When the number of identifiable stars was less than four, the pyramid algorithm cannot output a recognition result.	The recognition success rate decreased and the success rate may be related to the position of transient clusters in the recognition queue.
Pyramid algorithm		Compared to other algorithms, the recognition success rate was the highest.
Grid algorithm	The decrease in the number of identifiable stars led to a decrease in the recognition success rate.	The recognition success rate decreased the most severely and the erroneous output results may be related to the position of the transient clusters in the star map.

4. Discussion

As the proton irradiation fluence increases, the oxide trap charge, interface trap charge and bulk defects in CMOS image sensors increase. The oxide trap charge is mainly distributed in the shallow trench isolation (STI) region and gate oxide region, while the interface trap charge is mainly distributed at the Si-SiO$_2$ interface. These charges will increase the dark current of the image sensor, resulting in an increase in the background gray value in star maps. Bulk defects caused by displacement damage effects can generate new energy levels [17]. Some defect levels act as generation–recombination centers, thereby increasing the dark signal value. Some pixels have a higher dark current value (hot pixels), resulting in bright spots that occupy one pixel in the image [18]. The difference in dark signals between different pixels will reduce pixel uniformity, resulting in an increase in the standard deviation of the star map. Since the threshold segmentation value used in the pre-processing is the average grayscale value of the star map plus three times the standard deviation, it increases as the irradiation fluences increase. Moreover, the decrease in quantum efficiency of the CMOS image sensor after proton irradiation can lead to a decrease in the grayscale value of star areas. If the grayscale value of a star is less than the segmentation threshold, the star will not be selected. This is the main reason for the decrease in identifiable star caused by proton irradiation.

The reduction in the number of identifiable stars only affects the final output count of stars for the triangle algorithm and pyramid algorithm, without generating any incorrect output results. When the proton irradiation fluence was 3.6×10^{10} p/cm^2, there were only three identifiable stars in the star map, which did not meet the minimum requirement of the pyramid algorithm, so the pyramid algorithm did not output any results. The recognition success rate of the grid algorithm was affected by the pattern vectors built by the star map. The decrease in the number of identifiable stars will result in changes to the pattern vectors, which will affect the recognition success rate. If the nearest star S2 is not the one used to build the grid pattern database, it will result in errors in the rotation of the star map, generate completely wrong pattern vectors, and output incorrect recognition results.

The reason for the wrong identification results for some of the triangular and pyramidal structures at a proton irradiation fluence of 1.47×10^{10} p/cm^2 was related to the hot pixels. In previous research, it was discovered that the noise caused by the total ionizing dose effect can lead to an increase in star centroid positioning error through the diagonal distance relationship between star pairs [19]. However, the hot pixels generated by the displacement damage can have a greater impact on the star centroid positioning accuracy. In the star maps taken by the CMOS image sensor with a proton irradiation fluence of 1.47×10^{10} p/cm^2, the initial position of a star in relation to a hot pixel is depicted in Figure 5a. Here, the red box is the hot pixel, which is far away from the star and does not affect the accuracy of star centroid positioning; however, the position of the star in the star map was changed by the rotation of the Earth, and thus the hot pixel gradually came closer to the star (Figure 5b–d). The distribution of the grayscale value of the star was also changed.

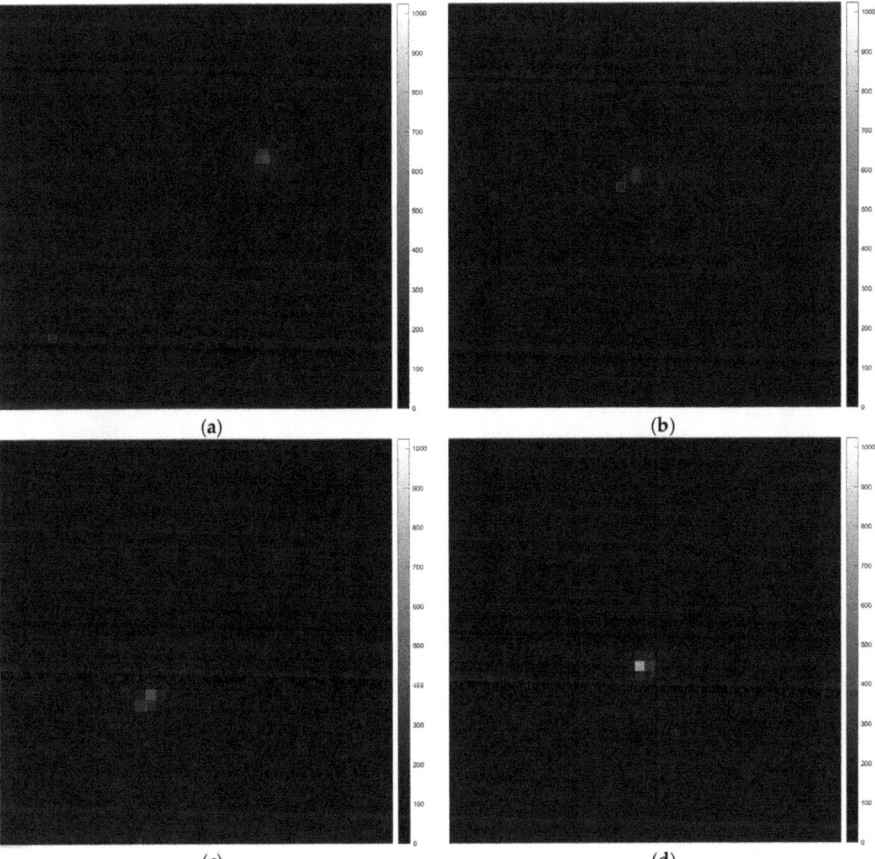

Figure 5. Different positional relationships of the hot pixel and the star: (**a**) the hot pixel is far away from the star; (**b**–**d**) the distance between the hot pixel and the center of the star is gradually approaching.

Without the influence of hot pixels, the energy distribution of a star approximately obeys a two-dimensional Gaussian distribution, as shown in Equation (4).

$$Ix, y = \frac{I_0}{2\pi\sigma_{PSF}^2} exp\left[-\frac{(x-x_0)^2}{2\sigma_{PSF}^2}\right] exp\left[-\frac{(y-y_0)^2}{2\sigma_{PSF}^2}\right] \qquad (4)$$

In Equation (4), $I(x, y)$ represents the energy of the star whose center is (x_0, y_0) at (x, y), I_0 is the total energy of the star, and σ_{PSF} is the Gaussian dispersion radius. The σ_{PSF} of this star sensor is 0.8. In order to prevent the saturation of the grayscale value of the center pixel of the star from affecting the centroid positioning, the exposure time of the image sensor was adjusted so that the center gray value of the lowest magnitude star was about 80% of the saturation gray value of the image sensor. This paper further analyzed the influence of a hot pixel on the centroid positioning of star from the different positional relationships and the grayscale value between a hot pixel and a star.

The grayscale value of each position within the 3 × 3 window of the star point is calculated according to Equation (4). Figure 6a shows the model of the impact of a hot pixel on it. When obtaining the centroid position of a star in the image, it will first locate the pixel position with the largest grayscale value in the area; select a 3 × 3 window with it as the center; and then use different centroid positioning algorithms to extract the centroid coordinates of the star point as needed, such as the centroid algorithm, square centroid algorithm, Gaussian centroid algorithm, and other algorithms. Due to the different calculation methods of these centroid positioning algorithms, the different centroid positioning algorithms also lead to different centroid positioning errors of a star. Due to the different calculation methods of these centroid positioning algorithms, the resulting centroid positioning errors will also be different.

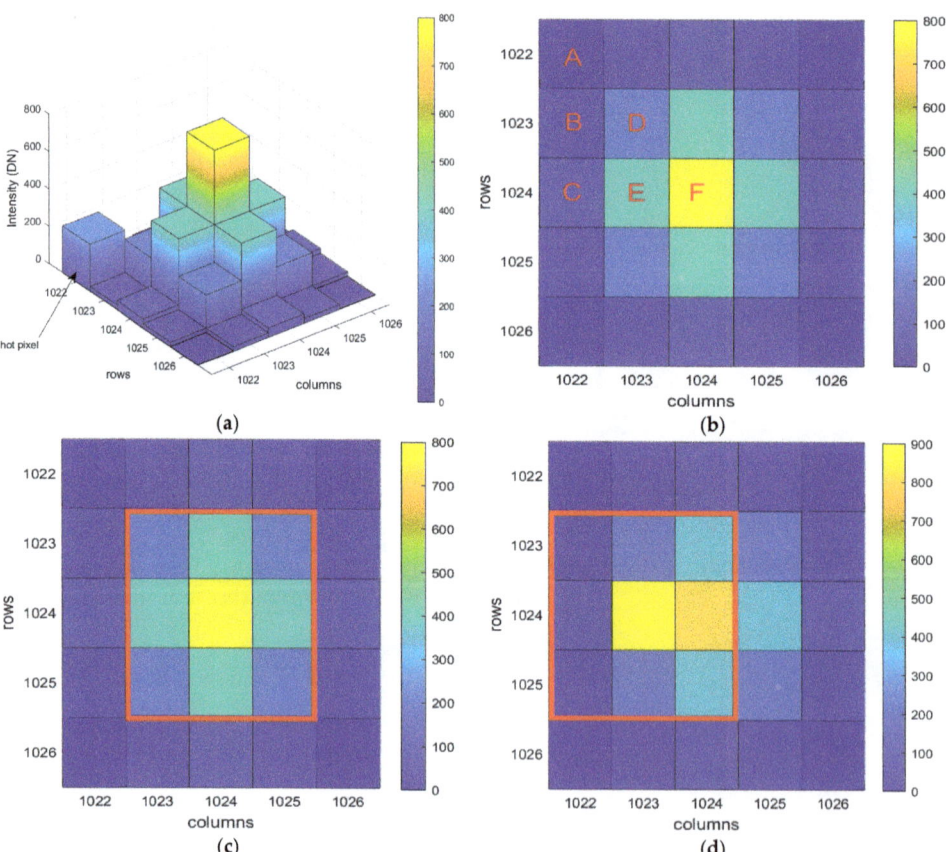

Figure 6. Schematic diagram of star gray value distribution and positioning window: (**a**) star gray value distribution; (**b**) six situations that affect star centroid positioning; (**c**) original centroid positioning window; (**d**) centroid positioning window shift.

According to the symmetry of the distribution of the star grayscale value and the size of the centroid positioning window, there are six cases of the influence of the hot pixels on the distribution of the star, as shown in Figure 6b. Among them, the F case mainly changed the Gaussian dispersion radius. For the cases D and E where the hot pixel was within the original window, the change of the grayscale value in the window led to an error of the centroid positioning of the star. When the grayscale value of a hot pixel was lower than that at the center of the star, the centroid positioning window remains the same as shown in the red box in Figure 6c. However, if the grayscale value of the hot pixel was higher than that of the center of the star, the center of the positioning window shifted to the location of the hot pixel. This is shown in Figure 6d. For cases A, B, and C outside the window, centroid positioning error will occur only when the grayscale value of the hot pixel is higher than the grayscale value of the star center.

The centroid position (x_1, y_1) is calculated using different centroid positioning methods for the five cases, and the error of centroid positioning $\Delta_{x,y}$ is calculated according to Equation (5). The result is shown in Figure 7.

$$\Delta_{x,y} = \sqrt{(x_1 - x_0)^2 + (y_1 - y_0)^2} \tag{5}$$

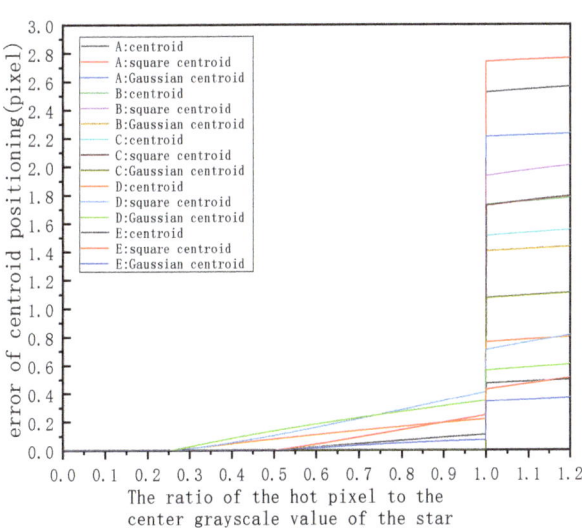

Figure 7. The error of centroid positioning calculated by using different centroid positioning methods.

It can be seen from Figure 7 that in the case of D and E, the error of centroid positioning increased with the increase in the grayscale value of the hot pixel. After the grayscale value of the hot pixel exceeded the grayscale value of the center of the star, the error of centroid positioning increased sharply due to the movement of the centroid positioning window, but in these cases, the centroid error calculated by the centroid positioning algorithm was within one pixel. In the case of A, B, and C, since the change of the grayscale value of the hot pixel did not affect the grayscale distribution of the star in the original window, only after the gray-scale value exceeds the grayscale value of the center of the star will the centroid positioning window move, and the error of centroid positioning occur. In these cases, the error of centroid positioning of the star was larger than one pixel, and the maximum error of centroid positioning calculated by the square centroid method exceeded 2.7 pixels. In all cases, the higher the grayscale value of the hot pixel and the farther the distance from the center of the star, the larger the error of centroid positioning; under the same influence conditions, the error of centroid positioning calculated by the Gaussian centroid algorithm is the smallest. Moreover, high-magnitude stars in the field of view have a lower grayscale value, and are more likely to be affected by hot pixels.

Taking the star Alnilam (SAO 132346) affected by a hot pixel in Figure 5 as an example, the right ascension and declination of another star Saiph in the field of view and this star were $(\alpha_A, \delta_A) = (86.93914, -9.66968)$, $(\alpha_B, \delta_B) = (84.05338, -1.20196)$, and the theoretical cosine value of the diagonal distance θ_{The} between the two stars can be calculated by Equation (2).

$$\theta_{The} = \begin{pmatrix} 0.052638 \\ 0.984386 \\ -0.167968 \end{pmatrix} \cdot \begin{pmatrix} 0.103579 \\ 0.994399 \\ -0.020976 \end{pmatrix} = 0.987849$$

The focal length of the star sensor test system was 24.048 mm, and the coordinates of the principal point obtained by calibration were $(x_0, y_0) = (1093.379, 1207.317)$. The coordinates of these two stars in the map were calculated to be (1505.018, 1410.990), (915.000, 1066.081), and then the cosine value of measured diagonal distance θ_{Mea} between the two stars was calculated by Equation (3).

$$\theta_{Mea} = \begin{pmatrix} -0.093631 \\ -0.046327 \\ 0.994528 \end{pmatrix} \cdot \begin{pmatrix} -0.040742 \\ -0.032258 \\ -0.998648 \end{pmatrix} = 0.987876$$

The difference between the θ_{Mea} and θ_{The} was 0.000027, exceeding the allowable error (0.000025), and thus it is impossible for this pair of stars to be correctly identified in the formed triangle and pyramid structure. In the star map not affected by the hot pixel, the coordinates of the center of the two stars were (1504.98, 1411.015) and (914.153, 1066.941), respectively. The calculated value from them is 98,786,586. The difference between the theoretical cosine values was 0.0000168. Therefore, these two stars can be correctly identified in the pyramid and triangle structures. On the basis of the matching results of Alnilam and all other stars, the error of the cosine value of the diagonal distance between the two stars would increase as the cosine value of the two stars decreased. That is, when the two stars in the map are far away from each other, a small positioning error generated by a hot pixel will have a large effect on the cosine value of the diagonal distance of the two stars. For the grid algorithm, it is difficult for a star to move from its original grid to other grids due to a small centroid offset. Therefore, the centroid error caused by hot pixels will not significantly affect the recognition success rate of the grid algorithm. Although annealing can reduce proton radiation damage to image sensors [20], it is difficult to install annealing devices in star sensors. The local threshold segmentation algorithm can be used instead of global thresholding in the pre-processing. This is effective for stars with lower magnitudes. For centroid positioning error caused by hot pixels, the allowable error ε used in recognition can be slightly increased based on the need to consider the speed of the recognition algorithm.

Incoming particles generate electron–hole pairs in the sensitive silicon volume through direct or indirect ionization [21]. Direct ionization means that the proton excites and releases off-core electrons of the Si atom through Coulomb interaction with Si, directly generating electron–hole pairs. Indirect ionization means that the proton generates new charged secondary particles through nuclear reactions, and these secondary particles continue to generate electron–hole pairs through either direct or indirect ionization. Part of the charge generated by the proton incident is collected by the depletion region of the pixel unit during the integration stage of the image sensor [22]. Due to the difference in the potential of the region before and after the integration time, the grayscale value of the pixel changes. Some of the charge is also collected by other pixel cells through diffusion, appearing as bright clusters in the image. Unlike the stars on a star map which generally remains at the same position for several consecutive frames, the transient cluster is related to the incident position of the proton and only exists in one frame.

After analyzing the output of the triangle algorithm, it was found that in a triangular structure composed of a single-event transient cluster and stars or other transient clusters, if the calculated diagonal distance is within allowable error ε and the result is unique, it will lead to an incorrect recognition result. If the transient clusters are at the head of the identification queue, they will form a triangular structure with other stars and transient clusters in the recognition queue multiple times for recognition, leading to a higher probability of outputting incorrect recognition results. While the transient cluster is at the end of the recognition queue, the transient cluster has less influence on the result because the recognition has been completed between the stars in the star map. The recognition success rate of the pyramid algorithm is improved because it needs to judge three more sets of diagonal distances than the triangle algorithm, and this algorithm will introduce other stars to validate the recognition results. After analyzing the results of all pyramid structure outputs, it was found that if there is one transient cluster in the pyramid structure, the recognition result is usually not output because the condition of six sets of diagonal distance matching cannot be satisfied. However, incorrect matching results may be output in pyramid structures composed of multiple transient clusters, but these incorrect results will be discarded in the subsequent verification process, and a small number of transient clusters have no effect on the final recognition results. Transient clusters can cause the grid algorithm to output incorrect results in two ways. The first way is that transient clusters are incorrectly selected as the reference star S1 or closest-neighbor star S2, which will cause the star map to be incorrectly rotated and generate completely incorrect mode vectors. The second way is that even if both the reference star and closest-neighbor star are real stars, but there are transient clusters in some grids. It will also generate incorrect pattern vectors. The algorithm of increasing the judgment conditions to verify clusters multiple times is effective when there are few clusters. But the time consumed by this algorithm will significantly increase as the number of clusters increases. By utilizing the characteristic that transient clusters exist in only one frame at the same position in the image, it is possible to distinguish them from stars [23]. However, since this algorithm requires the collection of multiple sets of star maps, the posture output frequency may be reduced.

5. Conclusions

Star map recognition is a key step in the output of attitude information from a star sensor. However, the success rate of star map recognition is affected by proton radiation. In this study, we investigated the mechanism of the cumulative irradiation effect and single-event transients of protons on the image sensor on star map recognition. The results of the study showed that increased dark current and non-uniformity of CMOS image sensor pixels led to a decrease in the number of identifiable star points. The hot pixels generated by displacement damage increased the centroid positioning error of the star when it was near the star. Transient clusters were found to be similar to a star and they affected the success rate of the star map recognition as "false stars".

For the triangle algorithm and the pyramid algorithm, the reduction in the number of stars to be recognized did not affect the recognition accuracy, but the increase in the centroid positioning error of the star points caused by hot pixels and the interference of "false stars" affected the recognition accuracy. For the grid algorithm, the decrease in the number of identifiable stars and the presence of "false stars" affected the mode vector constructed by the grid algorithm, resulting in a significant decrease in the identification success rate. The pyramid algorithm was found to have the highest stability among these three algorithms, and the grid algorithm was the worst.

Author Contributions: Conceptualization, J.F. and Y.C.; methodology, Y.C. and J.F.; software, Y.C.; validation, Y.C. and Y.L.; formal analysis, J.F.; investigation, J.F.; data curation, Y.C.; writing—original draft preparation, Y.C.; writing—review and editing, J.F.; visualization, Y.C.; supervision, Y.L.; project administration, L.W. and Q.G.; funding acquisition, J.F. and Y.L. All authors have read and agreed to the published version of the manuscript.

Funding: This research was funded by the National Natural Science Foundation of China under grant No. 12175307, the West Light Talent Training Plan of the Chinese Academy of Sciences under grant No. 2022-XBQNXZ-010, the Youth Science and Technology Talents Project of Xinjiang Uygur Autonomous Region No. 2022TSYCCX0094, and the Tianshan Innovation Team Program of Xinjiang Uygur Autonomous Region No. 2022D14003.

Institutional Review Board Statement: Not applicable.

Informed Consent Statement: Not applicable.

Data Availability Statement: Not applicable.

Conflicts of Interest: The authors declare no conflict of interest.

References

1. He, L.; Ma, Y.; Zhao, R.; Hou, Y.; Zhu, Z. High update rate attitude measurement method of star sensors based on star point correction of rolling shutter exposure. *Sensors* **2021**, *21*, 5724. [CrossRef] [PubMed]
2. Liebe, C.C. Star trackers for attitude determination. *IEEE Aerosp. Electron. Syst. Mag.* **1995**, *10*, 10–16. [CrossRef]
3. Saint-Pe, O.; Tulet, M.; Davancens, R.; Larnaudie, F.; Magnan, P.; Martin-Gonthier, P.; Corbiere, F.; Belliot, P.; Estribeau, M. Research-grade CMOS image sensors for remote sensing applications. In Proceedings of the Sensors, Systems, and Next-Generation Satellites VIII, Canary Islands, Spain, 13–16 September 2004; Volume 5570, pp. 549–556.
4. Sukhavasi, S.B.; Sukhavasi, S.B.; Elleithy, K.; Abuzneid, S.; Elleithy, A. CMOS image sensors in surveillance system applications. *Sensors* **2021**, *21*, 488. [CrossRef] [PubMed]
5. Li, J.; Liu, J.; Li, X.; Liu, Y.; Hao, Z. CMOS APS imaging system application in star tracker. In Proceedings of the Advanced Materials and Devices for Sensing and Imaging II, Beijing, China, 8–11 November 2004; Volume 5633, pp. 536–542.
6. Virmontois, C.; Toulemont, A.; Rolland, G.; Materne, A.; Lalucaa, V.; Goiffon, V.; Codreanu, C.; Durnez, C.; Bardoux, A. Radiation-induced dose and single event effects in digital CMOS image sensors. *IEEE Trans. Nucl. Sci.* **2014**, *61*, 3331–3340. [CrossRef]
7. Virmontois, C.; Goiffon, V.; Corbière, F.; Magnan, P.; Girard, S.; Bardoux, A. Displacement damage effects in pinned photodiode CMOS image sensors. *IEEE Trans. Nucl. Sci.* **2014**, *59*, 2872–2877. [CrossRef]
8. Rizzolo, S.; Goiffon, V.; Estribeau, M.; Paillet, P.; Marcandella, C.; Durnez, C.; Magnan, P. Total-ionizing dose effects on charge transfer efficiency and image lag in pinned photodiode CMOS image sensors. *IEEE Trans. Nucl. Sci.* **2017**, *65*, 84–91. [CrossRef]
9. Fu, J.; Feng, J.; Li, Y.D.; Wen, L.; Zhou, D.; Guo, Q. Effect of proton beam irradiation on the tracking efficiency of CMOS image sensors. *Radiat. Eff. Defects Solids* **2022**, *177*, 590–603. [CrossRef]
10. Ecoffet, R. Overview of in-orbit radiation induced spacecraft anomalies. *IEEE Trans. Nucl. Sci.* **2013**, *60*, 1791–1815. [CrossRef]
11. Minec-Dube, J.; Jacob, P.; Guillon, D.; Temperanza, D. Protons robustness improvement for the SED 26 star tracker. In Proceedings of the Guidance, Navigation and Control Systems, Loutraki, Greece, 17–20 October 2005; Volume 606.
12. Li, J.; Wei, X.; Wang, G.; Zhou, S. Improved grid algorithm based on star pair pattern and two-dimensional angular distances for full-sky star identification. *IEEE Access* **2013**, *8*, 1010–1020. [CrossRef]
13. Liebe, C.C. Pattern recognition of star constellations for spacecraft applications. *IEEE Aerosp. Electron. Syst. Mag.* **1993**, *8*, 31–39. [CrossRef]
14. Mortari, D. Search-less algorithm for star pattern recognition. *J. Astronaut. Sci.* **1993**, *45*, 179–194. [CrossRef]
15. Padgett, C.; Kreutz-Delgado, K. A grid algorithm for autonomous star identification. *IEEE Trans. Aerosp. Electron. Syst.* **1997**, *33*, 202–213. [CrossRef]
16. Mingqian, L.; Hong, T.; Kaili, L. An On-orbit Correction Method for Star Sensor Under Single Event Effect Based on PSF Reference Model. *Semicond. Optoelectron.* **2022**, *43*, 986–991.
17. Le Roch, A.; Virmontois, C.; Goiffon, V.; Tauzière, L.; Belloir, J.M.; Durnez, C.; Magnan, P. Radiation-induced defects in 8T-CMOS global shutter image sensor for space applications. *IEEE Trans. Nucl. Sci.* **2018**, *65*, 1645–1653. [CrossRef]
18. Virmontois, C.; Goiffon, V.; Magnan, P.; Girard, S.; Saint-Pe, O.; Petit, S.; Rolland, G.; Bardoux, A. Similarities between proton and neutron induced dark current distribution in CMOS image sensors. *IEEE Trans. Nucl. Sci.* **2012**, *59*, 927–936. [CrossRef]
19. Feng, J.; Wang, H.C.; Cui, Y.H.; Li, Y.D.; Guo, Q.; Wen, L.; Fu, J. Effects of gamma radiation on the performance of star sensors for star map recognition. *Radiat. Phys. Chem.* **2018**, *203*, 110607. [CrossRef]
20. Virmontois, C.; Goiffon, V.; Magnan, P.; Girard, S.; Inguimbert, C.; Petit, S.; Rolland, G.; Saint-Pé, O. Displacement damage effects due to neutron and proton irradiations on CMOS image sensors manufactured in deep submicron technology. *IEEE Trans. Nucl. Sci.* **2010**, *57*, 3101–3108. [CrossRef]
21. Goiffon, V. Radiation Effects on CMOS Active Pixel Image Sensors. In *Ionizing Radiation Effects in Electronics*; CRC Press: Boca Raton, FL, USA, 2015; pp. 295–332.

22. Beaumel, M.; Hervé, D.; van Aken, D.; Pourrouquet, P.; Poizat, M. Proton, electron, and heavy ion single event effects on the HAS2 CMOS image sensor. *IEEE Trans. Nucl. Sci.* **2014**, *61*, 1909–1917. [CrossRef]
23. Blarre, L.; Piot, D.; Jacob, P.; Minec, J.; Piriou, V.; Ouaknine, J. SED16 Autonomous Star Sensor Product Line in Flight Results, New Developments and Improvements in Progress. In Proceedings of the AIAA Guidance, Navigation, and Control Conference and Exhibit, San Francisco, CA, USA, 15–18 August 2005; pp. 2005–5930.

Disclaimer/Publisher's Note: The statements, opinions and data contained in all publications are solely those of the individual author(s) and contributor(s) and not of MDPI and/or the editor(s). MDPI and/or the editor(s) disclaim responsibility for any injury to people or property resulting from any ideas, methods, instructions or products referred to in the content.

Article

Study of the Within-Batch TID Response Variability on Silicon-Based VDMOS Devices

Xiao Li [1,2,3], Jiangwei Cui [1,2,3,*], Qiwen Zheng [1,2,3,*], Pengwei Li [4], Xu Cui [1,2,3], Yudong Li [1,2,3] and Qi Guo [1,2,3]

1. Key Laboratory of Functional Materials and Devices for Special Environments, Xinjiang Technical Institute of Physics and Chemistry, Chinese Academy of Sciences, Urumqi 830011, China
2. Xinjiang Key Laboratory of Electronic Information Material and Device, Xinjiang Technical Institute of Physics and Chemistry, Chinese Academy of Sciences, Urumqi 830011, China
3. University of Chinese Academy of Sciences, Beijing 100049, China
4. China Academy of Space Technology, Beijing 100049, China
* Correspondence: cuijw@ms.xjb.ac.cn (J.C.); qwzheng@ms.xjb.ac.cn (Q.Z.)

Abstract: Silicon-based vertical double-diffused MOSFET (VDMOS) devices are important components of the power system of spacecraft. However, VDMOS is sensitive to the total ionizing dose (TID) effect and may have TID response variability. The within-batch TID response variability on silicon-based VDMOS devices is studied by the ^{60}Co gamma-ray irradiation experiment in this paper. The variations in device parameters after irradiation is obtained, and the damage mechanism is revealed. Experimental results show that the standard deviations of threshold voltage, subthreshold swing, output capacitance, and diode forward voltage increase, while the standard deviation of maximum transconductance decreases after irradiation. The standard deviation of on-state resistance is basically unchanged before and after irradiation. By separating the trapped charges generated by TID irradiation, it is found that the deviation of the oxide trapped charges and the interface traps increase with the increase in the total dose. The reasons for the variation in device parameters after irradiation are revealed by establishing the relationship between the trapped charges and the electrical parameters before and after irradiation.

Keywords: VDMOS; total ionizing dose (TID); variability; oxide trapped charges; interface traps

1. Introduction

Silicon-based vertical double-diffused MOSFET (VDMOS) devices are widely used in the power system of spacecraft due to the high input impedance, large current gain, excellent noise margin, and small conduction loss, as well as a negative temperature coefficient and no secondary breakdown effect [1,2]. However, VDMOS is sensitive to the total ionizing dose (TID) effect since there is a parasitic NPN transistor in the VDMOS structure [3]. Moreover, process variation in VDMOS manufacturing occurs with different temperature distributions, impurity diffusion, and injection. The TID response is sensitive to process variations, as evidenced by the different TID responses of devices produced from the same wafer (within-wafer) or devices produced from the same batch (within-batch).

The TID response variability has been studied in previous works. Within-wafer TID response variability of NMOSFET and PMOSFET was measured by Hu et al. and Gerardin et al. [4,5]. They attributed the within-wafer TID response variability to the process variation in shallow trench isolation (STI) and random doping. The TID response variability of 25 nm single-level cell non-volatile memory device (NAND) flash memories from two different lots was studied by Bagatin et al. [6]. The statistical parameters such as mean value, standard deviation, and shapes of the error distributions were studied. Part-to-part and lot-to-lot variability of TID response in bipolar linear devices was studied by Guillermin et al. [7]. The three-sigma method and one-sided tolerance limit method were commonly used to take the variability into account. Within-wafer TID response variability on the buried oxide (BOX) layer of silicon-on-insulator (SOI) technology was investigated by Zheng et al. [8,9]. The larger standard deviation of threshold voltage and off-state leakage distribution for irradiated devices than un-irradiated devices were observed. They attributed it to the evolution of net trapped charges induced by TID in BOX affecting by positively charged silicon nanoclusters introduced by silicon ion implantation. The device variability induced by the TID effects was investigated by Ma et al. in commercial 16 nm bulk nFinFETs with a small number of samples [10]. It was found that transistors characterized by higher drain currents exhibit the worst TID degradation. They attributed this phenomenon to the impact of random dopant fluctuations on the TID effects and/or to variations in the hydrogen concentration responsible for the TID-induced interface traps. The above studies confirmed the fluctuation of the TID response of the devices in the wafer, as well as the variability of the radiation damage of device modules from different batches and within-batch. There have been studies on the TID effect of VDMOS [11–15]. However, to our best knowledge, there is no report on the within-batch TID response variability on silicon-based VDMOS devices.

Within-batch TID response variability on silicon-based VDMOS devices is investigated by irradiation experiment in this paper. The variability of threshold voltage, subthreshold swing, maximum transconductance, output capacitance, diode forward voltage, and on-state resistance before and after irradiation is analyzed. The reasons for the variation in device parameters under irradiation are revealed by analyzing the trapped charge induced by irradiation.

2. Experiment Details

The devices under test (DUT) are N-channel enhanced VDMOS within the same batch. When the device is turned on, the maximum drain current is 120A. The maximum gate-source voltage is 20 V. The device has three terminals, gate, drain, and source. The device is packaged with TO247. The numbers of DUT are from 1 to 68. The experiments were conducted by ^{60}Co gamma-ray at room temperature in the Xinjiang Technical Institute of Physics and Chemistry, Chinese Academy of Sciences. The dose rate was 50.24 rad(Si)/s and the dose levels were 5 krad(Si), 10 krad(Si), 15 krad(Si), 20 krad(Si), and 25 krad(Si). It was found that the ON bias condition ($V_{DS} = 0$ V, $V_{GS} = 20$ V) can induce greater radiation damage than other bias conditions. All devices were kept ON bias condition during the irradiation process.

The transfer characteristic curves ($I_{DS} - V_{GS}$) of the devices were measured by Keysight B1500A semiconductor parameter analyzer at room temperature, while the drain voltage was set to 0.1 V, the gate voltage swept between -0.5 V and 5 V, and the source was grounded. The threshold voltage (V_{TH}) of the device was extracted by the constant current method. V_{TH} is equal to the gate-source voltage (V_{GS}) when the drain current is equal to 250 µA. The subthreshold swing (SS) of the device was calculated by $SS = \frac{dV_{GS}}{d(\log I_{DS})}$. The transconductance was calculated by $G_M = \frac{dI_{DS}}{dV_{GS}}$. G_{MMAX} is the max transconductance. The output capacitance (C_{OSS}) was measured, while the frequency was 1.0 MHz, the gate-source voltage was 0 V, and the drain-source voltage was 25 V. The on-state resistance ($R_{DS(ON)}$) and diode forward voltage (V_{SD}) were measured by BC3193 Semiconductor Discrete Device Test System at room temperature. The specific parameters and test conditions are shown in Table 1.

Table 1. Test parameters of silicon-based VDMOS.

Test Equipment	Test Parameter	Test Conditions
B1500A	V_{TH}	$V_{DS} = V_{GS}, I_{DS} = 250\ \mu A$
B1500A	SS	$V_{DS} = 0.1\ V, I_{DS} - V_{GS}$ Curve
B1500A	G_{MMAX}	$V_{DS} = 0.1\ V, I_{DS} - V_{GS}$ Curve
B1500A	C_{OSS}	$V_{DS} = 25\ V, V_{GS} = 0\ V, f = 1\ MHz$
BC3193	V_{SD}	$V_{GS} = 0\ V, I_S = 75\ A$
BC3193	$R_{DS(ON)}$	$V_{GS} = 10\ V, I_D = 75\ A$

3. Experimental Results

The $I_{DS} - V_{GS}$ curves of 68 devices before and after TID irradiation are shown in Figure 1. It can be seen that the $I_{DS} - V_{GS}$ curves of the devices shift negatively as the dose increases. The variability between devices increases after irradiation.

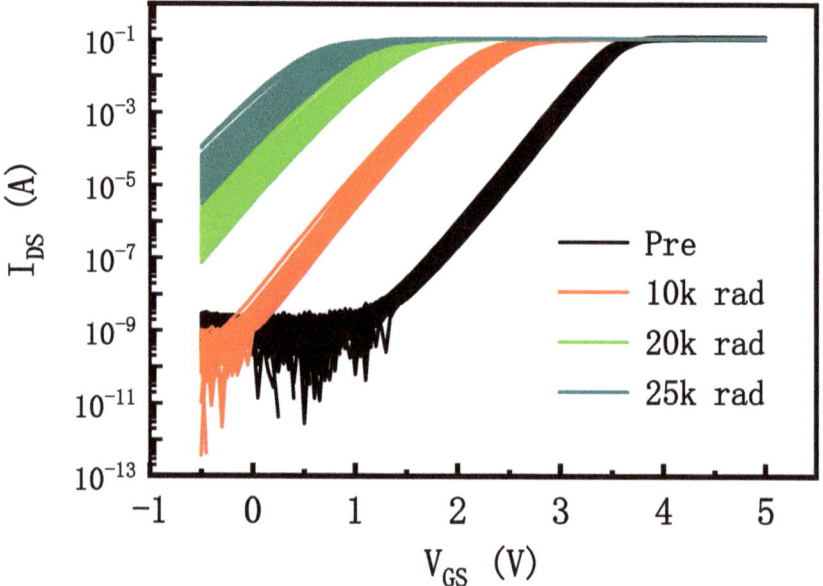

Figure 1. The shift of transfer characteristic curves of VDMOS devices before and after irradiation.

The variation in threshold voltage, subthreshold swing, and maximum transconductance extracted by the $I_{DS} - V_{GS}$ curves after irradiation is shown in Figure 2a, Figure 2b, and Figure 2c respectively. The mean value and standard deviation (σ) of the electrical parameters are calculated. With the increase in the total dose, the standard deviation of threshold voltage and subthreshold swing increase, while the standard deviation of maximum transconductance decreases. The experiment results verify the within-batch TID response variability on the VDMOS device since standard deviation measures the dispersion of a dataset relative to its mean value [8].

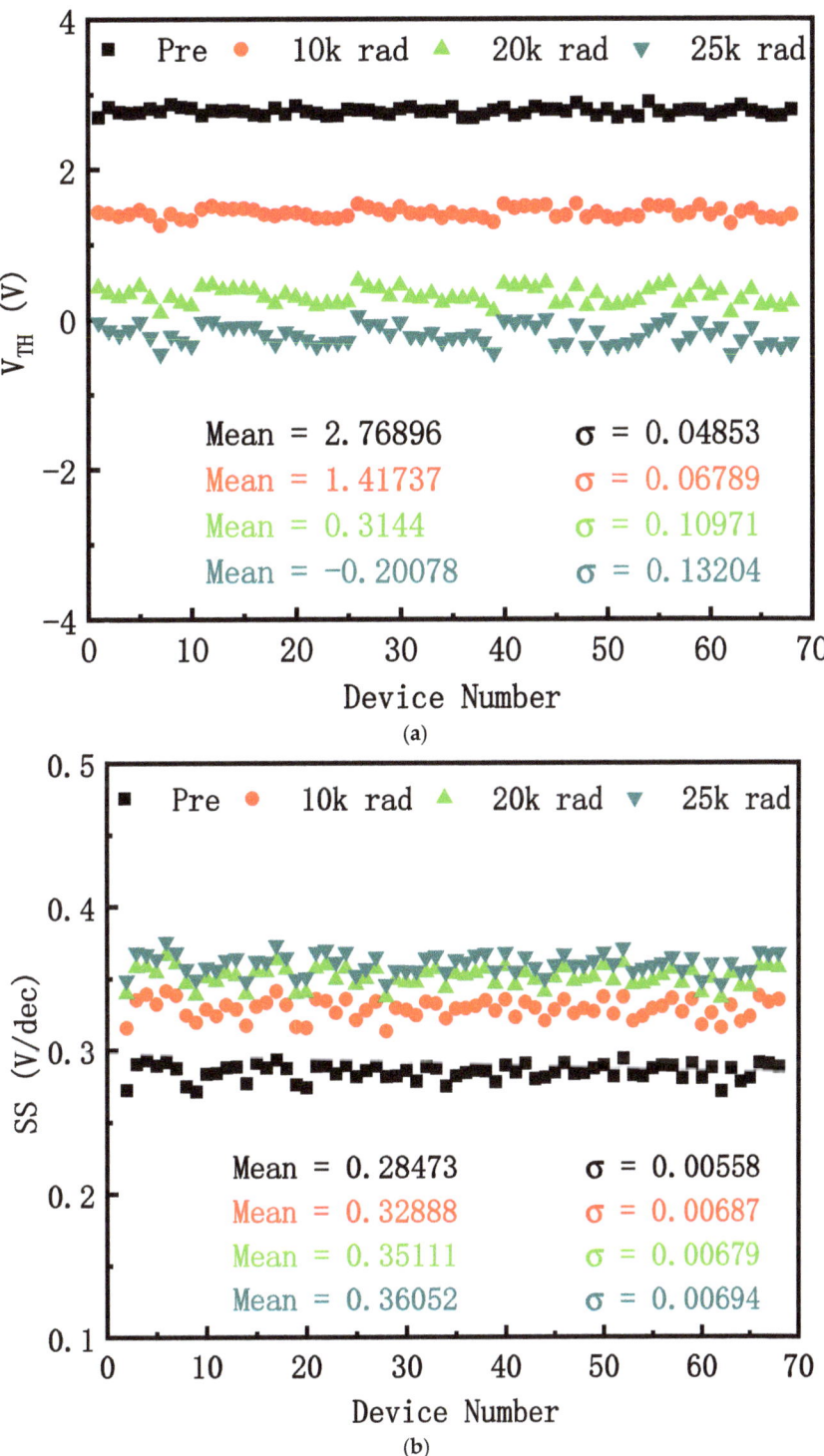

Figure 2. *Cont.*

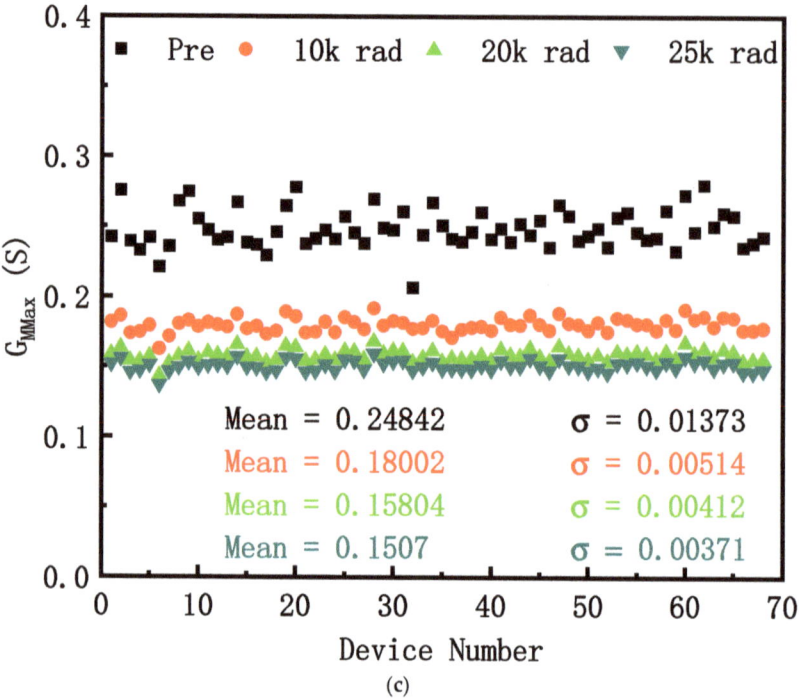

Figure 2. The variation in (**a**) V_{TH}, (**b**) SS, and (**c**) G_{MMAX} of the devices before and after irradiation.

The variation in output capacitance after irradiation is shown in Figure 3. The mean value and standard deviation of the electrical parameters increase after irradiation, which indicates that output capacitance variability also increases after irradiation.

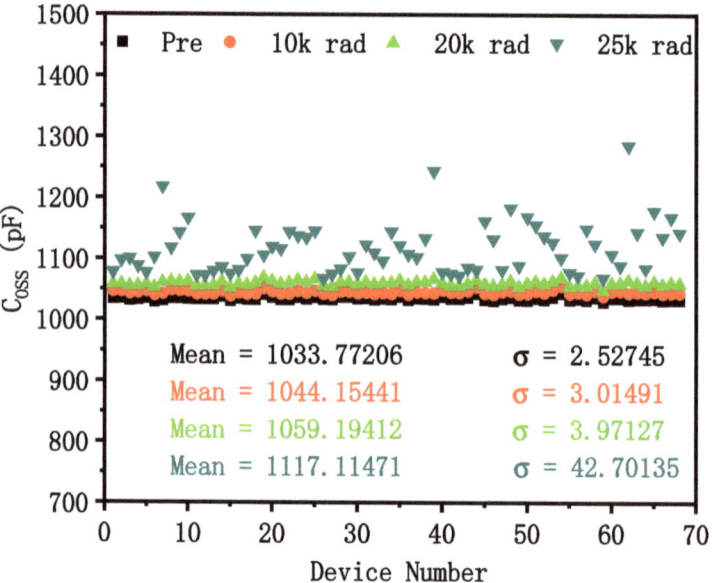

Figure 3. The variation in C_{OSS} of the devices before and after irradiation.

The variation in diode forward voltage and the on-state resistance after irradiation is shown in Figures 4 and 5, respectively. It can be seen that the variability of diode forward voltage increases as the total dose increases. The variability of on-state resistance is basically unchanged before and after irradiation.

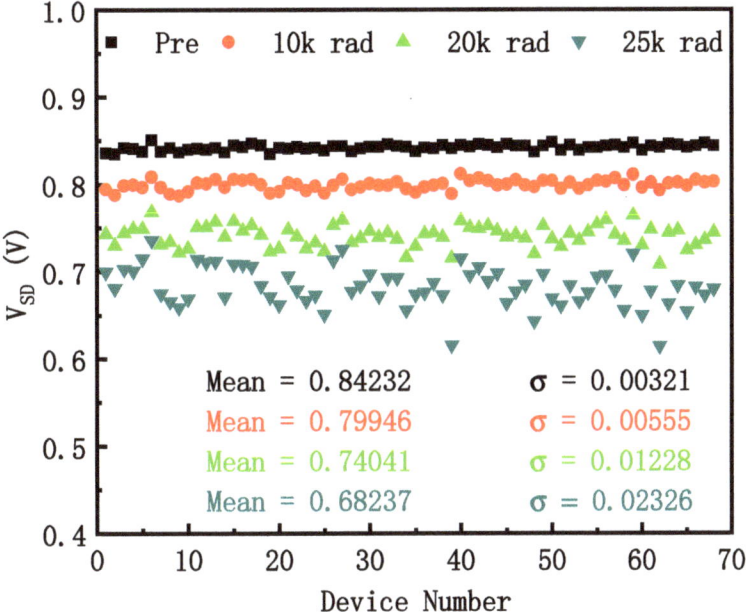

Figure 4. The variation in V_{SD} of the devices before and after irradiation.

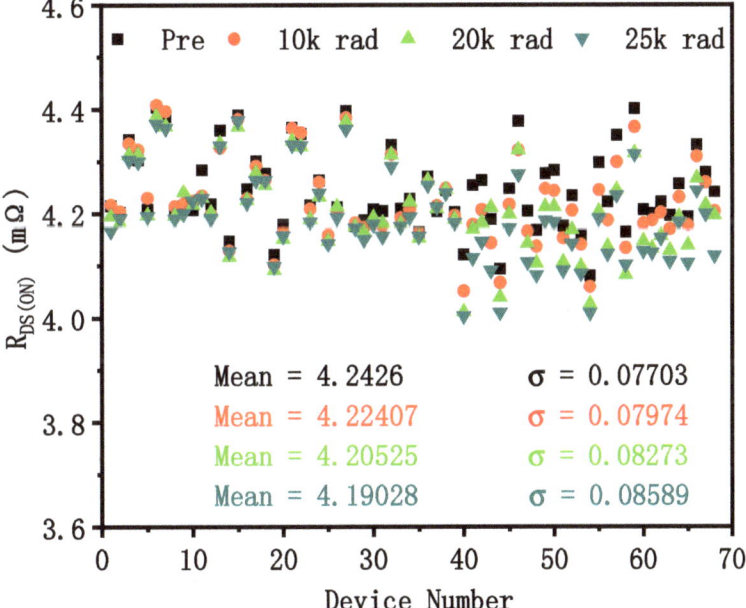

Figure 5. The variation in $R_{DS(ON)}$ of the devices before and after irradiation.

It can be seen from the above results that there is within-batch TID response variability of VDMOS devices in this paper. The changes in parameter variability after irradiation are listed in Table 2.

Table 2. The variation in test parameters of the devices after irradiation.

Test Parameter	Standard Deviation
V_{TH}	Increase
SS	Increase
G_{MMAX}	Decrease
C_{OSS}	Increase
V_{SD}	Increase
$R_{DS(ON)}$	Not Obvious

4. Discussion

For MOS devices, the initial electron–hole pairs generated by TID irradiation will eventually affect the electrical parameters of devices through a series of evolution. Four physical processes illustrate the evolution of electron–hole pairs generated by ionizing radiation at the interface of the SiO_2 gate and Si substrate, as shown in Figure 6 [16].

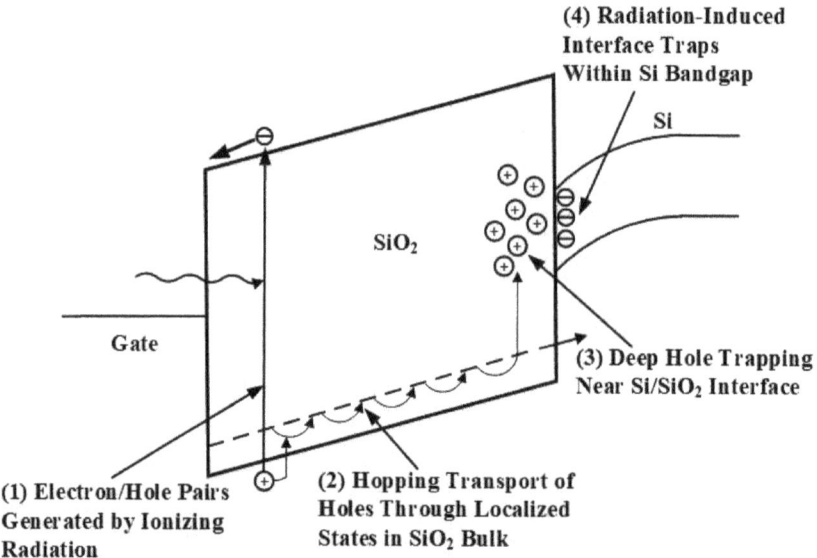

Figure 6. Band diagram of a MOS capacitor with a positive gate bias. Illustrated are the main processes for radiation-induced charge generation [16].

The specific process is as follows [16–19]:

1. The ionization of radiation particles in SiO_2 produces electron–hole pairs, the number of which is related to the ionization dose.
2. When the positive bias is applied to the gate, the drift motion of the electron–hole pairs in the oxide layer is the most significant. The electrons are removed by a fast drift (ps magnitude) towards the anodic ohmic contact, and the holes are relatively slow (s magnitude) to the cathodic ohmic contact.
3. In the drift process, some holes are captured to form the trap center. In the shallow-level trap center located in the gap of SiO_2, about 1 eV is distributed in the whole SiO_2 body. The holes can be transported in a jump mode. The center of the deep-level trap

with more than 3 eV in the gap of SiO_2 is mainly distributed near the SiO_2-Si interface, which is the relatively stable positive charge (N_{ot}) trapped by the oxide layer.
4. In the transition layer of the SiO_2-Si interface, the holes captured by the oxide layer are exchanged with the electrons in the substrate Si through the tunneling effect and finally captured by the defects at the interface to form the interface traps (N_{it}).
5. Therefore, the main reason for the variation in device parameters after irradiation is that the ionizing radiation destroys the energy band equilibrium, generates electron–hole pairs, and forms oxide-trapped charges and interface traps. The oxide layer is the most sensitive part to TID radiation in the MOS system [19].

A technique for separating the density of oxide-trapped charges (N_{ot}) and the density of interface traps (N_{it}) in MOS transistors through $I-V$ curves is proposed by McWhorter and Winokur [20,21]. In order to reveal the mechanism of the TID variability response of devices within-batch, the N_{ot} and N_{it} of devices used in this paper were extracted, as shown in Figure 7. The variability of N_{ot} and N_{it} of the devices increases as the total dose increases. It shows that the trapped charges induced by total dose irradiation have fluctuation, which causes the variability of relevant sensitive electrical parameters. The density of N_{ot} is much higher than the density of N_{it}, while N_{ot} is about $10^{12}/cm^2$, and N_{it} is about $10^{11}/cm^2$ at 25 krad(Si).

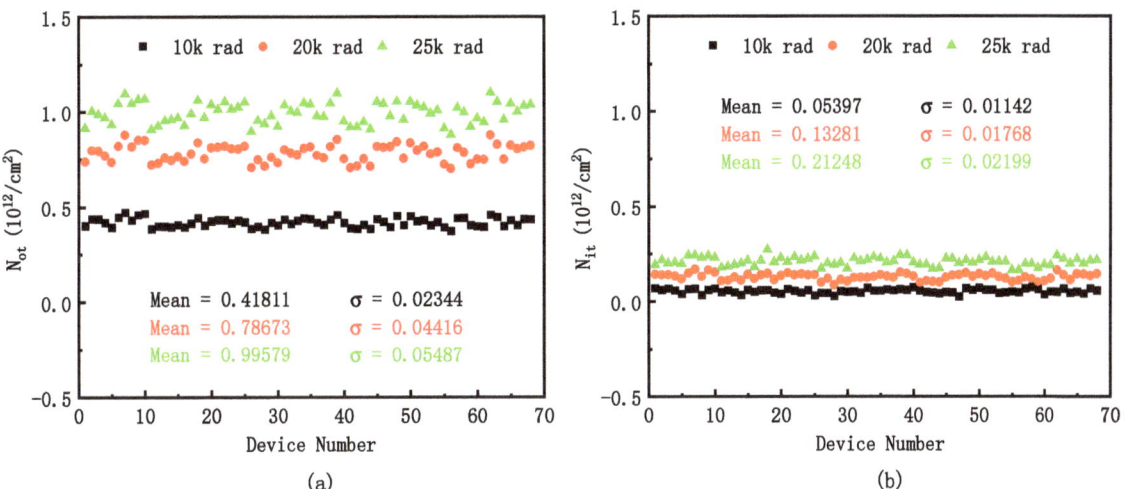

Figure 7. The variation in (**a**) oxide trapped charge (N_{ot}) density, (**b**) interface trap charge (N_{it}) density of the devices after irradiation.

Figure 8 shows the relation between the shift of threshold voltage (ΔV_{TH}) and the trapped charges ($|N_{ot}| - |N_{it}|$) of the within-batch devices after irradiation. As seen from the fit curve in Figure 8, the ΔV_{TH} and the $|N_{ot}| - |N_{it}|$ present the linear growth trend. For NMOSFET, the oxide-trapped charges cause a negative shift of the threshold voltage, while the interface traps cause a positive shift of the threshold voltage [3,19]. The calculation formula between ΔV_{TH} and trapped charges is [13,22–24]:

$$\Delta V_{TH} = -\frac{q|N_{ot}| - |N_{it}|}{C_{ox}} \quad (1)$$

where q (amount of charge) and C_{ox} (the gate oxide capacitance per unit area) are constant values. Since the density of N_{ot} is much higher than the density of N_{it}, the variability of the threshold voltage is mainly affected by the variability of N_{ot}.

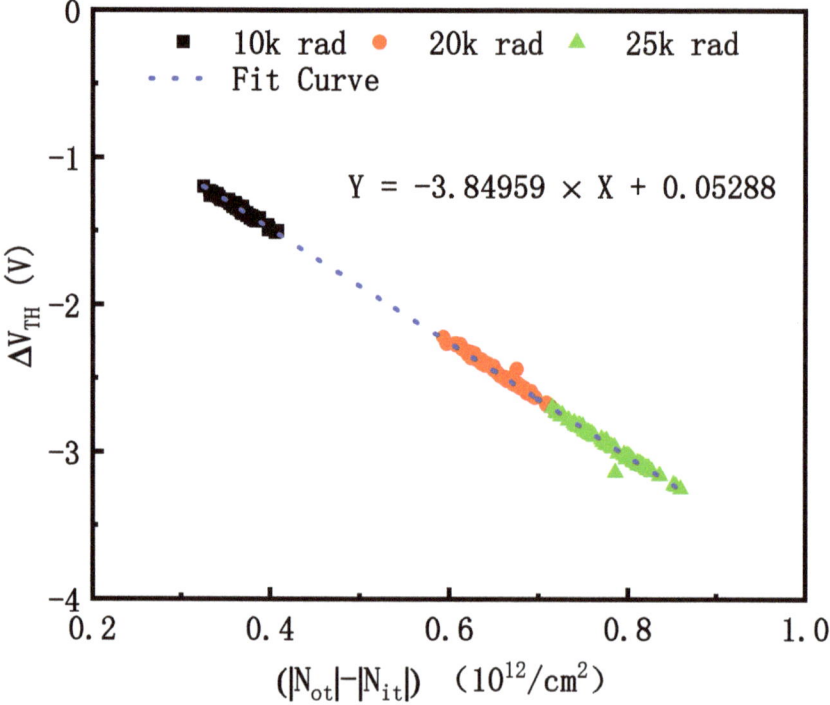

Figure 8. Relation between the shift of threshold voltage (ΔV_{TH}) and the trapped charges ($|N_{ot}| - |N_{it}|$) of the devices after irradiation.

V_{SD} is the forward voltage of the diode between the source and the drain. The threshold voltage reduces significantly or even becomes negative after irradiation, resulting in a conductive channel. Additionally, V_{SD} is across the source, channel, and drain [18,25]. V_{SD} is mainly affected by the threshold voltage, so the trend of within-batch variability is the same as the threshold voltage after irradiation.

Subthreshold swing and maximum transconductance are mainly affected by radiation-induced interface traps [26–29]. Interface traps are formed by TID irradiation at the interface between the device gate dielectric and the silicon substrate. The increase in the interface traps degrades the subthreshold swing of the devices [26]. The formula for SS is [21,30]:

$$SS = (\ln 10)\left(\frac{KT}{q}\right)\left(\frac{C_{ox} + C_D + C_{it}}{C_{ox}}\right) \quad (2)$$

$$C_{it} = q^2 D_{it} = \beta N_{it} \quad (3)$$

where K is the Boltzmann constant, T is the absolute temperature, C_D is the depletion layer capacitance, C_{it} is the interface traps capacitance, D_{it} is the interface traps density, and β is the correlation coefficient. Therefore, the within-batch variability of subthreshold swing increases after irradiation.

The variability of maximum transconductance is negatively correlated with the interface traps variability, while the variability trends of maximum transconductance and subthreshold swing are opposite. The formula for G_{MMAX} and N_{it} is [2,31,32]:

$$G_{MMAX} = G_{M0} \frac{1}{1 + \alpha N_{it}} \quad (4)$$

where G_{M0} and G_{MMAX} are the maximum transconductance values before and after irradiation, and α is the process fluctuation constants of the devices.

Figure 9 shows the relation between the shift of output capacitance (ΔC_{oss}) and N_{ot} of the within-batch devices after irradiation. As seen from the fit curve in Figure 9, the variation in ΔC_{OSS} and the N_{ot} presents the exponential growth trend. Therefore, the variability of output capacitance increases sharply at 25 krad(Si) in Figure 3. The output capacitance is equal to the sum of the drain-source capacitance and the gate-drain capacitance. The drain-source capacitance is the junction capacitance, which is not changed by the increase in radiation dose [19,33]. Therefore, the variation in output capacitance induced by TID is mainly affected by the gate-drain capacitance, which is a function of oxide capacitance, reverse capacitance, and depleted capacitance [33,34]. The output capacitance is sensitive to N_{ot}, so as to characterize the correlation between C_{OSS} and N_{ot}.

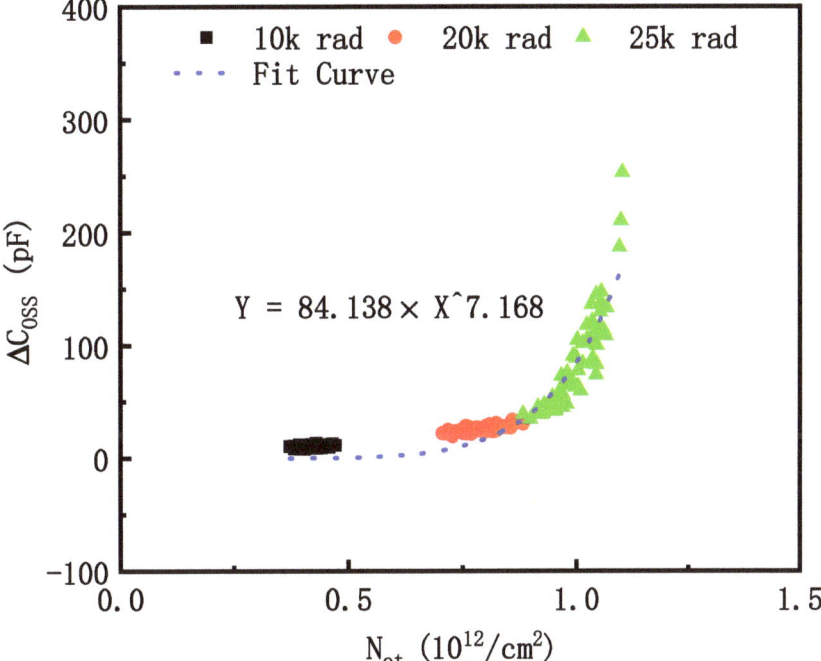

Figure 9. Relation between the shift of output capacitance (ΔC_{OSS}) and the oxide-trapped charge (N_{ot}) of the devices after irradiation.

The on-state resistance is regulated by the channel reverse voltage ($V_{GS} - V_{TH}$), which is closely related to the N_{ot} [25,35]. Because V_{GS} is much larger than the variation in V_{TH}, the variability of oxide trapped charges have no obvious effect on the variability of $R_{DS(ON)}$.

In general, the accumulation of trapped charges after irradiation magnifies the process differences of devices in the same batch and leads to differential variability in threshold voltage, subthreshold swing, maximum transconductance, output capacitance, and diode forward voltage. The correlations between the variations in electrical parameters and the trapped charges after irradiation are shown in Table 3. However, the variation in electrical parameters of devices in the same batch is a disadvantage to the stability and reliability of the spacecraft, which would lead to thermal failure, an unreasonable dead zone, or gate resonance problems in the circuit module. The differences in threshold voltage and output capacitance must be considered in the circuit design of space equipment, and the variability of on-state resistance in the same batch can be tolerated after irradiation.

Table 3. Correlation between the variation in electrical parameters and the trapped charges generated by irradiation.

Parameter	Elements	Major Impacts
V_{TH}	The oxide-trapped charges and the interface traps work together	Positively correlated with N_{ot}
SS	$SS = (\ln 10)\left(\frac{KT}{q}\right)\left(\frac{C_{ox}+C_D+\beta N_{it}}{C_{ox}}\right)$	Positively correlated with N_{it}
G_{MMAX}	$G_M = G_{M0}\frac{1}{1+\alpha N_{it}}$, Contrary to the trend of SS changes	Negatively correlated with N_{it}
C_{OSS}	It is related to the increase in the number of charges in the space charge region and the decrease in the width of the depletion layer	Exponential Relationship with N_{ot}
V_{SD}	After irradiation, a conductive channel is formed, and the changing trend is the same as the threshold voltage	Positively correlated with N_{ot}
$R_{DS(ON)}$	Regulation of channel reverse voltage (V_{GS}-V_{TH})	No Obvious

5. Conclusions

Silicon-based VDMOS devices are important components of the power system of spacecraft. However, the VDMOS is sensitive to the TID effect. Moreover, the TID response is sensitive to process variation, behaving as within-batch TID response variability. The within-batch TID response variability on silicon-based VDMOS devices is investigated by the ^{60}Co gamma-ray irradiation experiment in this paper. Experimental results show that with the increase in total dose, the variability of the within-batch devices parameters changes. The variability of threshold voltage, subthreshold swing, output capacitance, and diode forward voltage increases after irradiation. Furthermore, the variability of maximum transconductance decreases after irradiation, and the variability of on-state resistance is basically unchanged before and after irradiation. By extracting N_{ot} and N_{it} induced by TID irradiation in the devices, the relationship between the parameters variation and trapped charges are established, and the reasons for within-batch TID response variability are clarified.

Among them, it should be noted that the variability of threshold voltage and output capacitance within-batch shows different functional trends with the increase in radiation dose. The differences in threshold voltage and output capacitance must be considered in some new spacecraft. The new generation of spacecraft requires higher reliability and better performance of electronic devices. The electrical parameter margins are very small in some circuits. Considering the variability of electrical parameters between the same batch of devices caused by TID can reduce the loss and protect the circuit more accurately. The variability of on-state resistance in the same batch can be tolerated after irradiation in the circuit design of space equipment. This study provides a foundation for the establishment of scientific and reasonable TID effect evaluation and screening methods on the within-batch devices so as to ensure the stability and reliability of the power system of spacecraft.

Author Contributions: Conceptualization, J.C. and Q.Z.; methodology, writing—original draft preparation X.L.; writing—review and editing, J.C.; supervision, Y.L. and Q.G.; data curation, X.C.; formal analysis, P.L. All authors have read and agreed to the published version of the manuscript.

Funding: This work was supported in part by the Youth Innovation Promotion Association CAS (2020430), the West Light Foundation of the Chinese Academy of Science under Grant No. 2019-XBQNXZ-A-003, the National Natural Science Foundation of China under Grant 12275352, and the project under Grant No. 2022D14003.

Data Availability Statement: The data presented in this study are available on reasonable request from the corresponding authors.

Conflicts of Interest: The authors declare no conflict of interest.

References

1. Grant, D.A.; Gowar, J. *Power MOSFETS: Theory and Applications*, 1st ed.; John Wiley & Sons, Incorporated: New York, NY, USA, 1989; pp. 17–21.
2. Singh, G.; Galloway, K.F.; Russell, T.J. Radiation-Induced Interface Traps in Power Mosfets. *IEEE Trans. Nucl. Sci.* **1986**, *33*, 1454–1459. [CrossRef]
3. Wang, H. Research on the Radiation Effects of VDMOS Power Devices in Space Instruments. Master's Thesis, National University of Defense Technology, Changsha, China, 2017.
4. Hu, Z.; Liu, Z.; Shao, H.; Zhang, Z.; Ning, B.; Chen, M.; Bi, D.; Zou, S. Impact of within-wafer process variability on radiation response. *Microelectron. J.* **2011**, *42*, 883–888. [CrossRef]
5. Gerardin, S.; Bagatin, M.; Cornale, D.; Ding, L.; Mattiazzo, S.; Paccagnella, A.; Faccio, F.; Michelis, S. Enhancement of Transistor-to-Transistor Variability Due to Total Dose Effects in 65-nm MOSFETs. *IEEE Trans. Nucl. Sci.* **2015**, *62*, 2398–2403. [CrossRef]
6. Bagatin, M.; Gerardin, S.; Ferrarese, F.; Paccagnella, A.; Ferlet-Cavrois, V.; Costantino, A.; Muschitiello, M.; Visconti, A.; Wang, P.X. Sample-to-Sample Variability and Bit Errors Induced by Total Dose in Advanced NAND Flash Memories. *IEEE Trans. Nucl. Sci.* **2014**, *61*, 2889–2895. [CrossRef]
7. Guillermin, J.; Sukhaseum, N.; Varotsou, A.; Privat, A.; Garcia, P.; Vaillé, M.; Thomas, J.C.; Chatry, N.; Poivey, C. Part-to-part and lot-to-lot variability study of TID effects in bipolar linear devices. In Proceedings of the 2016 16th European Conference on Radiation and Its Effects on Components and Systems (RADECS), Bremen, Germany, 19–23 September 2016; pp. 1–8.
8. Zheng, Q.; Cui, J.; Yu, X.; Li, Y.; Lu, W.; He, C.; Guo, Q. Measurement and Evaluation of the Within-Wafer TID Response Variability on BOX Layer of SOI Technology. *IEEE Trans. Nucl. Sci.* **2021**, *68*, 2516–2523. [CrossRef]
9. Zheng, Q.; Cui, J.; Yu, X.; Li, Y.; Lu, W.; He, C.; Guo, Q. Impact of TID on Within-Wafer Variability of Radiation-Hardened SOI Wafers. *IEEE Trans. Nucl. Sci.* **2021**, *68*, 1423–1429. [CrossRef]
10. Ma, T.; Bonaldo, S.; Mattiazzo, S.; Baschirotto, A.; Enz, C.; Paccagnella, A.; Gerardin, S. Increased Device Variability Induced by Total Ionizing Dose in 16-nm Bulk nFinFETs. *IEEE Trans. Nucl. Sci.* **2022**, *69*, 1437–1443. [CrossRef]
11. Mo, J.J.; Chen, H.; Wang, L.P.; Yu, F.X. Total Ionizing Dose Effect and Single Event Burnout of VDMOS with Different Inter Layer Dielectric and Passivation. *J. Electron. Test.-Theory Appl.* **2017**, *33*, 255–259. [CrossRef]
12. Li, X.; Jia, Y.; Zhou, X.; Zhao, Y.; Tang, Y.; Li, Y.; Liu, G.; Jia, G. Degradation of Radiation-Hardened Vertical Double-Diffused Metal-Oxide-Semiconductor Field-Effect Transistor During Gamma Ray Irradiation Performed After Heavy Ion Striking. *IEEE Electron Device Lett.* **2020**, *41*, 216–219. [CrossRef]
13. Wang, R.; Li, Z.; Qiao, M.; Zhou, X.; Wang, T.; Zhang, B. Total Ionizing Dose Effects in 30-V Split-Gate Trench VDMOS. *IEEE Trans. Nucl. Sci.* **2020**, *67*, 2009–2014. [CrossRef]
14. Sun, Y.; Wang, T.; Liu, Z.; Xu, J. Investigation of irradiation effects and model parameter extraction for VDMOS field effect transistor exposed to gamma rays. *Radiat. Phys. Chem.* **2021**, *185*, 109478. [CrossRef]
15. Qin, Z.; Yang, J.; Li, X. Displacement damage on P-channel VDMOS caused by different energy protons. *Nucl. Instrum. Methods Phys. Res. Sect. B* **2019**, *461*, 232–236. [CrossRef]
16. Schwank, J.R.; Shaneyfelt, M.R.; Fleetwood, D.M.; Felix, J.A.; Dodd, P.E.; Paillet, P.; Ferlet-Cavrois, V. Radiation Effects in MOS Oxides. *IEEE Trans. Nucl. Sci.* **2008**, *55*, 1833–1853. [CrossRef]
17. Han, Z.; Zhao, Y. *Introduction to Radiation Hardened Integrated Circuit*, 1st ed.; Tsinghua University Press: Beijing, China, 2011; pp. 13–14.
18. Liu, W. *Radiation Effects and Reinforcement Techniques of Silicon Semiconductor Devices*, 1st ed.; Science Press: Beijing, China, 2013; pp. 10–19.
19. Oldham, T.R.; McLean, F.B. Total ionizing dose effects in MOS oxides and devices. *IEEE Trans. Nucl. Sci.* **2003**, *50*, 483–499. [CrossRef]
20. ASTM F996-11; Standard Test Method for Separating an Ionizing Radiation-Induced MOSFET Threshold Voltage Shift Into Components Due to Oxide Trapped Holes and Interface States Using the Subthreshold Current-Voltage Characteristics. ASTM International: West Conshohocken, PA, USA, 2011. [CrossRef]
21. He, Y.; Shi, Q.; Li, B.; Luo, H.; Lin, L. Oxide-trap and Interface-trap Charge Separation Analysis Techniques on MOSFET. *Reliab. Environ. Test. Electron. Prod.* **2006**, *24*, 26–29. [CrossRef]
22. Wu, H.; Huang, W. A 60V Radiation Hardened VDMOS Power Device. *Reliab. Environ. Test. Electron. Prod.* **2021**, *39*, 33–37.
23. Fan, P. Research on the Radiation and Thermal Stress Reliability of Typical Domestic VDMOS for Satellites Application. Master's Thesis, Shanghai Jiao Tong University, Shanghai, China, 2018.
24. Yang, G.; Wu, W.; Zhang, X.; Tang, P.; Yang, J.; Zhang, L.; Liu, S.; Sun, W. Experimental investigation on total-ionizing-dose radiation effects on the electrical properties of SOI-LIGBT. *Solid-State Electron.* **2021**, *175*, 107952. [CrossRef]
25. Liu, S.; DiCienzo, C.; Bliss, M.; Zafrani, M.; Boden, M.; Titus, J.L. Analysis of Commercial Trench Power MOSFETs' Responses to Co60 Irradiation. *IEEE Trans. Nucl. Sci.* **2008**, *55*, 3231–3236. [CrossRef]
26. Liu, Y.; Chen, H.; He, Y.; Wang, X.; Yue, L.; En, Y.; Liu, M. Radiation effects on the low frequency noise in partially depleted silicon on insulator transistors. *Acta Phys. Sin.* **2015**, *64*, 078501. [CrossRef]
27. Huang, J. Research on Current Model Induced by Total Dose Effects in NMOS Transistors. Master's Thesis, University of Electronic Science and Technology of China, Chendu, China, 2016.

28. Bonaldo, S.; Zhang, E.X.; Zhao, S.E.; Putcha, V.; Parvais, B.; Linten, D.; Gerardin, S.; Paccagnella, A.; Reed, R.A.; Schrimpf, R.D.; et al. Total-Ionizing-Dose Effects in InGaAs MOSFETs With High-k Gate Dielectrics and InP Substrates. *IEEE Trans. Nucl. Sci.* **2020**, *67*, 1312–1319. [CrossRef]
29. Liu, G.Z.; Li, B.; Xiao, Z.Q.; Sun, J.H.; Yu, Z.G.; Wei, J.H.; Wang, H.B.; Hong, G.S.; Shi, J.W. The TID Characteristics of a Radiation Hardened Sense-Switch pFLASH Cell. *IEEE Trans. Device Mater. Reliab.* **2020**, *20*, 358–365. [CrossRef]
30. Shi, M.; Wu, G.; Geng, L.; Zhang, R. *Semiconductor Device Physics*, 3rd ed.; Xi'an Jiaotong University Press: Xi'an, China, 2008; pp. 240–241.
31. Galloway, K.F.; Gaitan, M.; Russell, T.J. A Simple Model for Separating Interface and Oxide Charge Effects in MOS Device Characteristics. *IEEE Trans. Nucl. Sci.* **1984**, *31*, 1497–1501. [CrossRef]
32. Zheng, S.; Zeng, Y.; Chen, Z. Investigation of Total-Ionizing Dose Effects on the Two-Dimensional Transition Metal Dichalcogenide Field-Effect Transistors. *IEEE Access* **2019**, *7*, 79989–79996. [CrossRef]
33. Soliman, F.A.S.; Al-Kabbani, A.S.S.; Rageh, M.S.I.; Sharshar, K.A.A. Effects of electron-hole generation, transport and trapping in MOSFETs due to γ-ray exposure. *Appl. Radiat. Isot.* **1995**, *46*, 1337–1343. [CrossRef]
34. Xie, T.T.; Ge, H.; Lv, Y.H.; Chen, J. The impact of total ionizing dose on RF performance of 130 nm PD SOI I/O nMOSFETs. *Microelectron. Reliab.* **2021**, *116*, 114001. [CrossRef]
35. Sun, Y.B.; Wan, X.; Liu, Z.Y.; Jin, H.; Yan, J.Z.; Li, X.J.; Shi, Y.L. Investigation of total ionizing dose effects in 4H-SiC power MOSFET under gamma ray radiation. *Radiat. Phys. Chem.* **2022**, *197*, 110219. [CrossRef]

Disclaimer/Publisher's Note: The statements, opinions and data contained in all publications are solely those of the individual author(s) and contributor(s) and not of MDPI and/or the editor(s). MDPI and/or the editor(s) disclaim responsibility for any injury to people or property resulting from any ideas, methods, instructions or products referred to in the content.

Article

Effects of Different Factors on Single Event Effects Introduced by Heavy Ions in SiGe Heterojunction Bipolar Transistor: A TCAD Simulation

Zheng Zhang [1], Gang Guo [1,*], Futang Li [1], Haohan Sun [1], Qiming Chen [1], Shuyong Zhao [1], Jiancheng Liu [1] and Xiaoping Ouyang [1,2]

1. Department of Nuclear Physics, China Institute of Atomic Energy, Beijing 102413, China
2. Northwest Institute of Nuclear Technology, Xi'an 710024, China
* Correspondence: ggg@ciae.ac.cn

Abstract: In this paper, the effects of different factors, including the heavy ions striking location, incident angle, linear energy transfer (LET) value, projected range, ambient temperature and bias state, on the single event transient introduced by heavy ions irradiation in the SiGe heterojunction bipolar transistor (HBT) were investigated by the TCAD simulation. The results show that the current transient peak value, collected charge and carrier type of each terminal are changed by the striking location, incident angle and bias state. The current transient peak value and collected charge increase with the LET value, while they decrease with the ambient temperature. When heavy ions vertically irradiate the collector and substrate, the current transient peak value and collected charge increase with the projected range; therefore, the species of heavy ions should be considered in studying the single event effects of the SiGe HBT induced by heavy ions irradiation. The microphysical mechanism of these factors influencing the single event effects of the SiGe HBT is discussed in this work.

Keywords: SiGe heterojunction bipolar transistor; single event effect; single event transient; charge collection; TCAD simulation

1. Introduction

The space radiation environment is filled with a large number of high-energy charged particles, which will inevitably affect the electronic components of space missions [1–3]. The high-energy charged particles mainly come from galactic cosmic rays, solar cosmic rays and the Van Allen radiation belt, and they deposit energy into the aerospace devices to cause radiation effects, resulting in the functional failure of the aerospace devices and even the failure of the related space missions. The radiation effects of the aerospace devices can be divided into the ionizing effect and the non-ionizing effect. The ionizing effect mainly includes single event effects [4–6] and the total ionizing dose effect [7–9], while the non-ionizing effect is mainly the displacement damage effect [10,11]. These radiation effects occur simultaneously in the aerospace devices and interact with each other [12]. Since the launch of the first man-made satellite, 46% of spacecrafts and satellites suffered functional failures due to these radiation effects [13], ultimately resulting in mission failure.

In addition to the radiation effects caused by high-energy charged particles in the space radiation environment, the aerospace devices also face the challenge of an extremely low-temperature environment during the space mission. There are a lot of extremely low temperatures in the space environment, for example, the temperature on the Mars surface usually ranges from −133 °C to 27 °C, the temperature range of the lunar rover during its mission is usually from −180 °C to 120 °C and the lowest temperature at the polar craters on the lunar surface reach −230 °C. Therefore, it is of great significance to develop an aerospace device with excellent radiation resistance and extremely low-temperature characteristics for the future aerospace industry. Once the research is successful, a large amount of thermal

insulation equipment can be removed, which would not only reduce the cost of launching spacecraft and satellites but also enhance their deep-space exploration capabilities.

Since the early 2000s, NASA has been concerned about the use of electronic systems in the extreme space environments. Previous studies have indicated that the germanium silicon heterojunction bipolar transistor (SiGe HBT) has excellent total dose radiation resistance [14,15] and excellent low-temperature characteristics [16,17] due to the advantages of the silicon-based energy band engineering materials, semiconductor process and device structures. The SiGe HBT can operate normally in the temperature range of −180 °C to 125 °C, with the total ionizing dose effect resistance up to an Mrad(Si) magnitude [18,19] and the displacement damage resistance up to a 10^{15} cm^{-2} magnitude (equivalent fluence of 1 MeV neutron) [20]. Therefore, the SiGe HBT has an attractive application prospect in the field of the extreme space environment. However, a large number of studies have found that the SiGe HBT is very sensitive to the single event effects [21,22] and has a complex charge collection mechanism different from traditional bulk silicon devices. Thus, the study of the single event effects has always been a hot topic in the research of SiGe HBT radiation effects.

In recent years, many scholars have conducted a significant amount of research on the single event effects of the SiGe HBT. Many famous research institutions such as the Georgia Institute of Technology, Auburn University, Vanderbilt University, and the Boeing company have carried out a lot of research works on the single event effects of the commercial SiGe HBT produced by IMB, National Semiconductor, Jazz Semiconductor and other companies [23], in which a lot of the research works have been carried out based on the fourth-generation SiGe HBT produced by the IBM company. Since 2005, the Georgia Institute of Technology and Auburn University have studied the charge collection mechanism, key influencing factors and anti-radiation reinforcement design of the SiGe HBT single event effects with the help of proton, heavy ion and laser microbeam irradiation experiments and the TCAD numerical simulation [24,25]. The results show that the sensitive area of single event effects in the SiGe HBT with a deep trench isolation (DTI) structure is the DTI region. The DTI structure can not only prevent the external excess carriers from diffusing to the inner collection node but also limit the excess carriers from diffusing to the outside, resulting in a significant increase in the charge collection. The transient current amplitude and integral charge collection induced by the single event effects are closely related to the linear energy transfer (LET) value of the incident ions, but also to the projected range of the incident ions, indicating that the light-doping substrate in the bulk silicon process has an important effect on the sensitivity of the SiGe HBT single event effects. In 2015, Li et al. used laser microbeams to study the single event effects of the SiGe HBT produced by a Chinese company and IBM, respectively [26,27]. Due to the similar doping concentration in their collector region, the peak current values of the two SiGe HBTs collectors were close. The SiGe HBT produced by the Chinese company has a large C/S junction, and its collector has a strong charge collection capacity, so the transient current pulse width is large. From 2017 to 2019, Wei et al. studied the single event effects of the SiGe HBT produced by the Chinese company through the heavy ion microbeam and proton irradiation experiments [28]. The results show that the collector transient current peak value caused by heavy ions is significantly higher than that caused by a proton under the same bias state. The transient current pulse width of the SiGe HBT collector caused by heavy ions is also wider than that caused by protons. When the SiGe circuit works at a higher frequency and its working period is shortened to a time scale of ps, the collector transient current pulse induced by heavy ions covers more working periods than that of the collector transient pulse induced by protons, thus inducing a more serious multi-bit upset effect.

Through a TCAD simulation, Zhang et al. found that the incident angle of heavy ions would significantly change the ionization track length in the SiGe HBT [29], which would lead to the difference in the charge deposition and ultimately the difference in the charge collection. In the SiGe HBT, the emission direction of secondary particles produced by the intermediate and high-energy proton through a nuclear reaction is relatively random. Changing the incident angle of the proton can not uniquely determine the direction of the secondary particles; therefore, the single event effects in the SiGe HBT can not be changed significantly by changing the incident angle. Through a Monte Carlo simulation, Wei et al. found that when the SiGe HBT was irradiated by the proton in different incident angles [30], the main body of the collector transient current pulse waveform distribution shows the same basic characteristics, a fast rising edge and a relatively slow falling edge. However, as the incident angle of the proton increases, the falling edge becomes very slow, and the distribution range of the collector transient current pulse duration expands.

Above all, the single event effects in the SiGe HBT can be well reproduced by a TCAD simulation, which is not affected by the running time compared with the experimental study. At the same time, a TCAD simulation can complete the research content that is difficult to achieve in the experiment, so as to provide a theoretical basis for the practical application of the SiGe HBT in a space environment. Based on the process and structure of the SiGe HBT produced by the Chinese company, the effects of the striking location, incident angle and LET value of heavy ions, ambient temperature, bias state and other factors on the single event effects of the SiGe HBT were carried out by the TCAD simulation in this paper. The sensitive area of the charge collection, the effects of these factors on the current transient pulse peak value and width of each terminal and the charge collection amount were determined, which provides further theoretical support for the radiation-hardening technique of the SiGe HBT produced by the Chinese company.

2. Materials and Methods

In this paper, a domestic SiGe HBT is selected as the research object, whose inner structure is similar to that of the traditional bulk silicon npn vertical bipolar transistor, as shown in Figure 1. The base region is composed of SiGe material with gradual change in components. The introduction of Ge in the base region forms a slow mutation heterojunction at the emitter/base pole junction (E/B junction) and base/collector junction (B/C junction), as shown in Figure 2. The built-in electric field formed in the base region effectively improves the carrier transit time in the base region. Current gain increases exponentially with the band-gap variation $\triangle E_g$ after the introduction of Ge, as shown in the following equation:

$$h_{fe} = \frac{N_e}{N_b} \frac{V_{nb}}{V_{pe}} exp(\triangle E_g / kT) \tag{1}$$

where h_{fe} is the current gain, N_e is concentration in emitter region, N_b is the concentration in base region, V_{nb} is the electrons' velocity in base region, V_{pe} is the holes' velocity in emitter region, $\triangle E_g$ is the band-gap change, k is the Boltzmann's constant and T is the temperature. The base region thickness of the SiGe HBT is 0.08 µm, and the doping concentration is up to 10^{19} cm^{-3}, which effectively reduces the resistance of the base region and enables the SiGe HBT to simultaneously meet the requirements of high frequency and high gain. A shallow trough isolation (STI) was used to form an active region in the region from the base to the collector. Above the isolated oxide layer, a polysilicon layer doped boron and germanium exported the base, which was epitaxial by the dual polysilicon self-alignment process. Heavy doping epitaxial region can reduce resistance of the base region and the B/C junction. n$^+$ buried layer leads to the collector. The emitter region is manufactured using polysilicon and leads to the emitter contact at the top. Near the edge of the SiGe HBT, boron ions are injected by ion implantation process to form a P-type isolation wall and export the substrate.

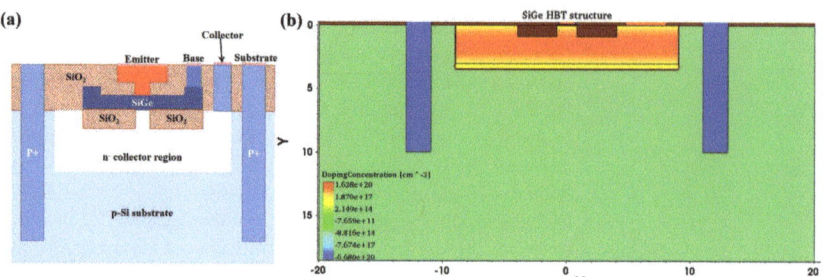

Figure 1. Schematic diagram (**a**) and the internal structure simulation profile (**b**) of the SiGe HBT.

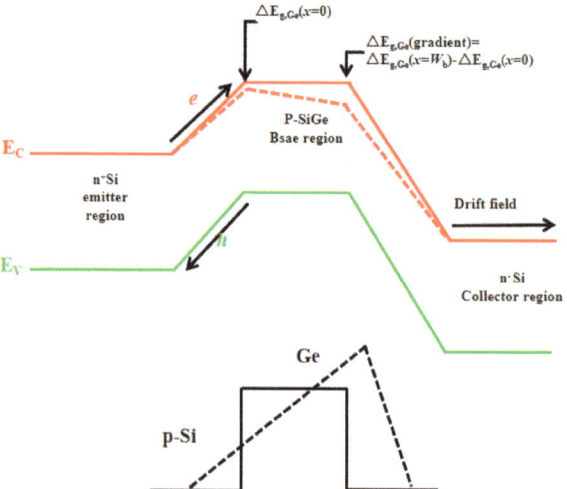

Figure 2. Schematic diagram of the band structure of the SiGe HBT.

Philips unified mobility model, SRH recombination model, Auger recombination model, velocity saturation model and band-gap narrowing model were used as physical models in TCAD simulation. The majority and minority carrier mobility for bipolar transistor can be accurately simulated by the Philips unified mobility model. The high concentrations of electron and hole in the SiGe HBT can be described by the SRH and Auger recombination models. The velocity saturation model is used due to the presence of high carrier density gradient. The band-gap narrowing model is used because germanium doping will cause gradual change in the band structure.

When heavy ions are striking on the SiGe HBT, a large number of electron–hole pairs are generated by ionization along the ions track, which distorts the potential in the depletion layer and forms a funnel potential toward the substrate. Under the action of the funnel electric field and concentration gradient, carriers are rapidly collected by each terminal through drift and diffusion, and such a large amount of charge collection will cause changes in the current of each terminal in a short time (a few nanoseconds). In TCAD simulation, the calculation of carrier generation rate caused by heavy ions irradiation is the key. The number of electron–hole pairs before the initial heavy ions striking is added to the carrier density at the beginning of the simulation, and the carrier generation rate after heavy ion incident is given by the following formula:

$$G(l, \omega, t) = G_{LET}(l) R(\omega, l) T(t) \tag{2}$$

where $R(\omega, l)$ and $T(t)$ represent the carrier generation rate as a function of space and time, respectively. Changing carrier with time is Gaussian distribution, that is, $T(t)$ can be expressed by Equation (3):

$$T(t) = \frac{2 \cdot exp(-(\frac{t-t_0}{\sqrt{2} \cdot S_{hi}})^2)}{\sqrt{2} \cdot S_{hi} \sqrt{\pi}(1 + erf(\frac{t_0}{\sqrt{2} \cdot S_{hi}}))} \quad (3)$$

where t_0 is the moment when heavy ions enter the SiGe HBT, S_{hi} is the Gaussian characteristic value. Changing carrier with space can follow either exponential function or Gaussian function. Gaussian distribution is used in this paper, and Equation (4) is a function representation of $R(\omega, l)$.

$$R(w, l) = exp(-(\frac{\omega}{\omega_t(l)})^2) \quad (4)$$

where ω is the vertical distance to the ion track, and $\omega_t(l)$ is the characteristic length. The striking locations of heavy ions on the SiGe HBT are, respectively, set in the center of emitter, base, collector and substrate.

Figure 3 shows the Gummel characteristic curve of the SiGe HBT obtained by TCAD simulation in this paper; it is in good agreement with the simulation results obtained by others and the tested values by the semiconductor parameter tester KETHLEY4200. This indicates that the SiGe HBT structure model constructed in this work can accurately reflect the actual performance of the SiGe HBT. In this paper, the substrate is biased at -5 V and all other terminals are biased at 0 V to achieve the worst bias, except where it is specifically stated, for example, in the study of the effects of bias state on the single event effects of the SiGe HBT induced by heavy ions.

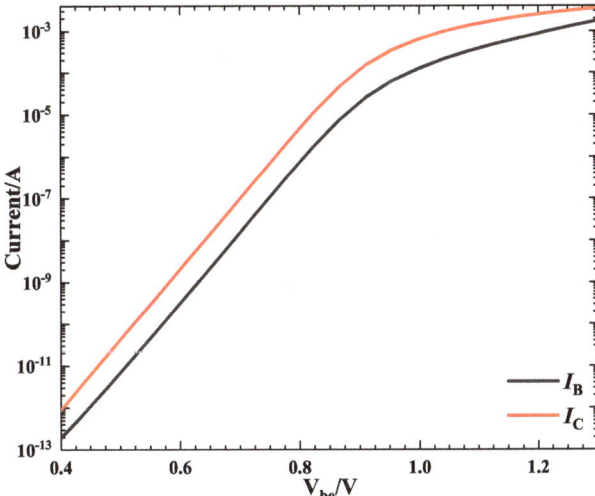

Figure 3. Gummel characteristic curve of the SiGe HBT.

3. Results and Discussion

3.1. Striking Location

The single event effects' sensitive area of the SiGe HBT can be obtained by analyzing the device structure and the simulation results of heavy ions striking at different locations. The charge collection quantity at each terminal is closely related to the striking location of the heavy ions. Figures 4 and 5 show the current transients and charge collection quantity of each terminal introduced by heavy ions striking at the center of the emitter, base, collector and substrate, respectively. When heavy ions are striking at the center of the emitter and base, electrons are collected by the emitter and collector, holes are collected by the base and the current of the substrate is not changed significantly by the heavy ions striking. When heavy ions are striking at the center of the collector and substrate, the electrons are collected by the collector, the holes are collected by the base and substrate and the current of the base and emitter is not changed significantly. The current transient peak value of each terminal is strongly dependent on the heavy ions' striking location. The current transient peak value caused by the irradiation of heavy ions on the emitter and base is much higher than that caused by the irradiation of heavy ions on the collector and substrate. When heavy ions are irradiating the collector, the current transient peak value is the minimum, which indicates that the emitter and base are the sensitive area of the single event transient. Electron–hole pairs are generated by ionization when the heavy ions are incident on the sensitive area of the SiGe HBT, the electrons are collected at the high potential region and the holes flow in the direction of decreasing potential. When heavy ions are striking the center of different terminals, different carrier transport modes result in the different response of the single event transient. When heavy ions irradiate the emitter and base of the SiGe HBT, the total charge quantity collected by the emitter and collector is equal to the charge quantity collected by the base. While heavy ions irradiate the collector and substrate of the SiGe HBT, the charge collected by the collector is equal to the charge quantity collected by the substrate. All of the results show that the emitter is the sensitive area of the SiGe HBT.

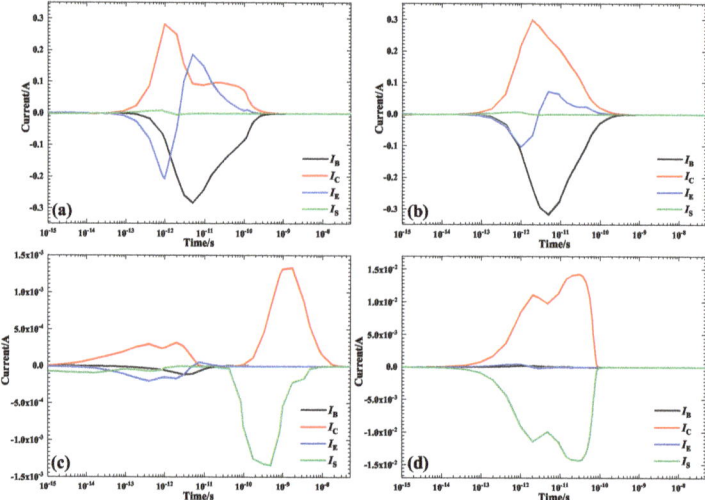

Figure 4. The current change in each terminal with time as heavy ions irradiate the center of emitter (**a**), base (**b**), collector (**c**) and substrate (**d**) of the SiGe HBT.

3.2. Incident Angle

The effective LET value of the heavy ions incident on the SiGe HBT surface changes with the incident angle. Figure 6 shows the current transients and collected charge of the base as the heavy ions irradiate the emitter and base of the SiGe HBT with different incident angles. With the increase in the incident angle, the current transient peak value increases first and then decreases, and the current transient pulse width does not change obviously. The collected charge quantity of the base changed with the incident angle, the collected charge quantity of the base is the maximum when the heavy ions irradiate the emitter with an angle of 60°, while the collected charge quantity of the base is the maximum when the heavy ions irradiate the base with an angle of 90°. As shown in Figure 1, the incident angle increases in a clockwise direction. The base current changes with the incident angle caused by the change in the distance from the carrier to the base, and it decreases first and then increases with the incident angle when the heavy ions irradiate the emitter of the SiGe HBT.

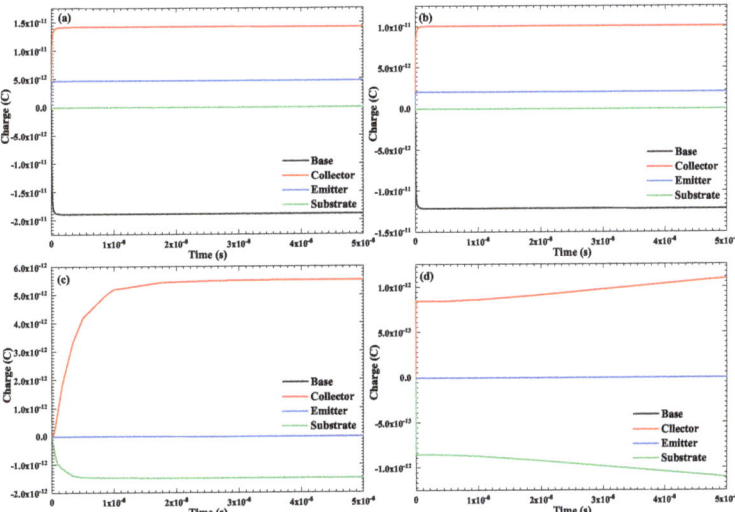

Figure 5. The collected charge change in each terminal with time as heavy ions irradiates the center of emitter (**a**), base (**b**), collector (**c**) and substrate (**d**) of the SiGe HBT.

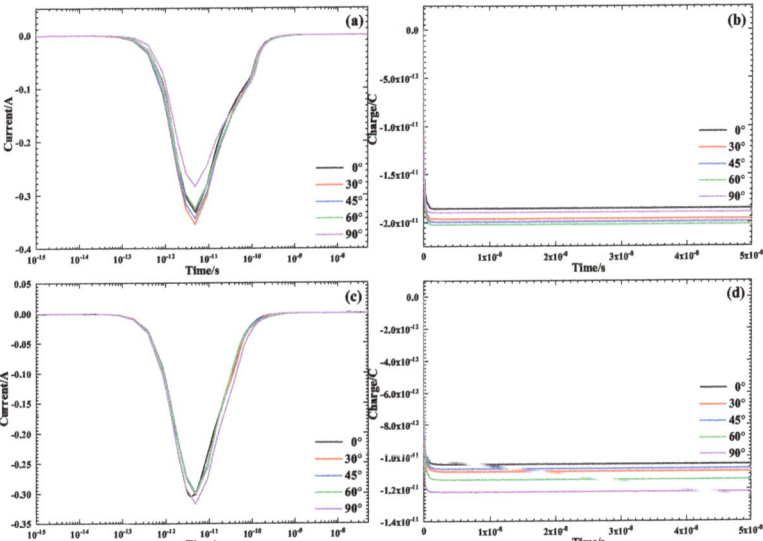

Figure 6. The current and collected charge change in the base with time as heavy ions irradiates the center of emitter (**a,b**) and base (**c,d**) of the SiGe HBT.

Figure 7 shows the current transients pulse and collected charge quantity of the collector as the heavy ions irradiate the collector and substrate of the SiGe HBT with different incident angles. When the heavy ions irradiate the collector, there are two transient peaks in the current of the collector caused by the drift and diffusion of electrons. The current transient peak value and collected charge quantity change with the incident angle. When the heavy ions irradiate the substrate, the current transient peak and pulse width of the collector are changed by the incident angle of the heavy ions. The current transient peak value and collected charge quantity of the collector caused by heavy ions irradiation with an angle of 0° is the lowest, while the values caused by heavy ions irradiation with an angle of 30° is the highest.

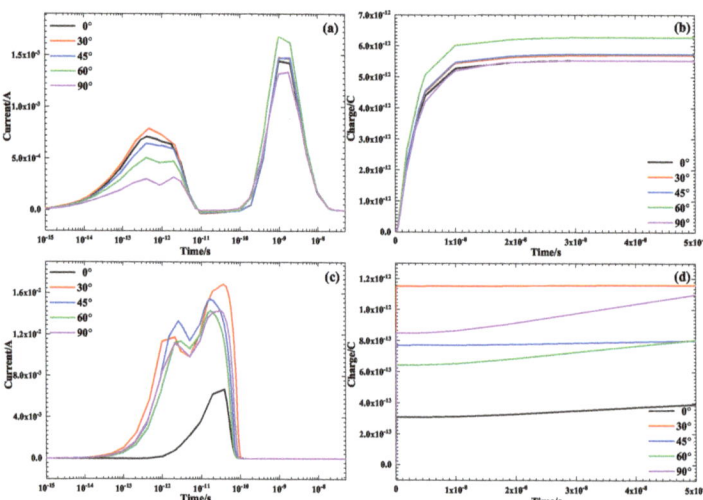

Figure 7. The current and collected charge change in the collector with time as heavy ions irradiate the center of collector (**a**,**b**) and substrate (**c**,**d**) of the SiGe HBT.

3.3. LET Values

The number of electron–hole pairs in the SiGe HBT produced by heavy ions irradiation with different LET values is different, and the number of the electrons and holes in SiGe HBT increases with the heavy ions' LET value. In the TCAD simulation, the LET value of the heavy ions is represented by the charge deposition quantity, and the charge deposition quantity of 0.1 pC/μm corresponds to the LET value of the heavy ions of 10 MeV·cm^2/mg. When studying the effect of the heavy ions' LET values on the single event effects of the SiGe HBT, the charge quantity deposited by the heavy ions in the SiGe HBT is set to vary in the range 0.1 to 1.5 pC/μm, and the corresponding heavy ion LET values vary in the range 10 to 150 MeV·cm^2/mg. Figures 8 and 9 show the change in the current and collected charge with time when the heavy ions with different LET values vertically irradiate the different terminals of the SiGe HBT. The current transient peak value, pulse width and collected charge increase with the heavy ions' LET values.

3.4. Projected Range

Different heavy ions with the same LET value have different projected ranges in the SiGe HBT, and lighter ions will have a longer projected range in the SiGe HBT. Whether the LET value can be used to characterize the single event effects induced by different heavy ions irradiation depends on the microstructure and the inner material of the device. Figures 10 and 11 show the current transients and collected charge quantity of the base and collector under the irradiation of heavy ions with the same LET value and different projected ranges, respectively. The projected range of heavy ions is set in the range of 1 to 10 μm. When heavy ions irradiate the emitter, the base current peak value and collected charge quantity increase with the projected range. When heavy ions irradiate the base, the base current peak value decreases with the projected range, but the collected charge is increased by the projected range because the transient pulse width increases with the projected range. When the heavy ions irradiate the collector and substrate, the current peak value, pulse width and collected charge quantity of the collector increase with the heavy ions' projected range. Therefore, the species of heavy ions should be considered when studying the single event effects of the SiGe HBT induced by heavy ions irradiation.

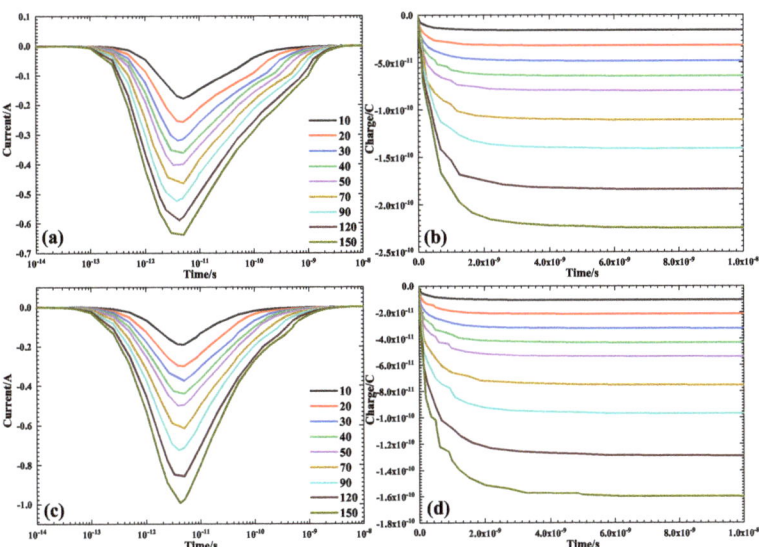

Figure 8. Change in base current and collected charge with time when heavy ions with different LET values vertically irradiate the emitter (**a**,**b**) and base (**c**,**d**) of the SiGe HBT.

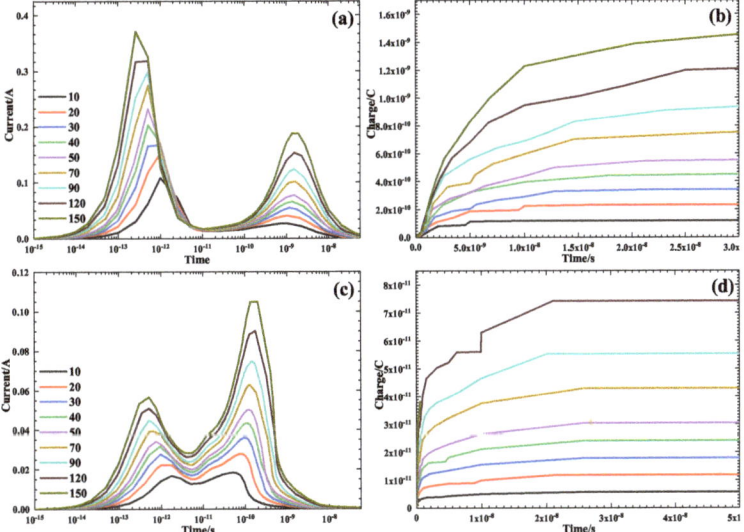

Figure 9. Change in collector current and collected charge with time when heavy ions with different LET values vertically irradiate the collector (**a**,**b**) and substrate (**c**,**d**) of the SiGe HBT.

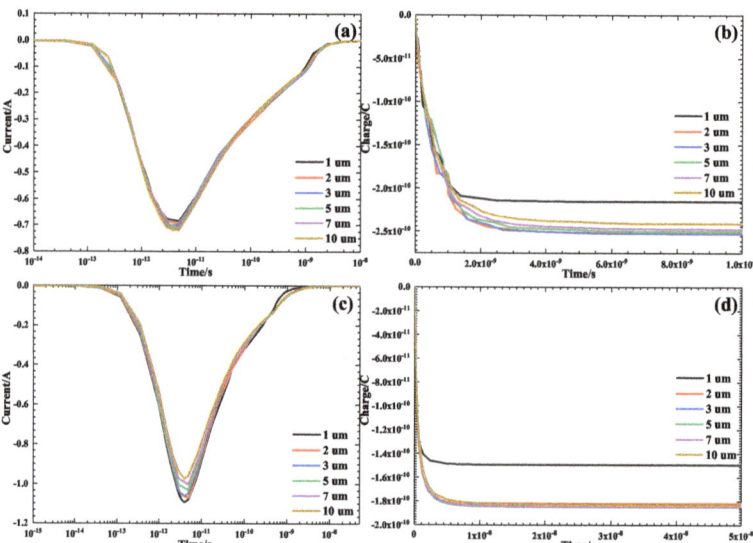

Figure 10. Change in base current and collected charge with time when heavy ions with different projected range vertically irradiate the emitter (**a**,**b**) and base (**c**,**d**) of the SiGe HBT.

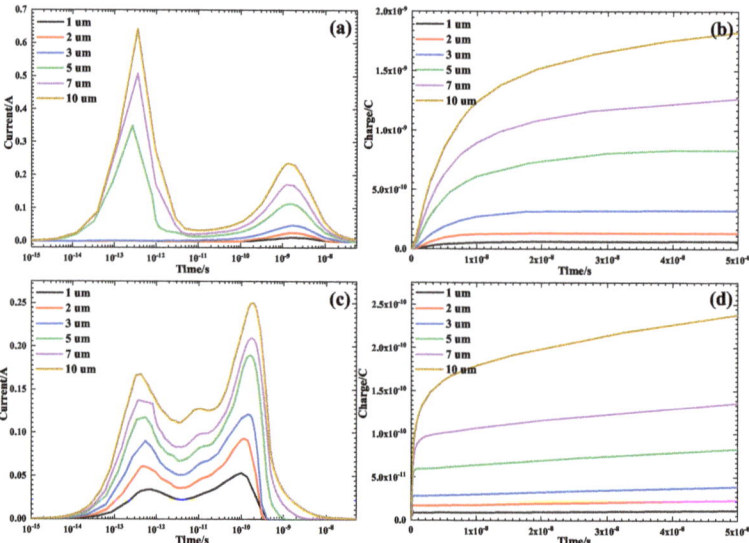

Figure 11. Change in collector current and collected charge with time when heavy ions with different projected range vertically irradiate the collector (**a**,**b**) and substrate (**c**,**d**) of the SiGe HBT.

3.5. Ambient Temperature

Under different temperatures, the mobility of the carrier in the SiGe HBT is different, and the mobility of the carrier decreases with the temperature. Therefore, the ambient temperature during the heavy ions irradiation has a significant influence on the single event effects of the SiGe HBT. Figures 12 and 13 show the current transients and collected charge quantity of the base and collector introduced by the heavy ions irradiation under different ambient temperatures. The current transient peak value and collected charge quantity decrease with the ambient temperature, while the pulse width is not significantly changed by the ambient temperature. We can conclude that increasing the ambient temperature of the SiGe HBT can effectively reduce the single event effects caused by the heavy ions irradiation.

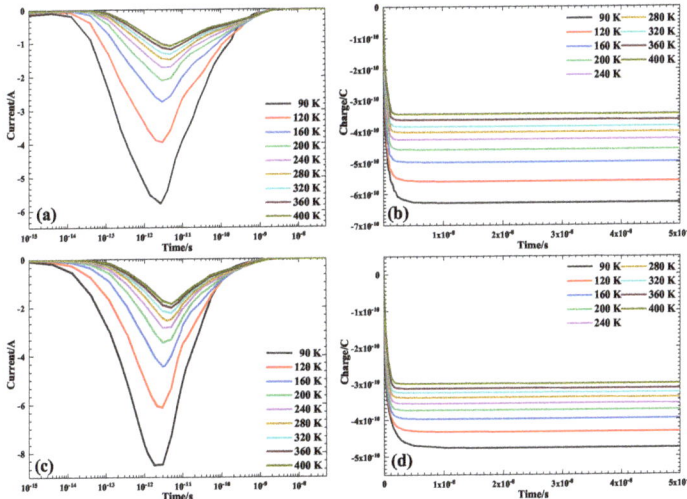

Figure 12. Change in base current and collected charge with time when heavy ions vertically irradiate the emitter (**a**,**b**) and base (**c**,**d**) of the SiGe HBT under different ambient temperatures.

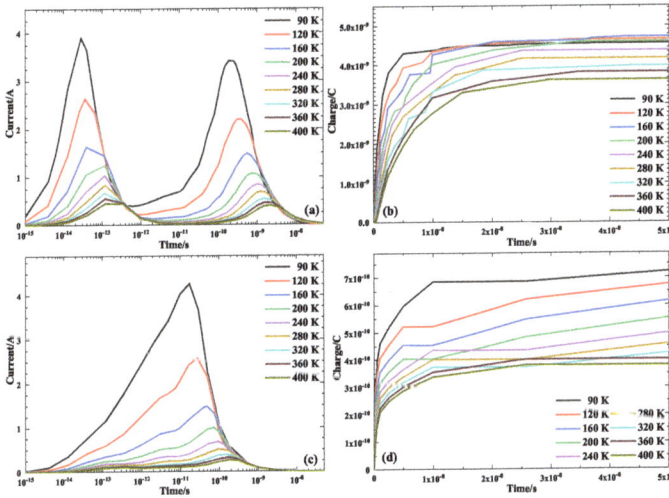

Figure 13. Change in collector current and collected charge with time when heavy ions vertically irradiate the collector (**a**,**b**) and substrate (**c**,**d**) of the SiGe HBT under different ambient temperatures.

3.6. Bias State

Previous studies have shown that the inverse bias of the large area C/S junction enhances the funnel effect, making the SiGe HBT sensitive to the single event effects. To compare the effects of different bias states on the SiGe HBT single event effects and considering the practical application in circuits, the positive bias (base = +1.2 V, collector = +3 V), off bias (emitter = +3 V, collector = +3 V), collector positive bias (collector = +3 V) and substrate inverse bias (substrate = −3 V), four kinds of work bias states that form the inverse bias C/S junction, were selected in this work. Figure 14 shows the current change in each terminal with time when heavy ions vertically irradiate the emitter of the SiGe

HBT under different bias states. When the SiGe HBT is in the substrate inverse bias state, electrons are collected by the collector and emitter, and holes are collected by the base and substrate. When the SiGe HBT is in the positive bias state, electrons are collected by the collector and base, and holes are collected by the emitter. When the SiGe HBT is in the off bias state, electrons are collected by the collector and emitter, and holes are collected by the base. When the SiGe HBT is in the substrate inverse bias state, electrons are collected by the collector and substrate, and holes are collected by the base and emitter. The current transient peak value when the SiGe HBT is in the collector positive bias state is the highest, followed by the SiGe HBT in the off bias and positive bias states, and the SiGe HBT in the substrate inverse bias state has the lowest current transient peak value. Therefore, the single event effect of the SiGe HBT is changed by the bias state during the heavy ions irradiation.

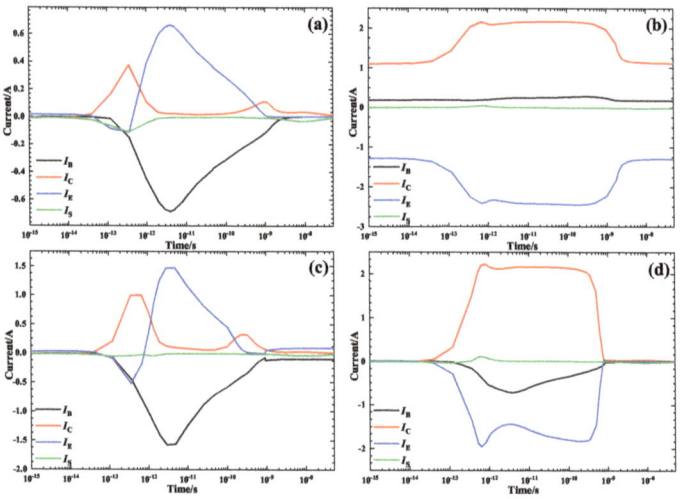

Figure 14. Change in current with time when heavy ions vertically irradiate the emitter of the SiGe HBT under the substrate inverse bias (**a**), positive bias (**b**), off bias (**c**) and collector positive bias (**d**) states.

Figure 15 shows the collected charge quantity change in each terminal with time when the heavy ions vertically irradiate the emitter of the SiGe HBT under different bias states. When the SiGe HBT is in the substrate inverse bias state, the collected charge quantity of the collector and substrate increases with time, while the collected charge quantity of the base and emitter rapidly reaches saturation after heavy ions irradiation, and the charge quantity of the electrons collected at the collector and emitter is equal to the charge quantity of the holes collected at the base and substrate. When the SiGe HBT is in the positive bias state, the collected charge of the base, collector and emitter increases with time, while the collected charge quantity of the substrate is not changed with time, and the charge quantity of the electrons collected at the base and collector is equal to the charge quantity of the holes collected at the emitter. When the SiGe HBT is in the off bias state, the collected charge quantity of the base, collector, emitter and substrate increases with time, and the charge quantity of the electrons collected at the collector and emitter is equal to the charge quantity of the holes collected at the base and substrate. When the SiGe HBT is in the collector positive bias state, the collected charge quantity of the collector and substrate increases slowly with time, while the collected charge quantity of the base and emitter rapidly reaches saturation, and the charge quantity of the electrons collected at the collector is equal to the charge quantity of the holes collected at the base, emitter and substrate. In the 5×10^{-8} s after the heavy ions irradiation, the collected charge quantity of the terminal of the SiGe HBT in the positive bias state is the highest, followed by the SiGe HBT in the off bias and collector positive bias states, and the least quantity of charge is collected at the terminal of the SiGe HBT in the substrate inverse bias state. Therefore, the quantity of the collected charge at each terminal of the SiGe HBT after the heavy ions irradiation depends on the bias state of the device.

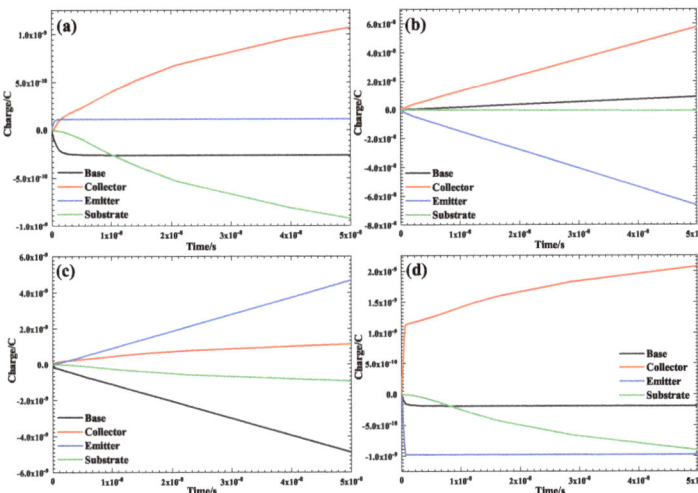

Figure 15. Change in collected charge with time when heavy ions vertically irradiate the emitter of the SiGe HBT under the substrate inverse bias (**a**), positive bias (**b**), off bias (**c**) and collector positive bias (**d**) states.

Figures A1–A3 show the change in the current with time when the heavy ions irradiate the base, collector and substrate of the SiGe HBT under the different bias states, respectively. When the heavy ions irradiate the base, the electrons are collected by the collector and emitter, and the holes are collected by the base and substrate under the substrate inverse bias and off bias. The collector and base collect the electrons under the positive bias state, while the emitter collects holes. Under the collector positive bias state, the electrons are collected by the collector and substrate, while the holes are collected by the base and emitter. The bias state not only changes the peak value and pulse shape of the current transients at each terminal but also changes the type of carriers collected at each terminal. When heavy ions irradiate the collector and substrate of the SiGe HBT, there are two peaks in the current of the collector and substrate, except for the positive bias state, and the peak values are not changed significantly by the bias states.

Figures A4–A6 show the collected charge quantity of each terminal when the heavy ions irradiate the base, collector and substrate under the positive bias, off bias and collector positive bias states, respectively. When the heavy ions irradiate the different locations of the SiGe HBT, the quantity of the collected charge at each terminal is changed by the bias state. The collected charge quantity increases linearly with time under the positive bias state and slowly with time under the other bias states. Therefore, the collected charge quantity at each terminal of the SiGe HBT is affected by both the irradiation position of the heavy ions and the bias state during the irradiation.

4. Conclusions

To understand the microphysical mechanism of single event effects in the SiGe HBT induced by heavy ion irradiation, the effects of the heavy ion striking location, incident angle, LET value, projected range, ambient temperature and bias state on the single event effects were investigated in this paper by using a TCAD simulation. The results show that the current transient peak value and collected carrier type of each terminal was changed by these factors. The current transient peak value increases with the LET and projected range of the heavy ions and decreases with the ambient temperature. The single event effects of the SiGe HBT are not only affected by the heavy ion irradiation parameters such as the incident angle, LET value and projected range, but they are also affected by the striking location, ambient temperature and bias state. The peak value of the current transient peak value increases with the projected range when the emitter, collector and substrate of the SiGe HBT are irradiated by heavy ions with the same LET value, while the current transient peak value decreases slowly with the projected range when the heavy ions irradiate the base, which indicates that the species of heavy ions should be taken into account when carrying out research on the single event effects of the SiGe HBT induced by heavy ions irradiation. The main reason for the change in the single event effects is the change in the carrier mobility and transport mode under different factors. According to the simulation results, we can conclude that the single-particle effect caused by heavy ion irradiation can be weakened by increasing the pseudo electrode to carry away the electron–hole pairs generated by heavy ions irradiation, increasing the isolation area of the insulating materials to prevent the electron–hole pairs being collected by the terminals, increasing the ambient temperature to reduce the mobility of carriers, and other methods.

Author Contributions: Conceptualization, Z.Z. and G.G.; methodology, H.S. and J.L.; validation, Q.C.; data analysis, F.L. and S.Z.; resources, G.G. and X.O.; writing—original draft preparation, Z.Z.; writing—review and editing, Z.Z. and Q.C. All authors have read and agreed to the published version of the manuscript.

Funding: This research received no external funding.

Data Availability Statement: The data used to support the findings of this study are available from the corresponding author upon request.

Conflicts of Interest: The authors declare no conflicts of interest.

Abbreviations

The following abbreviations are used in this manuscript:

TCAD Technology Computer-Aided Design
HBT   Heterojunction Bipolar Transistor
LET   Linear Energy Transfer

Appendix A

The following figures show the change in current and collected charge of each terminal with time when heavy ions irradiate the base, collector and substrate of the SiGe HBT under the different bias states, respectively.

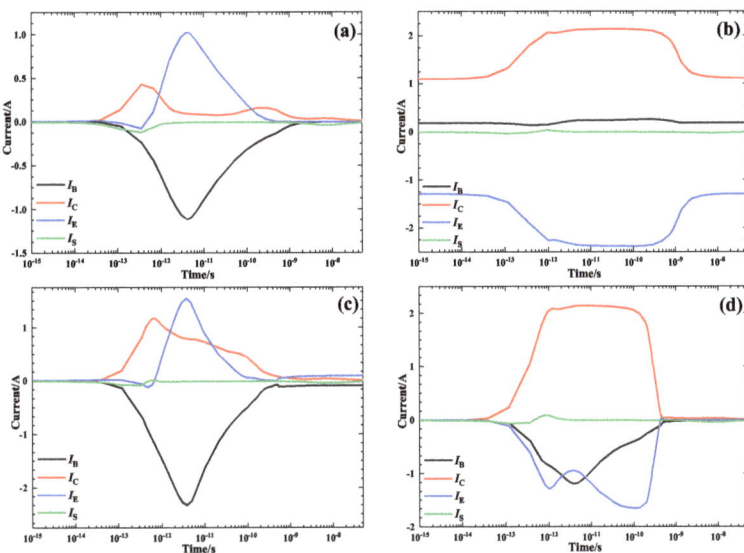

Figure A1. Change in current with time when heavy ions vertically irradiate the base of the SiGe HBT under the substrate inverse bias (**a**), positive bias (**b**), off bias (**c**) and collector positive bias (**d**) states.

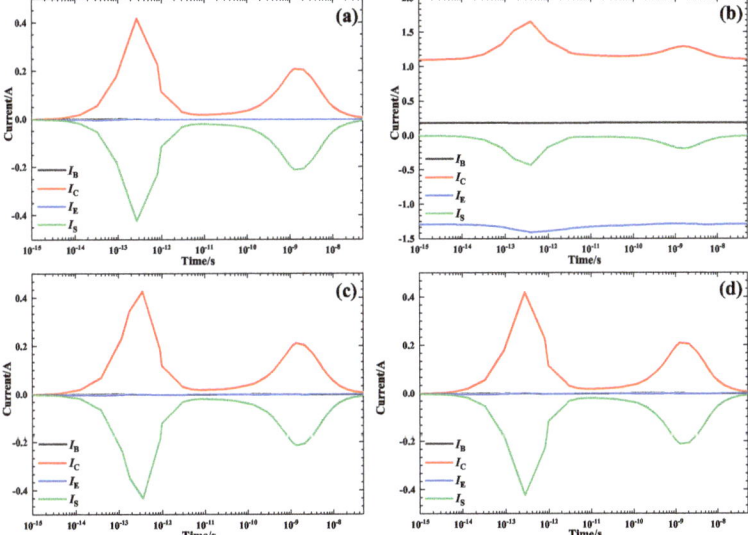

Figure A2. Change in current with time when heavy ions vertically irradiate the collector of the SiGe HBT under the substrate inverse bias (**a**), positive bias (**b**), off bias (**c**) and collector positive bias (**d**) states.

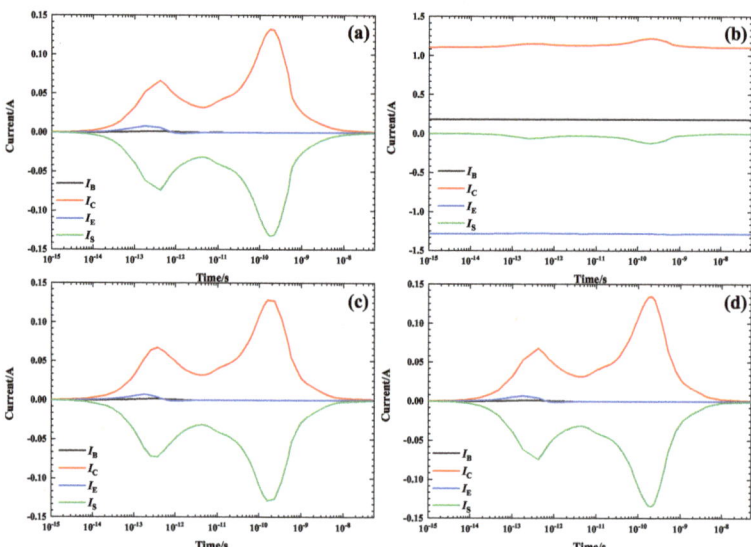

Figure A3. Change in current with time when heavy ions vertically irradiate the substrate of the SiGe HBT under the substrate inverse bias (**a**), positive bias (**b**), off bias (**c**) and collector positive bias (**d**) states.

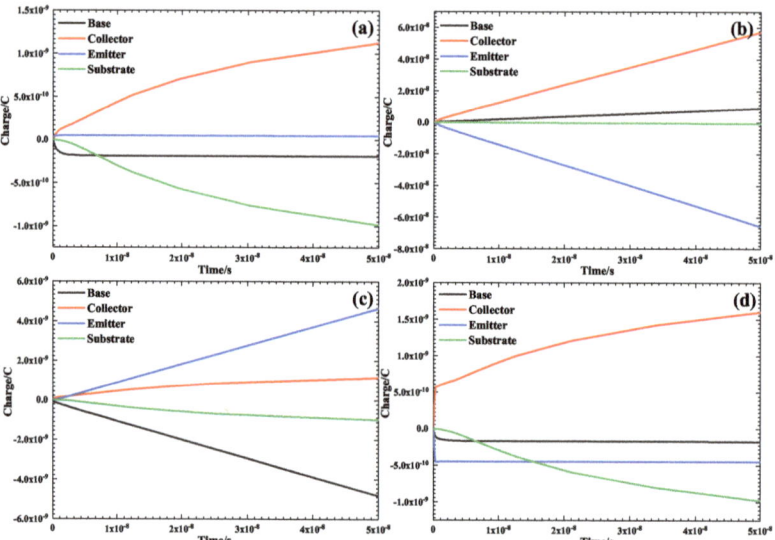

Figure A4. Change in collected charge with time when heavy ions vertically irradiate the emitter of the SiGe HBT under the substrate inverse bias (**a**), positive bias (**b**), off bias (**c**) and collector positive bias (**d**) states.

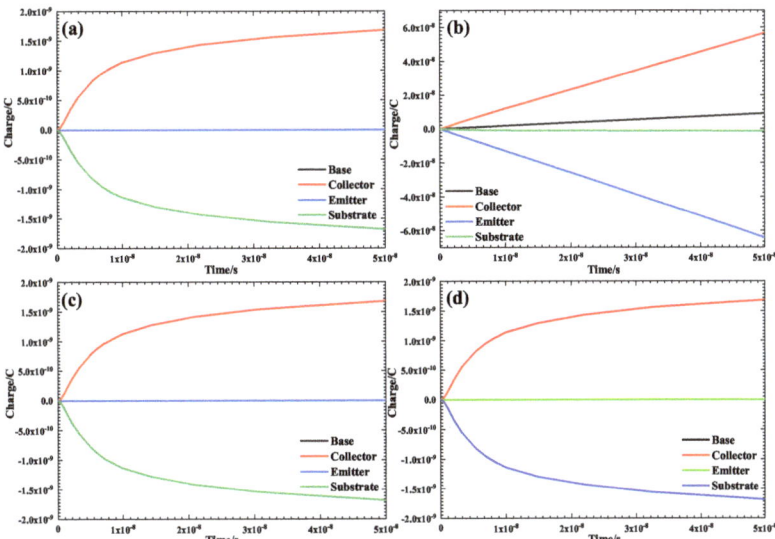

Figure A5. Change in collected charge with time when heavy ions vertically irradiate the emitter of the SiGe HBT under the substrate inverse bias (**a**), positive bias (**b**), off bias (**c**) and collector positive bias (**d**) states.

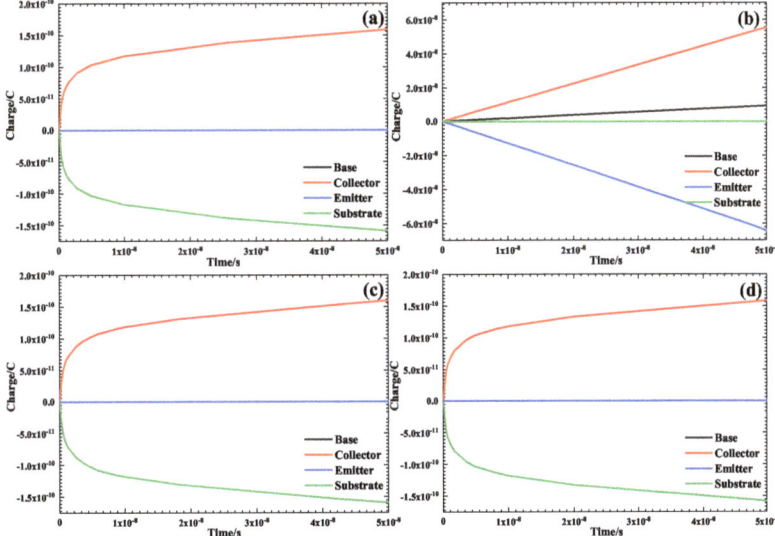

Figure A6. Change in collected charge with time when heavy ions vertically irradiate the emitter of the SiGe HBT under the substrate inverse bias (**a**), positive bias (**b**), off bias (**c**) and collector positive bias (**d**) states.

References

1. Li, X.; Yang, J.; Fleetwood, D.M.; Liu, C.; Wei, Y.; Barnaby, H.; Galloway, K. Hydrogen soaking, displacement damage effects, and charge yield in gated lateral bipolar junction transistors. *IEEE Trans. Nucl. Sci.* **2018**, *65*, 1271–1276. [CrossRef]
2. Witulski, A.F.; Ball, D.R.; Galloway, K.F.; Javanainen, A.; Lauenstein, J.M.; Sternberg, A.L.; Schrimpf, R.D. Single-event burnout mechanisms in SiC power MOSFETs. *IEEE Trans. Nucl. Sci.* **2018**, *65*, 1951–1955. [CrossRef]
3. Petrosyants, K.O.; Kozhukhov, M.V. Physical TCAD model for proton radiation effects in SiGe HBTs. *IEEE Trans. Nucl. Sci.* **2016**, *63*, 2016–2021. [CrossRef]

4. Dodd, P.E. Physics-based simulation of single-event effects. *IEEE Trans. Device Mater. Reliab.* **2005**, *5*, 343–357. [CrossRef]
5. Pickel, J.C. Single-event effects rate prediction. *IEEE Trans. Nucl. Sci.* **1996**, *43*, 483–495. [CrossRef]
6. Sun, H.; Guo, G.; Sun, R.; Zhao, W.; Zhang, F.; Liu, J.; Zhang, Z.; Chen, Y.; Zhao, Y. Study on Single Event Upsets in a 28 nm Technology Static Random Access Memory Device Based on Micro-Beam Irradiation. *Electronics* **2022**, *11*, 3413. [CrossRef]
7. Pease, R.L. Total ionizing dose effects in bipolar devices and circuits. *IEEE Trans. Nucl. Sci.* **2003**, *50*, 539–551. [CrossRef]
8. Fleetwood, D.M. Total ionizing dose effects in MOS and low-dose-rate-sensitive linear-bipolar devices. *IEEE Trans. Nucl. Sci.* **2013**, *60*, 1706–1730. [CrossRef]
9. Bernard, M.; Dusseau, L.; Buchner, S.; McMorrow, D.; Ecoffet, R.; Boch, J.; Vaillé, J.R.; Schrimpf, R.; LaBel, K. Impact of total ionizing dose on the analog single event transient sensitivity of a linear bipolar integrated circuit. *IEEE Trans. Nucl. Sci.* **2007**, *54*, 2534–2540. [CrossRef]
10. Srour, J.; Marshall, C.J.; Marshall, P.W. Review of displacement damage effects in silicon devices. *IEEE Trans. Nucl. Sci.* **2003**, *50*, 653–670. [CrossRef]
11. Srour, J.; Palko, J. Displacement damage effects in irradiated semiconductor devices. *IEEE Trans. Nucl. Sci.* **2013**, *60*, 1740–1766. [CrossRef]
12. Zhang, J.; Guo, H.; Pan, X.; Guo, Q.; Zhang, F.; Feng, J.; Wang, X.; Wei, Y.; Wu, X. Synergistic effect of total ionizing dose on single event effect induced by pulsed laser microbeam on SiGe heterojunction bipolar transistor. *Chin. Phys. B* **2018**, *27*, 108501. [CrossRef]
13. Ya'Acob, N.; Zainudin, A.; Magdugal, R.; Naim, N.F. Mitigation of space radiation effects on satellites at Low Earth Orbit (LEO). In Proceedings of the IEEE International Conference on Control System, Computing and Engineering (ICCSCE), Penang, Malaysia, 25–27 November 2017.
14. Zhang, J.X.; Guo, H.X.; Guo, Q.; Wen, L.; Cui, J.W.; Xi, S.B.; Wang, X.; Deng, W. 3D simulation of heavy ion induced charge collection of single event effects in SiGe heterojunction bipolar transistor. *Acta Phys. Sin.* **2013**, *62*, 221–229.
15. Zhang, J.X.; He, C.H.; Guo, H.X.; Du, T.; Xin, W. Three-dimensional simulation study of bias effect on single event effects of SiGe heterojunction bipolar transistor. *Acta Phys. Sin.* **2014**, *63*, 1071–1074.
16. Pruvost, S.; Delcourt, S.; Telliez, I.; Laurens, M.; Bourzgui, N.E.; Danneville, F.; Monroy, A.; Dambrine, G. Microwave and noise performance of SiGe BiCMOS HBT under cryogenic temperatures. *IEEE Electron Device Lett.* **2005**, *26*, 105–108. [CrossRef]
17. Cressler, J.D. SiGe BiCMOS Technology: An IC Design Platform for Extreme Environment Electronics Applications. In Proceedings of the IEEE International Reliability Physics Symposium, Phoenix, AZ, USA, 15–19 April 2007.
18. Inanlou, F.; Lourenco, N.E.; Fleetwood, Z.E.; Song, I.; Cressler, J.D. Impact of Total Ionizing Dose on a 4th Generation, 90 nm SiGe HBT Gaussian Pulse Generator. *IEEE Trans. Nucl. Sci.* **2014**, *61*, 3050–3054. [CrossRef]
19. Hegde, V.N.; Pradeep, T.M.; Pushpa, N.; Praveen, K.C.; Bhushan, K.G.; Cressler, J.D.; Prakash, A. A Comparison of Electron, Proton and Gamma Irradiation Effects on the I-V Characteristics of 200 GHz SiGe HBTs. *IEEE Trans. Device Mater. Reliab.* **2018**, *18*, 592–598. [CrossRef]
20. Metcalfe, J.; Dorfan, D.E.; Grillo, A.A.; Jones, A.; Lucia, D.; Martinez-McKinney, F.; Mendoza, M.; Rogers, M.; Sadrozinski, H.F.W.; Seiden, A.; et al. Evaluation of the Radiation Tolerance of SiGe Heterojunction Bipolar Transistors Under 24-GeV Proton Exposure. *IEEE Trans. Nucl. Sci.* **2006**, *53*, 3889–3893. [CrossRef]
21. Reed, R.A.; Marshall, P.W.; Pickel, J.C.; Carts, M.A.; Fodness, B.; Niu, G.; Fritz, K.; Vizkelethy, G.; Dodd, P.E.; Irwin, T.; et al. Heavy-ion broad-beam and microprobe studies of single-event upsets in 0.20-/spl mu/m SiGe heterojunction bipolar transistors and circuits. *IEEE Trans. Nucl. Sci.* **2003**, *50*, 2184–2190. [CrossRef]
22. Pellish, J.A.; Reed, R.A.; Schrimpf, R.D.; Alles, M.L.; Varadharajaperumal, M.; Niu, G.; Sutton, A.K.; Diestelhorst, R.M.; Espinel, G.; Krithivasan, R.; et al. Substrate engineering concepts to mitigate charge collection in deep trench isolation technologies. *IEEE Trans. Nucl. Sci.* **2006**, *53*, 3298–3305. [CrossRef]
23. Jiang, N.; Ma, Z.; Ma, P.; Racanelli, M. Impact of proton radiation on the large-signal power performance of SiGe power HBTs. *IEEE Trans. Nucl. Sci.* **2006**, *53*, 2361–2366. [CrossRef]
24. Lourenco, N.E.; Fleetwood, Z.E.; Ildefonso, A.; Wachter, M.T.; Roche, N.J.H.; Khachatrian, A.; McMorrow, D.; Buchner, S.P.; Warner, J.H.; Itsuji, H.; et al. The impact of technology scaling on the single-event transient response of SiGe HBTs. *IEEE Trans. Nucl. Sci.* **2016**, *64*, 406–414. [CrossRef]
25. Cressler, J.D. Radiation effects in SiGe technology. *IEEE Trans. Nucl. Sci.* **2013**, *60*, 1992–2014. [CrossRef]
26. Li, P.; Guo, H.X.; Guo, Q.; Zhang, J.X.; Wei, Y. Laser-induced single event transients in local oxidation of silicon and deep trench isolation silicon-germanium heterojunction bipolar transistors. *Chin. Phys. Lett.* **2015**, *32*, 088505. [CrossRef]
27. Li, P.; Guo, H.X.; Guo, Q.; Zhang, J.X.; Xiao, Y.; Wei, Y.; Cui, J.W.; Wen, L.; Liu, M.H.; Wang, X. Single-event response of the SiGe HBT in TCAD simulations and laser microbeam experiment. *Chin. Phys. B* **2015**, *24*, 088502. [CrossRef]
28. Zhang, J.; Guo, H.; Zhang, F.; He, C.; Li, P.; Yan, Y.; Wang, H.; Zhang, L. Heavy ion micro-beam study of single-event transient (SET) in SiGe heterojunction bipolar transistor. *Sci. China Inf. Sci.* **2017**, *60*, 1–3. [CrossRef]

29. Zhang, J.; Guo, H.; Wen, L.; Guo, Q.; Cui, J.; Wang, X.; Deng, W.; Zhen, Q.; Fan, X.; Xiao, Y. 3-D simulation of angled strike heavy-ion induced charge collection in silicon–germanium heterojunction bipolar transistors. *J. Semicond.* **2014**, *35*, 044003. [CrossRef]
30. Wei, J.; Li, Y.; Yang, W.; He, C.; Li, Y.; Zang, H.; Li, P.; Zhang, J.; Guo, G. Proton-induced current transient in SiGe HBT and charge collection model based on Monte Carlo simulation. *Sci. China Technol. Sci.* **2020**, *63*, 851–858. [CrossRef]

Disclaimer/Publisher's Note: The statements, opinions and data contained in all publications are solely those of the individual author(s) and contributor(s) and not of MDPI and/or the editor(s). MDPI and/or the editor(s) disclaim responsibility for any injury to people or property resulting from any ideas, methods, instructions or products referred to in the content.

Article

Effect of Trapped Charge Induced by Total Ionizing Dose Radiation on the Top-Gate Carbon Nanotube Field Effect Transistors

Hongyu Ding [1,2], Jiangwei Cui [1,2,*], Qiwen Zheng [1,2,*], Haitao Xu [3,4], Ningfei Gao [3], Mingzhu Xun [1,2], Gang Yu [1,2], Chengfa He [1,2], Yudong Li [1,2] and Qi Guo [1,2]

[1] XinJiang Technical Institute of Physics and Chemistry, Chinese Academy of Sciences, Urumqi 830011, China
[2] School of Electronic, Electrical and Communication Engineering, University of Chinese Academy of Sciences, Beijing 100049, China
[3] Beijing HuaTanYuanXin Electronics Technology Ltd. Co., Beijing 101399, China
[4] Beijing Institute of Carbon-Based Integrated Circuits, Beijing 100195, China
* Correspondence: cuijw@ms.xjb.ac.cn (J.C.); qwzheng@ms.xjb.ac.cn (Q.Z.)

Abstract: The excellent performance and radiation-hardness potential of carbon nanotube (CNT) field effect transistors (CNTFETs) have attracted wide attention. However, top-gate structure CNTFETs, which are often used to make high-performance devices, have not been studied enough. In this paper, the total ionizing dose (TID) effect of the top-gate structure CNTFETs and the influence of the substrate on top-gate during irradiation are studied. The parameter degradation caused by the irradiation- and radiation-damage mechanisms of the top-gate P-type CNTFET were obtained by performing a Co-60 γ-ray irradiation test. The results indicate that the transfer curves of the top-gate P-type CNTFETs shift negatively, the threshold voltage and the transconductance decrease when TID increases, and the subthreshold swing decreases first and then increases with the increase in TID. The back-gate transistor is constructed by using the substrate as a back-gate, and the influence of back-gate bias on the characteristics of the top-gate transistor is tested. We also test the influence of TID irradiation on the characteristics of back-gate transistors, and reveal the effect of trapped charge introduced by radiation on the characteristics of top-gate transistors. In addition, the CNTFETs that we used have obvious hysteresis characteristics. After irradiation, the radiation-induced trapped charges generated in oxide and the OH groups generated by ionization of the CNT adsorbates aggravate the hysteresis characteristics of CNTFET, and the hysteresis window increases with the increase in TID.

Keywords: carbon nanotube field effect transistor; total ionizing dose; radiation effect; trapped charge

1. Introduction

Metamaterial is an artificial material that does not exist in nature. It exhibits material properties not seen in nature, such as permittivity, a bulk modulus or refractive index, which surpasses substances found in nature [1]. In recent years, due to their unique properties, metamaterials have been applied to epsilon-negative metamaterial (ENM), biosensors, metamaterial platforms, and other aspects [2–6], which have developed rapidly in various fields and attracted wide attention.

The development of silicon-based semiconductor technology follows Moore's law, i.e., the physical size of transistors approaches the limit, and performance hits the bottleneck. Semiconductor devices have entered the post-Moore era, more noticeably than in the Moore era, as new materials and new technology have emerged. Carbon nanotube (CNT) has attracted attention as a metamaterial with excellent properties. A carbon-based device based on carbon nanotube materials has the characteristic of high energy efficiency, with ease of three-dimensional heterogeneous integration, it is low cost, and so on. It is an

Citation: Ding, H.; Cui, J.; Zheng, Q.; Xu, H.; Gao, N.; Xun, M.; Yu, G.; He, C.; Li, Y.; Guo, Q. Effect of Trapped Charge Induced by Total Ionizing Dose Radiation on the Top-Gate Carbon Nanotube Field Effect Transistors. *Electronics* **2023**, *12*, 1000. https://doi.org/10.3390/electronics12041000

Academic Editor: Paul Leroux

Received: 17 January 2023
Revised: 6 February 2023
Accepted: 12 February 2023
Published: 17 February 2023

Copyright: © 2023 by the authors. Licensee MDPI, Basel, Switzerland. This article is an open access article distributed under the terms and conditions of the Creative Commons Attribution (CC BY) license (https://creativecommons.org/licenses/by/4.0/).

important technical route to continue Moore's law, and it has been considered a promising technology in aerospace and other fields [7–13].

Devices with improved performance and reliability are needed to cope with the radiation effects in space applications. Carbon-based devices show better performance and radiation-hardness than silicon-based devices [14–16]. Carbon-based devices have great potential in space applications, such as space SRAM and radiation-hardness IC. Furthermore, due to its excellent electrical performance and special physical structure, CNTFET has undergone rapid development in flexible electronic devices [17], which can be applied to wearable electronic devices for the human body and space environment detection devices in space stations in the future.

The carbon nanotube field effect transistor (CNTFET) is the basic unit of a carbon-based integrated circuit. The channel region of CNTFET is composed of CNT. CNTFET can be divided into various types according to different processes and structures, including back-gate structure, top-gate structure, ion gel gate structure, etc. In space applications, CNTFET will face radiation damage effects caused by high-energy particles, resulting in the degradation of device performance. The total ionizing dose effect (TID) is one of the most important radiation damage effects.

Researchers have studied the TID effect on CNTFETs with different structures. Cory D. Cress et al. used a Co-60 γ-ray source to conduct a TID test on P-type back-gate CNTFET without a passivity layer [18]. The results indicated that irradiation did not generate significant damage to the CNT channel, and it was believed that the change of device characteristics mainly depended on materials around CNT, interface state, and air environment. Yudan Zhao et al. further investigated the influence of CNT absorbability and CNT arrangement on the TID effect of back-gate CNTFET. They demonstrated that the air molecules adsorbed by CNT in devices without a passivation layer, the fixed charge in materials of the passivation layer, and the improvement of the contact junction in network arrangement were the predominant factors influencing the TID effect of CNTFET [19].

In recent years, Lian-Mao Peng's team researched the radiation effect of ion gel gate CNTFET, and the results indicated that the structure of the ion gel gate could protect the channel of CNT, improved the radiation tolerance effectively [16,20,21], and could be recovered by annealing after irradiation. However, the author also mentioned that the structure suffers from low performance, difficulty in scaling down, and a narrow range of operating temperatures.

Compared with the first two structures, the top-gate device can not only protect the channel of CNT effectively and enhance the radiation tolerance of CNTFET, but can also reduce the size of the device more easily, which can be used for high-performance carbon nanotube devices [22–24]. In 2019, Xinyang Zhao et al. studied the TID effect of a flexible top-gate carbon nanotube thin-film transistor (CNTFT) using Co-60 γ-ray. The research showed that the radiation resistance of flexible CNTFETs was comparable to that of rigid CNTFETs, which were fabricated on the oxide substrate [25]. For the TID of rigid CNTFET, Maguang Zhu et al. conducted research on the TID of SRAM composed of top-gate CNTFET [26]. In addition, they tested TID on each part of the top-gate CNTFETs independently [15]. The results showed that CNTs are a kind of radiation-hardened semiconducting material, and the damage of substrate material and gate oxide material was the main factor of TID. However, it did not reveal the influence process of radiation-induced trapped charge in the substrate material on the characteristics of the top-gate transistor. Up to now, there has been little research on the sensitivity parameters, damage rules, and mechanisms of top-gate CNTFETs, and relevant research is urgently required for the design of CNT-based devices.

In this paper, we researched TID of P-type CNT-network FETs with a top-gate structure. The variation of threshold voltage, transconductance, subthreshold swing, and hysteresis characteristics of the device before and after irradiation by Co-60 γ-ray are analyzed. The influence mechanism of trapped charge introduced by TID irradiation on the characteristics of the top-gate CNTFETs is discussed. Additionally, the influence of back-gate pressure on front-gate transfer characteristic curves (I_{DS}-V_{GS}) and its control ability were tested. Finally, the possible reasons for CNTFETs hysteresis and the relationship of hysteresis characteristics with TID are explained.

2. Device Structures and Experimental Details

The structure of the test samples used in this paper is shown in Figure 1a. The CNTFET substrates are Si and SiO_2. The Si substrate is 500 μm thick, and the SiO_2 substrate is 280 nm thick. The carbon tube material comprises a reticulated film with a density of 15 pieces/um. The device image is shown in Figure 1b. The semiconducting CNT-network films are the channel regions of the devices. CNT films are undoped, and the width/length ratio is 30 μm/4 μm. The Pd layers serve as electrodes. The function of the P-type transistor is obtained according to the difference of work function between Pd and CNT. The contacts between CNTs and Pd electrodes are Ohmic contacts, which was detected by testing the I_{DS}-V_{DS} characteristics of the device. The gate oxide layers consisted of high-k materials Y_2O_3 and HfO_2, which are 10 nm in total. Samples are wafer-level unpackaged devices.

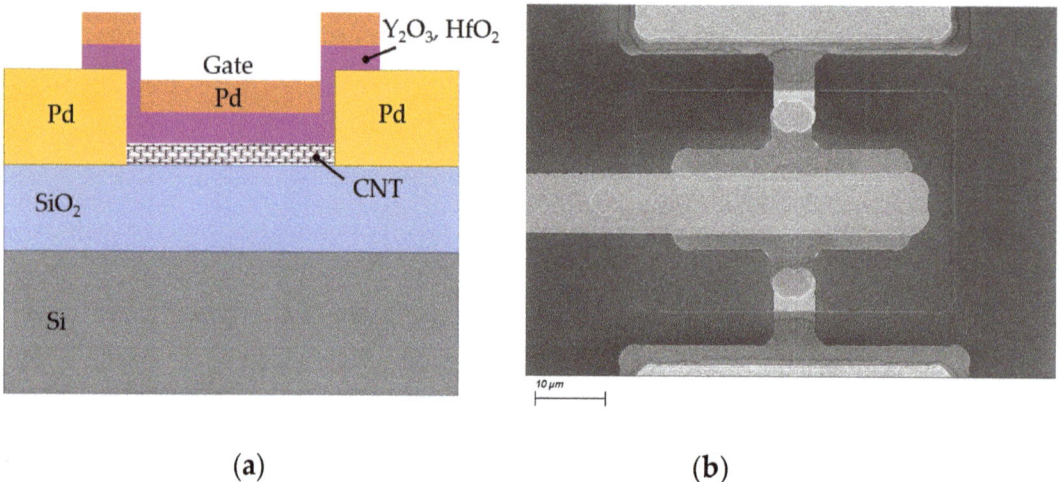

Figure 1. (a) Structure of the top-gate CNTFET; (b) SEM of the top-gate CNTFET.

The irradiation experiments were carried out on a Co-60 γ-ray source in Xinjiang Technical Institute of Physics and Chemistry, Chinese Academy Sciences. The average energy of γ ray is 1.25 MeV, and the devices are irradiated in atmospheric environment. The dose rate is 195.476 rad(Si)/s during irradiation. Dose levels are 200 k rad(Si), 500 k rad(Si), 1 M rad(Si), 2 M rad(Si), and 3 M rad(Si). The total irradiation cumulative time is 255 min 47 s. The accumulation time of each dose level was 17 min 3 s, 25 min 35 s, 42 min 38 s, 85 min 16 s, and 85 min 16 s, respectively. During irradiation, devices are floating. Co-60 γ-ray has a dose-building area when it propagates in the medium [27], and there might be a dose error in the sensitive area because the devices that we used were not packaged. In order to ensure uniformity of the absorbed dose during irradiation, a pre-balanced layer of about 5 mm was added in front of the devices, and a backscatter layer of about 2 cm was added behind the devices to ensure balance of secondary electrons and the absorbed dose in the sensitive area.

Devices were measured using the probe station and the Keithley 4200SCS parameter analyzer at room temperature. The measurement included the I_{DS}-V_{GS} characteristic curves of the linear region and saturated region. The order of scanning may affect the hysteresis characteristics as the gate voltage (V_{GS}) sweeps from 0.5 V to −2 V, with the −0.02 V step first, and from −2 V to 0.5 V. The test information is shown in Table 1. In order to avoid the effect of annealing on the devices, the test time of each dose level is less than 45 min.

Table 1. Measurement information in this paper.

TID	Dose Rate	Test Items
200 k rad(Si) 500 k rad(Si) 1 M rad(Si) 2 M rad(Si) 3 M rad(Si)	195.476 rad(Si)/s	1. $V_{DS} = -0.1$ V, $V_{GS} = (0.5\ V) - (-2\ V)$, bidirectional scanning 2. $V_{DS} = -2$ V, $V_{GS} = (0.5\ V) - (-2\ V)$, bidirectional scanning

3. Results and Discussion

3.1. Effect of TID Radiation on I_{DS}-V_{GS} Characteristics of Top-Gate CNTFET

Figure 2 shows the I_{DS}-V_{GS} curves under different irradiation doses. It can be seen from Figure 2 that the I_{DS}-V_{GS} characteristics shift negatively after TID radiation, and the saturation current in the linear region decreases with the increase in TID.

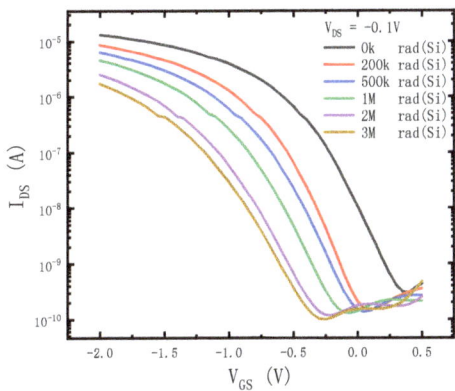

Figure 2. Transfer characteristic curves after irradiation.

The threshold voltage (V_{th}), maximum transconductance ($G_{M\ Max}$), and subthreshold swing (SS) of the devices before and after irradiation were extracted from the I_{DS}-V_{GS} curves, as shown in Figure 3a, 3b and 3c, respectively. As can be seen from the figure, with the increase in accumulated doses of irradiation, V_{th} shifts negatively and $G_{M\ Max}$ decreases monotonically, while the subthreshold swing (SS) first decreases and then increases.

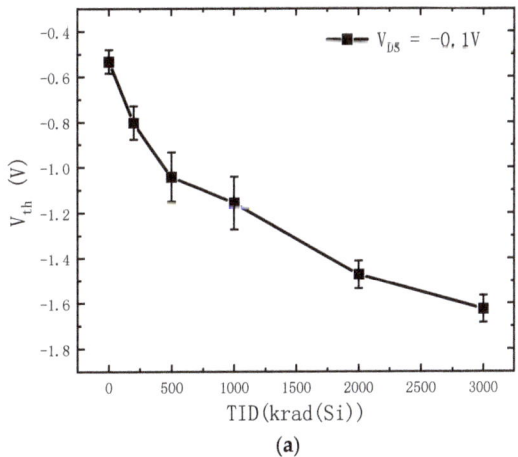

(a)

Figure 3. Cont.

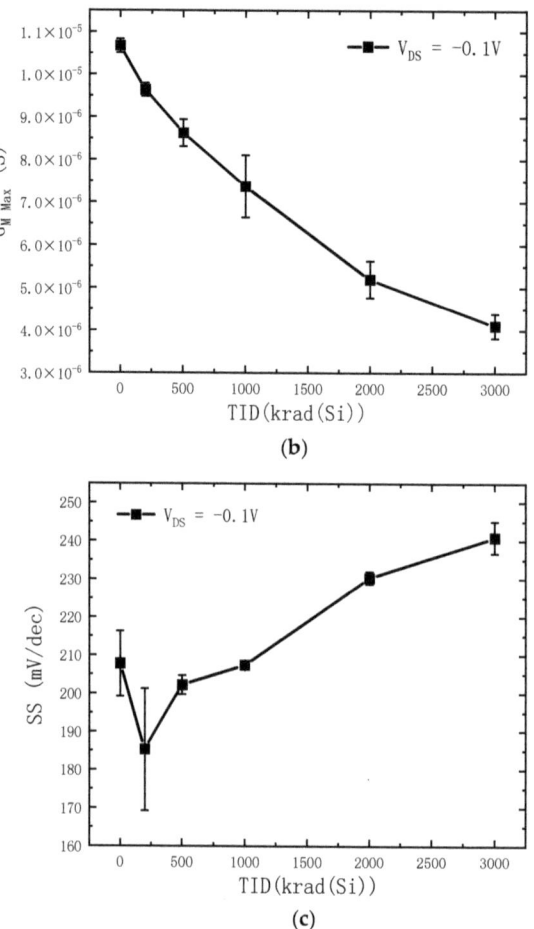

Figure 3. The changes of (**a**) threshold voltage (V_{th}), (**b**) maximum transconductance ($G_{M\ Max}$) and (**c**) subthreshold swing (SS) after irradiation.

The relation between threshold voltage and trapped charges induced by radiation is [28,29]:

$$V_{th} = V_{th0} + \frac{\Delta Q_{ot} + \Delta Q_{it}}{C_{ox}} \quad (1)$$

where V_{th0} is the threshold voltage before irradiation, ΔQ_{ot} is the charge variation of the top-gate oxide layer, ΔQ_{it} is the change in the interface-trapped charges, and C_{ox} is the capacitance of the top-gate oxide layer. Equation (1) demonstrated that both radiation-induced oxide-trapped charge and interface-trapped charge could lead to threshold voltage shift.

Equation (2) is the formula of transconductance:

$$G_m = \mu C_{ox} \frac{W}{L} V_{DS} \quad (2)$$

Equation (3) is the formula of subthreshold swing:

$$SS = \frac{kT}{q}\left(1 + \frac{C_{dep} + C_{it}}{C_{ox}}\right) ln10 \quad (3)$$

where μ is the carrier mobility, C_{dep} is the depletion capacitance, C_{it} is the interface trap capacitance, k is Boltzmann constant, T is temperature, and q is electronic charge. Carrier mobility μ is determined as follows:

$$\mu = \frac{\mu_0}{1 + \alpha \Delta N_{it}} \quad (4)$$

Interface trap capacitance C_{it} is obtained with Equation (5):

$$C_{it} = q^2 D_{it} \quad (5)$$

where μ_0 is the mobility pre-irradiation, ΔN_{it} is the radiation dependent densities of trapped charges in interface SiO_2/CNT, α is the parameter reflecting the technology change, and D_{it} is the density of interface traps.

The transconductance is proportional to the carrier mobility, while the carrier mobility is mainly affected by the interface traps. Therefore, an increase in irradiation dose increases interface traps, which then results in Coulomb scattering that in turn affects carrier mobility in the channel, and eventually causes a decrease in the maximum transconductance of the device.

Both depletion capacitance and interface trap capacitance could affect subthreshold swing. However, CNTs are ultrathin channel materials while C_{dep} can be neglected [15]. The subthreshold swing of CNTFETs is thus only affected by C_{it}, and the density of interface traps is the main influencing factor for C_{it} from Equation (5).

Because we used a CNT-network FET, CNTs in the channel region are staggered, and there is a large number of junctions between CNTs in the network films. According to previous research, high energy γ-ray can produce defects near CNT junctions. Through annealing under γ irradiation [30] or the oxidation of CNT due to the ionization effect [31,32], CNT junctions can recombine with nearby defects, improving the performance of CNT [19,33] that repairs the interface traps on the surface of CNT, and reduces the density of interface traps to a certain degree. Therefore, at the initial stage of radiation, that is from 0 k rad (Si) to about 200 k rad (Si), the recombination process of CNT junctions that can repair interfacial traps plays a major role, and the subthreshold swing of CNTFET decreases with the increase in TID. With the increase in TID, the number of junctions begins to decrease, and interface traps become the dominant factor, which increases the subthreshold swing with the increase in TID.

Considering the above reasons, the negative shift of I_{DS}-V_{GS} characteristics in Figure 2 can be confirmed to be caused by radiation. During irradiation, electron hole pairs are generated in the gate dielectric layer and/or oxide substrate, and the hole mobility is relatively low, which is easily captured by oxide and interface traps, forming positive trapped charges and requiring a higher negative gate voltage to control the channel. Therefore, as TID increases, the negative shift of the I–V curve occurs.

3.2. Influence of Trapped Charges in SiO_2 Substrate Induced by Irradiation on Top-Gate CNTFET Characteristics

With regard to the devices in this paper, theoretically, the TID can generate trapped charges in both the top-gate dielectric layer and the substrate oxide, leading to degradation of the device parameters. In order to determine where the trapped charges were the dominant factor, we measured the I_{DS}-V_{GS} characteristics of the devices under different substrate voltages.

As shown in Figure 4, positive voltages were applied to the substrate in order to simulate the effect of the substrate's accumulated oxide-trapped charges on top-gate transistors. With the increase in the substrate voltage, the I_{DS}-V_{GS} curve shifts negatively and the saturation current decreases gradually, which is consistent with the phenomenon observed in radiation. It is shown that the positive oxide-trapped charges are also generated in the substrate under γ-ray irradiation, which is consistent with the effect of trapped charges generated in gate on the I_{DS}-V_{GS} characteristics.

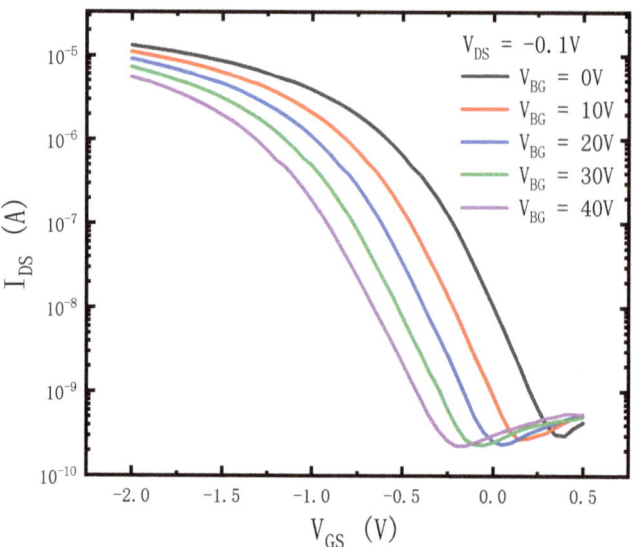

Figure 4. I_{DS}-V_{GS} characteristics of devices at different substrate voltages.

Figure 5a–c shows the threshold voltage, the maximum transconductance, and the subthreshold swing of top-gate transistor under various back-gate voltages, respectively. It may be assumed that positive oxide-trapped charges in the substrate bring holes closer to the surface of the CNT channel and increase the generation of interface traps. As a result, transconductance decreases and subthreshold swing increases when the substrate voltage is increased. On the other hand, Figure 5b also shows that the phenomenon of the subthreshold swing decreasing first and then increasing during irradiation, as shown in Figure 3c, derives from other factors rather than interface-trapped charges.

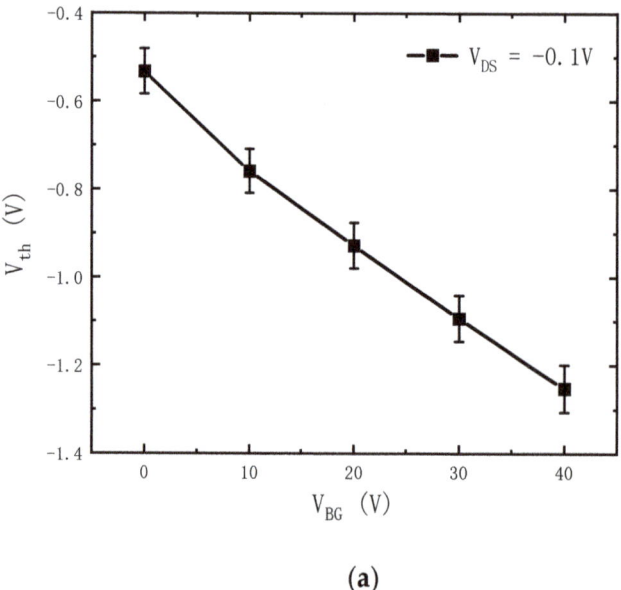

(a)

Figure 5. *Cont.*

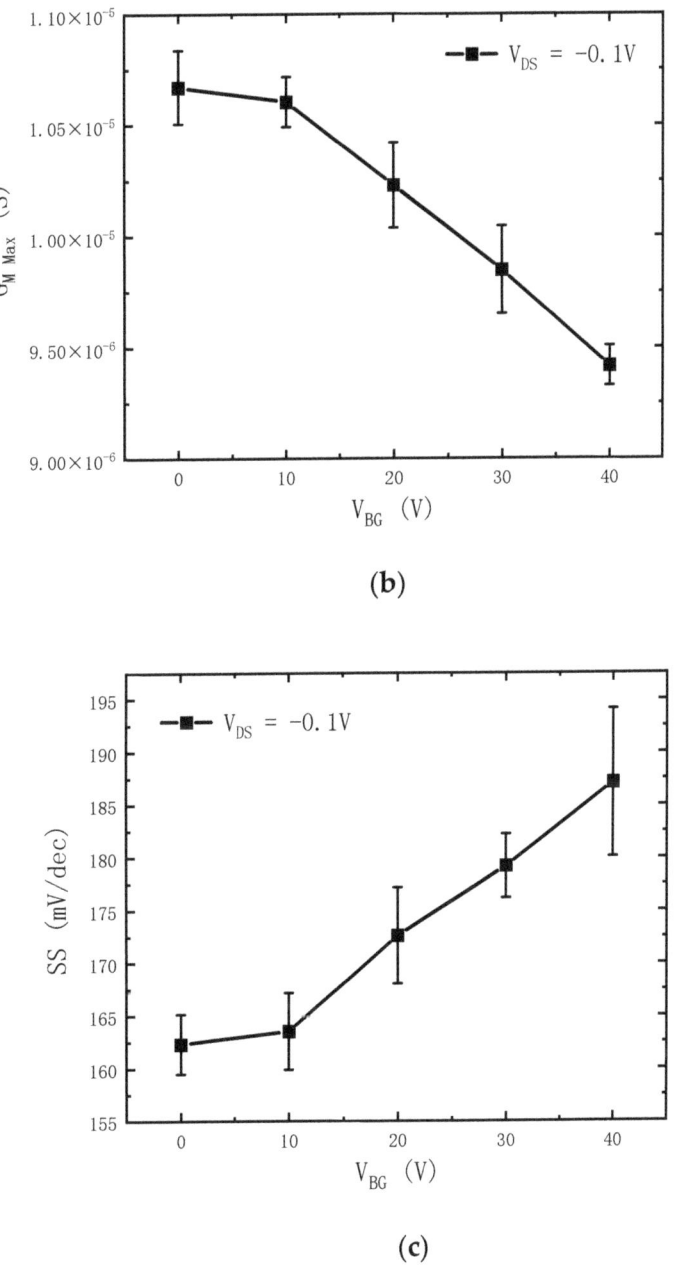

Figure 5. (**a**) The changes of threshold voltage under various substrate voltages. (**b**) Influence of different substrate voltages on transconductance. (**c**) The effect of different substrate voltages on subthreshold swing.

In addition, we used the SiO_2 substrate as the back-gate to test the TID radiation effect on the characteristics of the back-gate transistor. Figure 6 shows the I_{DS}-V_{BG} characteristic curves of the back-gate transistor at different TIDs. Figure 6 also shows that the I_{DS}-V_{BG} characteristic curves of the back-gate transistor shift negatively, and the saturation current decreases, which is consistent with the front-gate characteristic curves in Figure 2. The threshold voltage of the back-gate transistor is also extracted by the constant voltage method, as shown in Figure 7. In the same figure, it can be seen that the threshold voltage of the back-gate transistor changes obviously. It is supposed that there will be many oxide-trapped charges in the SiO_2 substrate that are induced by irradiation.

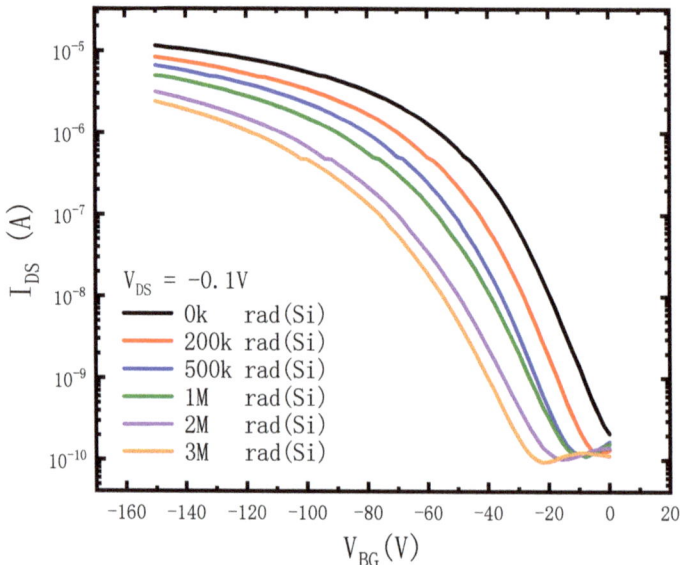

Figure 6. I_{DS}-V_{BG} characteristics of the back-gate after irradiation.

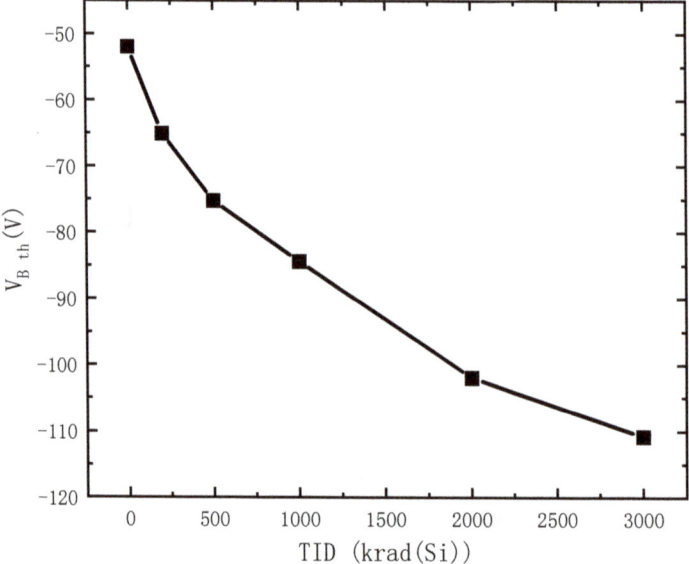

Figure 7. Threshold voltage of the back-gate transistor after irradiation.

Figure 8 shows the coupling coefficient between the back-gate transistor and the front-gate transistor. The device coupling coefficient is almost linear, indicating that the back-gate has better control over the front-gate, the density of trapped charges near the SiO$_2$ interface is low, and the TID effect has little influence on the coupling effect for the test samples in this paper.

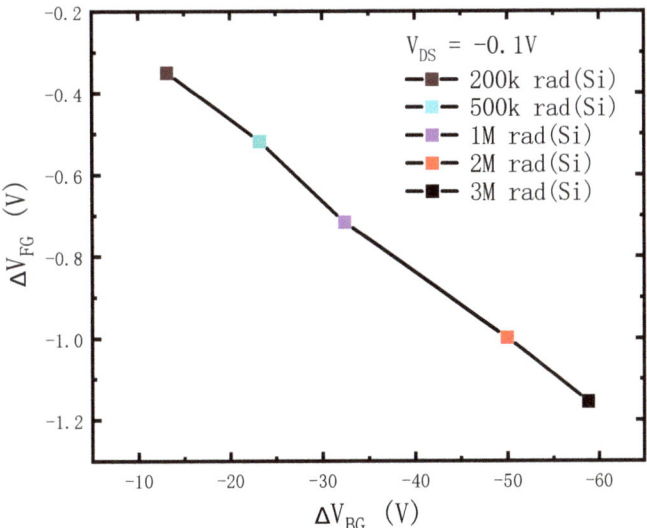

Figure 8. Changes of coupling coefficient.

3.3. TID Influence on Hysteresis Characteristics of Top-Gate CNTFET

For field effect transistors, the stable control of voltage to current is an important index of device performance. The hysteresis characteristics will affect the stability of the device, and it will pose a problem that should be discussed when CNTFET is used in real life.

The channel of CNTFET is composed of organic material CNTs, and the interface becomes complicated when it comes into contact with inorganic materials. Meanwhile, high-k materials such as Y$_2$O$_3$ and HfO$_2$ that are used as the gate dielectric have more defects than SiO$_2$. This will more likely generate current hysteresis characteristics [34]. Thus, we measured the hysteresis characteristics through bidirectional scanning for gate voltage, as shown in Figure 9. The maximum gate voltage difference under the same drain current I$_{DS}$ was used to define the hysteresis window V_M. The test sequence involves scanning of the gate voltage V$_{GS}$ from 0.5 V to −2 V, which is defined as negative scanning in this paper. Subsequently, it is scanned back from −2 V to 0.5 V, which is defined as positive scanning in this paper. It can be clearly seen from Figure 9 that the hysteresis window, V_M, increased with the increase in TID.

Previous research has shown that the reasons for the hysteresis of devices are mainly due to the following four kinds of charges [35–37]: (1) positive oxide fixed charge that is generally very close to the channel (about 2 nm); (2) oxide-trapped charge; (3) interface-trapped charge; and (4) moving charge in oxide. Due to the properties of high-k materials, there are more fixed positive charges in Y$_2$O$_3$ and HfO$_2$ than in SiO$_2$. As mentioned in the previous analysis, the generation of oxide-trapped charge and interfacial trapped charge are consistent with the change of hysteresis characteristics with the accumulation of TID. Therefore, the overall quantity of oxide-trapped charges and interfacial trapped charges increases with the increase in TID and could generate the change of hysteresis characteristics. The hysteresis characteristics are proportional to TID.

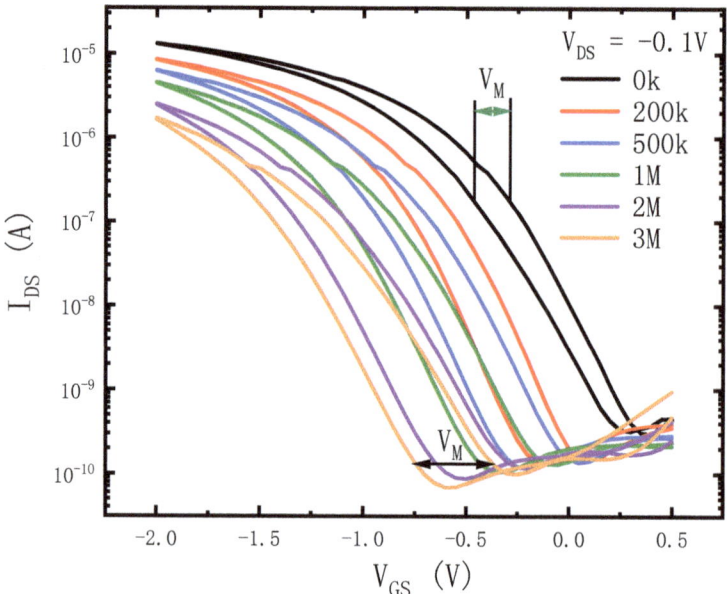

Figure 9. Hysteresis characteristics as a function of TID.

On the other hand, CNTFETs are different from traditional transistors. CNTs have strong adsorption, and it is difficult to avoid the adsorption of oxygen, molecules of water, and other impurities in the fabrication process. At present, it is widely believed that water molecules adsorbed by CNT and the presence of OH group in the form of silanol are the main sources of hysteresis characteristics in CNTFETs [38]. Water molecules and groups will trap electrons in CNTs and induce excess holes in the channel, resulting in the transistor current becoming larger than expected under the condition of the same V_{GS}. When the V_{GS} scans negatively in the measurement process, the negative gate voltage leads to the electrons captured by the water molecules and the groups being gradually released. Therefore, the transistor current becomes smaller under the same gate voltage V_{GS} condition when scanning back. This phenomenon will be exacerbated by the process of irradiation. The water molecules that are adsorbed on the surface of the CNTs are ionized during Co-60 γ-ray irradiation and produce more OH groups. In this way, the electron capture and release of adsorbates on the CNT surface becomes more intense, which further exacerbates the hysteresis characteristics of CNTFETs.

4. Conclusions

CNTFETs have been considered to be some of the most promising candidates for continuing Moore's law and have received much attention in the aerospace field. The TID effect of CNTFET has been the focus of much research. In different device structures, the top-gate structure is an important CNTFET structure. It is used to manufacture high-performance CNTFETs because it scales down easily. However, the lack of studies on TID of the top-gate structure CNTFET restricts the development of radiation-hardness top-gate CNTFETs.

In this paper, the effect of radiation-induced trapped charges on the top-gate CNT-FET characteristics is studied. Firstly, we measure the change of I_{DS}-V_{GS} characteristics in the top-gate transistors with the total dose increases, and show that radiation could induce oxide-trapped charges and interface-trapped charges in the CNTFET. They lead to a negative shift in the transistor I_{DS}-V_{GS} characteristics. Both oxide-trapped charges and interface-trapped charges lead to the threshold voltage shifting negatively. The interface-trapped charges not only affect the threshold voltage, but also reduce the mobility of carriers. This results in the transconductance reduction with the increase in TID. In addition to the influence of interface-trapped charges, the subthreshold swing is also affected by CNT junctions in the channel. During irradiation, a large number of CNT junctions recover, and some interface traps are repaired, which is the main factor affecting the subthreshold swing at the initial stage of irradiation. With the reduction in CNT junctions, the accumulation of interface-trapped charges has become the main factor affecting the subthreshold swing. Thus, the subthreshold swing decreases first and then increases. In order to confirm the source and influence of radiation-induced trapped charges, we construct back-gate transistors, test the effect of different back-gate voltages on the front-gate I_{DS}-V_{GS} characteristics, and obtain the TID influence on the back-gate transistor characteristics. We also extract the coupling coefficient between the back-gate transistor and the front-gate transistor. The results indicate that the TID has little effect on the coupling coefficient, and the back-gate has better control ability on the front-gate of the samples. Finally, in the research of hysteresis characteristics, the results show that the hysteresis characteristics of CNTFETs increased with the increase in oxide-trapped charges and interface-trapped charges after radiation. Furthermore, the hysteresis characteristics of CNTFETs significantly increased following adsorbates ionization on the surface of the CNT channel. In general, the hysteresis characteristics of CNTFETs are proportional to TID.

The ultimate purpose of the TID study of CNTFETs is to apply this to the space environment, where the particle situation is complex and the radiation dose rate is low. Different particles and dose rates may cause different phenomena. To estimate the impact of the actual environment as accurately as possible, radiation conditions that are closer to the actual application environment could be considered in future research, namely different radiation sources and low dose rates of radiation exposure. In addition, a quantitative analysis of trap charge is also necessary. As the technology of CNTFETs develops, trap charge can be extracted by devices with better subthreshold swing characteristics [39,40], which is helpful in the development of anti-irradiation technology of CNTFET.

Author Contributions: Conceptualization, J.C., Q.Z.; methodology, H.D., J.C., Q.Z. and C.H.; validation, H.D., J.C. and Q.Z.; formal analysis, H.D.; investigation, H.D., J.C., M.X. and G.Y.; resources, H.X. and N.G.; data curation, H.D.; writing—original draft preparation, H.D.; writing—review and editing, H.D. and J.C.; visualization, H.D.; supervision, Y.L., Q.G.; project administration, J.C.; funding acquisition, J.C. All authors have read and agreed to the published version of the manuscript.

Funding: This research was supported in part by the Youth Innovation Promotion Association CAS (2020430), the West Light Foundation of the Chinese Academy of Science under Grant No. 2019-XBQNXZ-A-003, and the National Natural Science Foundation of China under Grant 12275352.

Institutional Review Board Statement: Not applicable

Informed Consent Statement: Not applicable.

Data Availability Statement: Data are not available in a publicly accessible repository and they cannot be shared upon request.

Conflicts of Interest: The authors declare no conflict of interest.

References

1. Lincoln, R.L.; Scarpa, F.; Ting, V.P.; Trask, R.S. Multifunctional composites: A metamaterial perspective. *Multifunct. Mater.* **2019**, *2*, 043001. [CrossRef]
2. Mohammadi Estakhri, N.; Edwards, B.; Engheta, N. Inverse-designed metastructures that solve equations. *Science* **2019**, *363*, 1333–1338. [CrossRef] [PubMed]
3. Lalegani, Z.; Ebrahimi, S.S.; Hamawandi, B.; La Spada, L.; Batili, H.; Toprak, M.S. Targeted dielectric coating of silver nanoparticles with silica to manipulate optical properties for metasurface applications. *Mater. Chem. Phys.* **2022**, *287*, 126250. [CrossRef]
4. Pacheco-Peña, V.; Beruete, M.; Rodríguez-Ulibarri, P.; Engheta, N. On the performance of an ENZ-based sensor using transmission line theory and effective medium approach. *New J. Phys.* **2019**, *21*, 043056. [CrossRef]
5. Akbari, M.; Shahbazzadeh, M.J.; La Spada, L.; Khajehzadeh, A. The Graphene Field Effect Transistor Modeling Based on an Optimized Ambipolar Virtual Source Model for DNA Detection. *Appl. Sci.* **2021**, *11*, 8114. [CrossRef]

6. Greybush, N.J.; Pacheco-Peña, V.; Engheta, N.; Murray, C.B.; Kagan, C.R. Plasmonic Optical and Chiroptical Response of Self-Assembled Au Nanorod Equilateral Trimers. *ACS Nano* **2019**, *13*, 1617–1624. [CrossRef]
7. Javey, A.; Guo, J.; Wang, Q.; Lundstrom, M.; Dai, H. Ballistic carbon nanotube field-effect transistors. *Nature* **2003**, *424*, 654–657. [CrossRef]
8. Franklin, A.D.; Luisier, M.; Han, S.J.; Tulevski, G.; Breslin, C.M.; Gignac, L.; Lundstrom, M.S.; Haensch, W. Sub-10 nm Carbon Nanotube Transistor. *Nano Lett.* **2012**, *12*, 758–762. [CrossRef]
9. Chen, B.; Zhang, P.; Ding, L.; Han, J.; Qiu, S.; Li, Q.; Zhang, Z.; Peng, L.-M. Highly Uniform Carbon Nanotube Field-Effect Transistors and Medium Scale Integrated Circuits. *Nano Lett.* **2016**, *16*, 5120–5128. [CrossRef]
10. De Volder, M.F.; Tawfick, S.H.; Baughman, R.H.; Hart, A.J. Carbon Nanotubes: Present and Future Commercial Applications. *Science* **2013**, *339*, 535–539. [CrossRef]
11. Yang, Y.; Ding, L.; Chen, H.; Han, J.; Zhang, Z.; Peng, L.M. Carbon nanotube network film-based ring oscillators with sub 10-ns propagation time and their applications in radio-frequency signal transmission. *Nano Res.* **2018**, *11*, 300–310. [CrossRef]
12. Peng, L.M.; Zhang, Z.; Wang, S. Carbon nanotube electronics: Recent advances. *Mater. Today* **2014**, *17*, 433–442. [CrossRef]
13. Yang, L.; Wang, S.; Zeng, Q.; Zhang, Z.; Pei, T.; Li, Y.; Peng, L.M. Efficient photovoltage multiplication in carbon nanotubes. *Nat. Photonics* **2011**, *5*, 672–676. [CrossRef]
14. Liu, L.; Han, J.; Xu, L.; Zhou, J.; Zhao, C.; Ding, S.; Shi, H.; Xiao, M.; Ding, L.; Ma, Z.; et al. Aligned, high-density semiconducting carbon nanotube arrays for high-performance electronics. *Science* **2020**, *368*, 850–856. [CrossRef]
15. Zhu, M.; Zhou, J.; Sun, P.; Peng, L.M.; Zhang, Z. Analyzing Gamma-Ray Irradiation Effects on Carbon Nanotube Top-Gated Field-Effect Transistors. *ACS Appl. Mater. Interfaces* **2021**, *13*, 47756–47763. [CrossRef]
16. Zhu, M.; Xiao, H.; Yan, G.; Sun, P.; Jiang, J.; Cui, Z.; Zhao, J.; Zhang, Z.; Peng, L.-M. Radiation-Hardened and Repairable Integrated Circuits Based on Carbon Nanotube Transistors with Ion Gel Gates. *Nat. Electron.* **2020**, *3*, 622–629. [CrossRef]
17. Li, X.; Wang, X.; Deng, J.; Li, M.; Shao, S.; Zhao, J. Printed carbon nanotube thin film transistors based on perhydropolysilazane-derived dielectrics for low power flexible electronics. *Carbon* **2022**, *191*, 267–276. [CrossRef]
18. Cress, C.D.; McMorrow, J.J.; Robinson, J.T.; Friedman, A.L.; Landi, B.J. Radiation Effects in Single-Walled Carbon Nanotube Thin-Film-Transistors. *IEEE Trans. Nucl. Sci.* **2010**, *57*, 3040–3045. [CrossRef]
19. Zhao, Y.; Li, D.; Xiao, L.; Liu, J.; Xiao, X.; Li, G.; Jin, Y.; Jiang, K.; Wang, J.; Fan, S.; et al. Radiation effects and radiation hardness solutions for single-walled carbon nanotube-based thin film transistors and logic devices. *Carbon* **2016**, *108*, 363–371. [CrossRef]
20. Luo, M.; Zhu, M.; Wei, M.; Shao, S.; Robin, M.; Wei, C.; Cui, Z.; Zhao, J.; Zhang, Z. Radiation-Hard and Repairable Complementary Metal-Oxide-Semiconductor Circuits Integrating printed n-type Indium Oxide and p-type Carbon Nanotube Field-Effect Transistors. *ACS Appl. Mater. Interfaces* **2020**, *12*, 49963–49970. [CrossRef]
21. Wang, Y.; Xiao, L. Repairable Integrated Circuits for Space. *Nat. Electron.* **2020**, *3*, 586–587. [CrossRef]
22. Javey, A.; Kim, H.; Brink, M.; Wang, Q.; Ural, A.; Guo, J.; McIntyre, P.; McEuen, P.; Lundstrom, M.; Dai, H. High-κ dielectrics for advanced carbon-nanotube transistors and logic gates. *Nat. Publ. Group* **2002**, *1*, 241–246. [CrossRef] [PubMed]
23. Pitner, G.; Zhang, Z.; Lin, Q.; Su, S.K.; Gilardi, C.; Kuo, C.; Kashyap, H.; Weiss, T.; Yu, Z.; Chao, T.A.; et al. Sub-0.5 nm Interfacial Dielectric Enables Superior Electrostatics: 65 mV/dec Top-Gated Carbon Nanotube FETs at 15 nm Gate Length. In Proceedings of the 2020 IEEE International Electron Devices Meeting (IEDM), San Francisco, CA, USA, 12–18 December 2020; IEEE: New York, NY, USA, 2020; pp. 3.5.1–3.5.4.
24. Qiu, C.; Zhang, Z.; Xiao, M.; Yang, Y.; Zhong, D.; Peng, L.M. Scaling carbon nanotube complementary transistors to 5-nm gate lengths. *Science* **2017**, *355*, 271. [CrossRef] [PubMed]
25. Zhao, X.; Yu, M.; Cai, L.; Liu, J.; Wang, J.; Wan, M.; Wang, J.; Wang, C.; Fu, Y. Radiation effects in printed flexible single-walled carbon nanotube thin-film transistors. *AIP Adv.* **2019**, *9*, 105121. [CrossRef]
26. Zhu, M.G.; Zhang, Z.; Peng, L.M. High-Performance and Radiation-Hard Carbon Nanotube Complementary Static Random-Access Memory. *Adv. Electron. Mater.* **2019**, *5*, 1900313. [CrossRef]
27. Wang, Y.; Xiang, Z.Q.; Hu, H.F.; Cao, F. Feasibility Study of Semifloating Gate Transistor Gamma-Ray Dosimeter. *IEEE Electron Device Lett.* **2015**, *36*, 99–101. [CrossRef]
28. Petrosjanc, K.O.; Adonin, A.S.; Kharitonov, I.A.; Sicheva, M.V. SOI device parameter investigation and extraction for VLSI radiation hardness modeling with SPICE. In Proceedings of the 1994 IEEE International Conference on Microelectronic Test Structures, San Diego, CA, USA, 22–25 March 1994; pp. 126–129.
29. Galloway, K.F.; Gaitan, M.; Russell, T.J. A Simple Model for Separating Interface and Oxide Charge Effects in MOS Device Characteristics. *IEEE Trans. Nucl. Sci.* **1984**, *31*, 1497–1501. [CrossRef]
30. Li, B.; Feng, Y.; Ding, K.; Qian, G.; Zhang, X.; Zhang, J. The effect of gamma ray irradiation on the structure of graphite and multi-walled carbon nanotubes. *Carbon* **2013**, *60*, 186–192. [CrossRef]
31. Miao, M.; Hawkins, S.C.; Cai, J.Y.; Gengenbach, T.R.; Knott, R.; Huynh, C.P. Effect of gamma-irradiation on the mechanical properties of carbon nanotube yarns. *Carbon* **2011**, *49*, 4940–4947. [CrossRef]
32. Skakalova, V.; Hulman, M.; Fedorko, P.; Lukáč, P.; Roth, S. Effect of gammairradiation on single-wall carbon nanotube paper. *AIP Conf. Proc.* **2003**, *685*, 143–147. [CrossRef]
33. Vitusevich, S.A.; Sydoruk, V.A.; Petrychuk, M.V.; Danilchenko, B.A.; Klein, N.; Offenhäusser, A.; Bosman, G. Transport properties of single-walled carbon nanotube transistors after gamma radiation treatment. *J. Appl. Phys.* **2010**, *107*, 063701. [CrossRef]

34. Leroux, C.; Mitard, J.; Ghibaudo, G.; Garros, X.; Reimbold, G.; Guillaumot, B.; Martin, F. Characterization and modeling of hysteresis phenomena in high K dielectrics. In Proceedings of the IEEE International Electron Devices Meeting, IEDM Technical Digest, San Francisco, CA, USA, 13–15 December 2004; pp. 737–740.
35. Schroder, D.K. *Semiconductor Material and Device Characterization*, 3rd ed.; John Wiley & Sons, Inc.: Hoboken, NJ, USA, 2006.
36. Kessler, J.O.; Tompkins, B.E.; Blanc, J. Variable-characteristic p-n-junction devices based on reversible ion drift. *Solid-State Electron.* **1963**, *6*, 297–307. [CrossRef]
37. Snow, E.H.; Grove, A.S.; Deal, B.E.; Sah, C.T. Ion Transport Phenomena in Insulating Films. *J. Appl. Phys.* **1965**, *36*, 1664–1673. [CrossRef]
38. Kim, W.; Javey, A.; Vermesh, O.; Wang, Q.; Li, Y.; Dai, H. Hysteresis Caused by WaterMolecules in Carbon Nanotube Field-Effect Transistors. *Nano Lett.* **2003**, *3*, 193–198. [CrossRef]
39. Patil, P.D.; Ghosh, S.; Wasala, M.; Lei, S.; Vajtai, R.; Ajayan, P.M.; Ghosh, A.; Talapatra, S. Gate-Induced Metal–Insulator Transition in 2D van der Waals Layers of Copper Indium Selenide Based Field-Effect Transistors. *ACS Nano* **2019**, *13*, 13413–13420. [CrossRef] [PubMed]
40. *ASTM F996-11*; Standard Test Method for Separating an Ionizing Radiation-Induced Mosfet Threshold Voltage Shift into Components Due to Oxide Trapped Holes and Interface States Using the Subthreshold Current-Voltage Characteristics. ASTM International: West Conshohocken, PA, USA, 2011.

Disclaimer/Publisher's Note: The statements, opinions and data contained in all publications are solely those of the individual author(s) and contributor(s) and not of MDPI and/or the editor(s). MDPI and/or the editor(s) disclaim responsibility for any injury to people or property resulting from any ideas, methods, instructions or products referred to in the content.

Article

The Inflection Point of Single Event Transient in SiGe HBT at a Cryogenic Temperature

Xiaoyu Pan [1,2], Hongxia Guo [2,*], Chao Lu [1], Hong Zhang [3] and Yinong Liu [1]

[1] The Key Laboratory of Particle and Radiation Imaging, Ministry of Education, Department of Engineering Physics, Tsinghua University, Beijing 100084, China
[2] State Key Laboratory of Intense Pulsed Radiation Simulation and Effect, Northwest Institute of Nuclear Technology, Xi'an 710024, China
[3] The School of Material Science and Engineering, Xiangtan University, Xiangtan 411105, China
* Correspondence: guohxnint@126.com

Abstract: Basing our findings on our previous pulsed laser testing results, we have experimentally demonstrated that there is an inflection point of a single event transient (SET) in the silicon-germanium heterojunction bipolar transistors (SiGe HBTs) with a decreasing temperature from +20 °C to −180 °C. Additionally, the changes in the parasitic resistivity of the carrier collection pathway due to incomplete ionization could play a key role. In this paper, we found that the incident-heavy ion's parameters could also have an important impact on the SET inflection point by introducing the ion track structures generated by Geant4 simulation to the TCAD transient simulation. Heavy ion with a low linear energy transfer (LET) will not trigger the ion shunt effect of SiGe HBT and the inflection point will not occur until −200 °C. For high LET ions' incidence, the high-density electron-hole pairs (EHPs) could significantly affect the parasitic resistivity on the pathway and lead to an earlier inflection point. The present results and methods could provide a new reference for the effective evaluation of single-event effects in bipolar transistors and circuits at cryogenic temperatures and provide new evidence of the SiGe technology's potential for applications in extreme cryogenic environments.

Keywords: SiGe HBT; Geant4; TCAD simulation; single event transient; cryogenic temperature

1. Introduction

Today, NASA is preparing to go back to the moon with Artemis missions and will build an Artemis Base Camp on the lunar surface (−180 °C ~ +120 °C). SpaceX is also making continuous efforts to land human beings on Mars (−133 °C ~ +27 °C) in Starships. All of these great missions require the support of large thrust rockets and how to improve their payload is a concern. As we all know, there are usually bulky "warm boxes" to protect the electronic systems in an extreme environment which could cause additional consumption [1]. Fortunately, SiGe HBT could be a candidate to change this situation [2].

SiGe HBT has excellent RF performance and good compatibility with silicon-based technologies, and has been widely used in wireless communication, phased-array radars, etc. [3]. Furthermore, SiGe HBT has inherent resistance to a total ionizing dose (TID) effect [4]. Meanwhile, thanks to the introduction of Ge content to the intrinsic base region, it could work over a wide temperature range (especially cryogenic temperatures) [5]. Hence, electronic systems using SiGe technologies have the potential to operate well without the "warm boxes".

However, things do not always go smoothly. SiGe HBT is sensitive to a SET and this sensitivity increases as the device feature size decreases [5]. There have been many related studies at room temperature [6–8], which can help us to understand the underlying mechanism of SiGe HBT's SET. According to the existing research, there are few studies on the impact of temperature on the SiGe HBT's SET. And generally, researchers attribute the main cause of the SET's variation with temperature to the carrier mobility's variation such as the study on proton-induced SEU in SiGe digital logic at cryogenic temperatures in which the SET peaks increase as the temperature decreases [9]. The inflection point of the SET peaks was found by the TCAD simulation for the first time, which shows the impact of impurities' incomplete ionization (abbreviated as i.i.) at cryogenic temperature [10,11]. However, the heavy ion's LET is only 0.01 pC/μm in the simulation results that could not trigger the ion shunt effects [12] of the SiGe HBT's emitter/base/collector/substrate (E/B/C/S) stack.

As is shown in Figure 1, our previous study experimentally demonstrated the existence of the SET's inflection point for the first time by carrying out pulsed laser testing over a wide temperature range (−180 °C ~ +20 °C) [13]. We found that the change in parasitic resistance in the carrier collection pathway is an important reason for the peak inflection point. Additionally, the parasitic resistance depends on both the concentration and mobility of electrons and holes. The variation in the carrier mobility with the temperature has been studied extensively for a long time. Furthermore, we also discussed the ionization rate of intrinsic doping in our previous study. One more thing to mention so far is that we have not yet discussed the impact of the heavy-ion induced EHPs on the parasitic resistance.

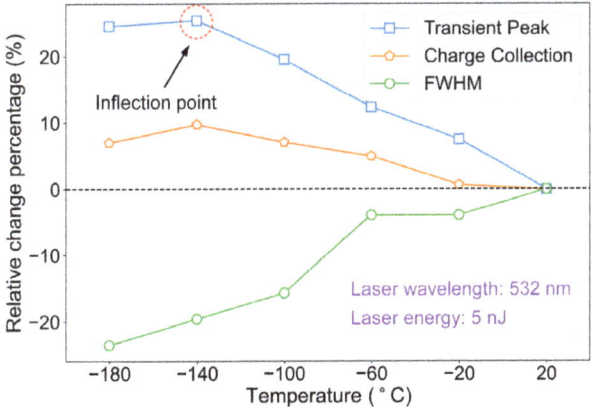

Figure 1. The relative percentage change (relative to 20 °C) on collector's transient peak, charge collection, and FWHM from pulsed laser testing.

In this paper, we focused on a study of the impact of the incident heavy ion's parameters on the inflection point of SET peaks. We built a simulation method which helped us to introduce the heavy-ion induced EHP's distribution generated by Geant4 calculation to the TCAD device simulation directly. When the LET value of the incident heavy ion is too low to trigger the ion shunt effect, the collector's transient current is mainly derived from collector/base (C/B) junction and collector/substrate (C/S) junction. In this case, heavy-ion induced EHPs are relatively low to the intrinsic doping and the temperature corresponding to the inflection point comes later (even up to −200 °C). As a comparison, when the LET value is relatively high and the ion shunt effect turns on at this time, then the heavy-ion induced EHPs can also have a significant impact on the total parasitic resistance of the charge collection pathway. In this case, the inflection point will come much earlier (about −160 °C).

2. TCAD 2-D Process Simulation

2.1. DUT Description

In this paper, the device under test (DUT) is a low-noise SiGe HBT (NPN transistor) provided by the School of Integrated Circuits, Tsinghua University. The DUT's lithographic node is 400 nm and the chip layout is configured as a 4E5B2C interdigital structure, shown in Figure 2. The peak Ge content in the base region is close to 14% and has a trapezoidal distribution. The detailed device information can be found in our previous study and will not be repeated here [14].

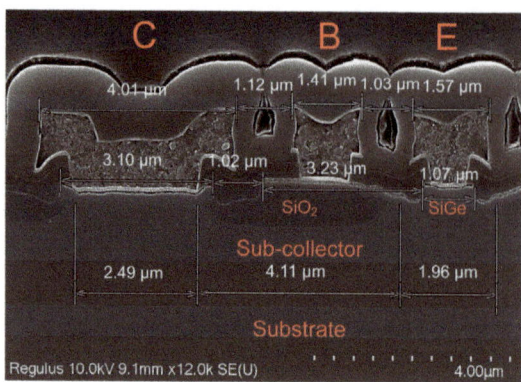

Figure 2. The SEM figure shows a cross-section of the DUT.

2.2. TCAD 2-D Process Model

In this section, we built a 2-D TCAD process model according to the DUT's production process, shown in Figure 3. In particular, this model is a simplified 1E2B2C structure to save the simulation time and achieve better simulation convergence at a cryogenic temperature. When the ion shunt effect is triggered, it can be simply understood that the EHPs generated by the incident heavy ion could build a bridge between the emitter and the collector. At this time, the total parasitic resistance of the collector's charge collection pathway includes the emitter resistance R_E, vertical base resistance R_{B-V}, selectively implanted collector (SIC) resistance R_C and lateral sub-collector resistance R_{SC}.

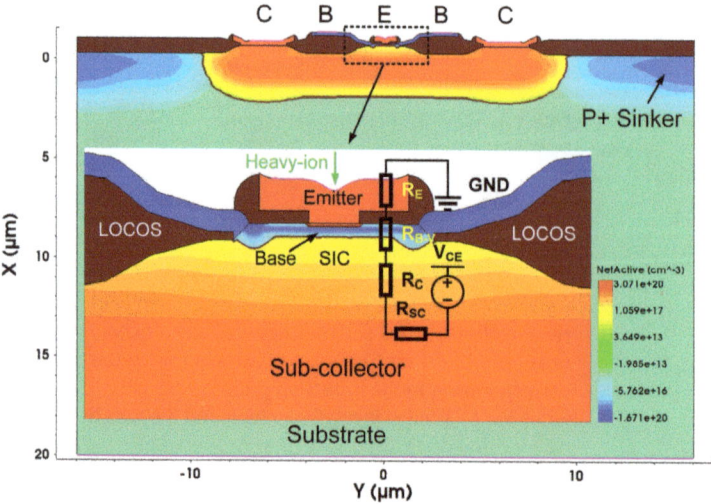

Figure 3. The 2-D TCAD process model with 1E2B2C structure and the inside zoom view shows the parasitic resistance of the E/B/C stack when the ion shunt effect is triggered.

2.3. Simulation Results

Because it is the most sensitive volume, we fixed the heavy ion's incident position at the emitter center during the whole simulation. The characteristic distance and incident depth of the heavy ion are set to 0.2 µm and 20 µm, respectively. In addition, the device bias is defined as the cut-off bias V_{CE} = 2 V, V_{BE} = 0 V (or the C/S junction reverse bias). In particular, the simulation physics models are consistent with our prior study, including the incomplete ionization model [13].

We chose four temperature points from −140 °C to −200 °C with intervals of 20 °C. The simulation results are shown in Figure 4, one can see that the transient current is quite different when the LET values of the incident heavy ion are 0.01 pC/µm, 0.05 pC/µm and 0.1 pC/µm respectively.

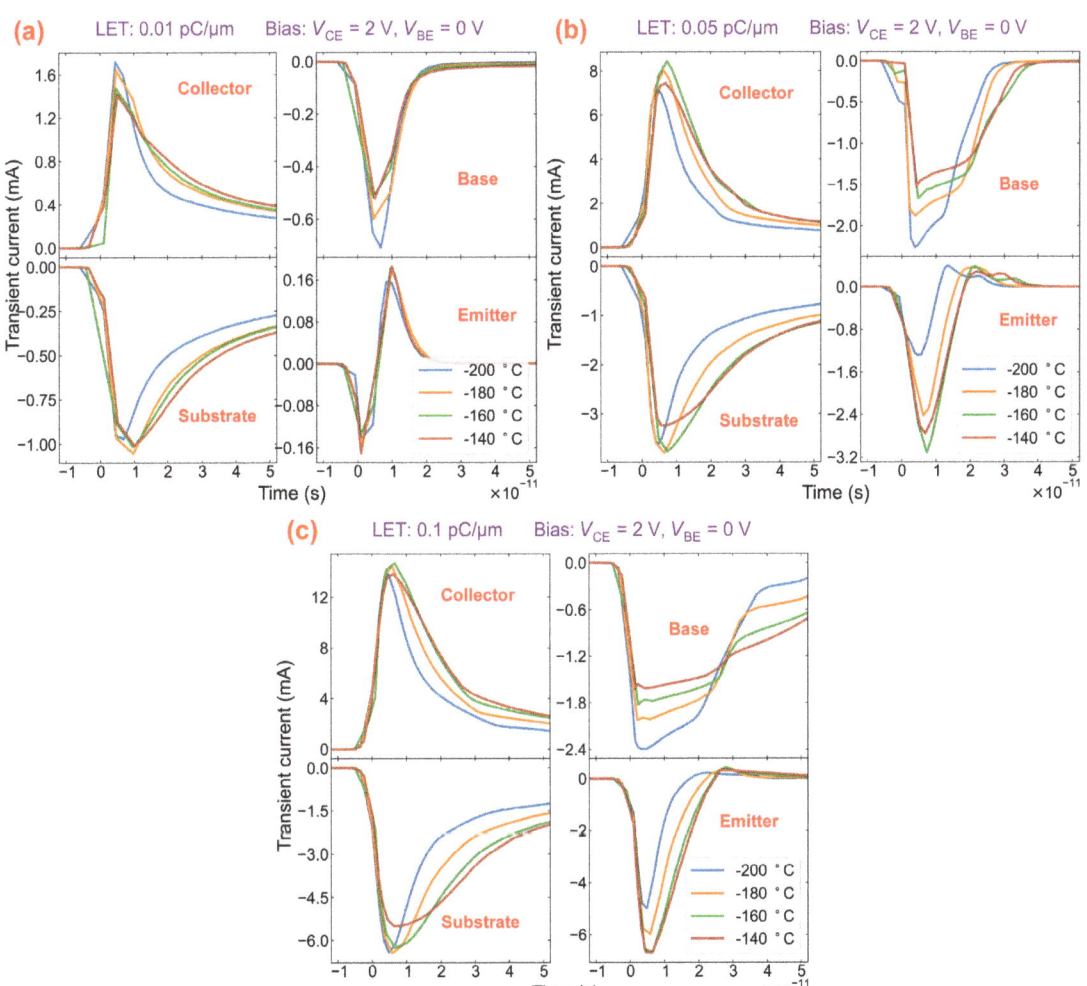

Figure 4. The transient current of the four device electrodes from the TCAD simulation at cryogenic temperatures when the LETs are (**a**) 0.01 pC/µm, (**b**) 0.05 pC/µm and (**c**) 0.1 pC/µm, respectively.

Firstly, when the LET is 0.01 pC/µm as in Figure 4a, the ion shunt effect will not be triggered which can be recognized by the weak transient peak of the emitter. At this point, the collector's transient peaks have continued to increase and not shown an inflection point with the temperature decreasing.

As a comparison, when the LET values are 0.05 pC/µm and 0.1 pC/µm, as in Figure 4b,c, the high-density ionized EHPs can connect the emitter and the collector, and there will be a lot of electrons transferred directly from the emitter to the collector (or the ion shunt effect is triggered on). At this time, the collector's transient peaks will have an obvious inflection point around −160 °C.

That is to say, when the initial ionized EHPs by incident heavy ions are high enough to trigger the ion shunt effect, the inflection point will come earlier.

For a clear analysis, we plotted the transient peaks of the four device electrodes at cryogenic temperatures as in Figure 5. As is generally known, the transient current of the collector could be the sum of the other electrodes, as in (1) [15],

$$i_{cn} = -(i_{bp} + i_{sp} + i_{en}) \tag{1}$$

where i_{cn}, i_{bp}, i_{sp}, and i_{en} represent the transient currents of the collector, base, substrate and emitter, respectively. In addition, the subscript n indicates "electron collection" and p indicates "hole collection". It is not surprising that the sum of all the electrodes' currents should be zero. From Figure 5, we can extract three key features:

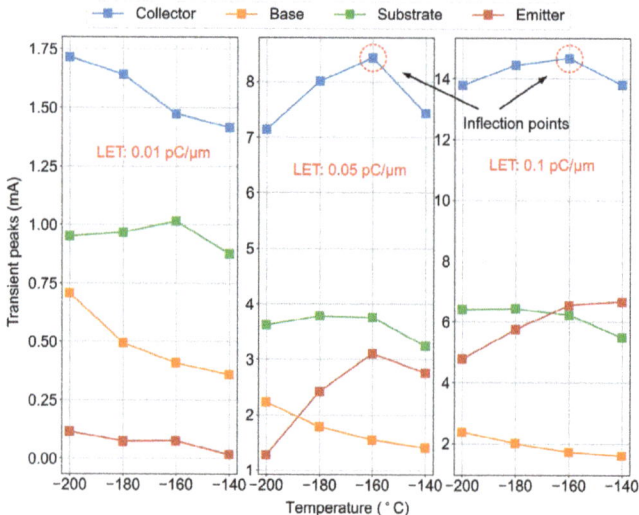

Figure 5. The transient peaks (absolute values) of the four device electrodes from the TCAD simulation at cryogenic temperatures with different LET values.

First, whether the ion shunt effect is on or not can be directly reflected by the share of the emitter transient peak (the red lines and squares), which will increase with the LET value. Additionally, when the ion shunt effect is on (LET values are 0.05 pC/μm and 0.1 pC/μm), there will be an inflection point (about −160 °C ~ −140 °C) of the emitter transient peaks. That is to say, the relatively high LET values lead to earlier inflection points. At this time, the total parasitic resistance on the emitter "electron collection" path includes the heavy doping R_E, moderate doping R_{B-V}, light doping R_C and heavy doping R_{SC} in Figure 3. At cryogenic temperatures, the R_{B-V} and R_C increase as the temperature decreases. In contrast, the R_E and R_{SC} decrease with the temperature. The presence of these two competitive mechanisms together leads to the inflection point. In the future, the total parasitic resistance at a specific temperature will need to be calculated by 3-D simulation.

Second, the base transient peak (the orange lines and squares) continues to increase as the temperature decreases, regardless of the LET value. This is due to the high doping concentration in the intrinsic and epitaxial base regions (R_{B-L}) which means that the impurities are almost completely ionized even at a cryogenic temperature. At this time, the parasitic resistance on the base "hole collection" path is mainly influenced by the carrier mobility.

Third, the substrate transient peaks (the green lines and squares) have shown a very different pattern from the emitter. As we can see, the relatively low LET values lead to earlier inflection points. The total parasitic resistance on the substrate "hole collection" path is dominated by the lightly doped substrate and C/S junction regions. When the LET value is 0.01 pC/μm, the heavy ion-induced EHPs' density is relatively low, and the parasitic resistance is controlled by the intrinsic impurity ionization rate and carrier mobility. At this time, the i.i. of the impurities at low temperatures will lead to the peak inflection point (about −160 °C). In contrast, when the LET value is relatively high (such as 0.05 pC/μm and 0.1 pC/μm), the heavy ion-induced EHPs' density will also modulate the parasitic resistance. In extreme cases, the total resistance will be completely taken over by the initial EHPs.

So far, we have found that the inflection point of the collector transient peaks occurs as a combined result of the temperature dependence of the parasitic resistance on the above three carrier collection paths. Furthermore, we can obtain the key conclusion that if we want to conduct a ground-based simulation experiment (typically high LET values), cooling down with the liquid nitrogen (−196 °C) can already meet the requirements.

3. Ion Track Simulation by Geant4

3.1. Initial Ion Track Structures

In the space radiation environment, heavy ions' energy could even reach hundreds of GeV per nucleon (GeV/amu) and the peak flux is around hundreds of MeV/amu. However, for the ground SEE testing facilities in the world, the heavy ion beam's energy could not exceed 100 MeV/amu [16,17]. As is generally known, the same ion at different energies will have different LET values or different ion track structures; thus, the heavy ion-induced initial EHPs' distribution will be different.

In this section, we will take the typical heavy ion (Fe) in space as an example and study the impact of ion energy on the SiGe HBT's SET inflection point.

The ion track structure was obtained by Geant4 (version 10.7) Monte Carlo simulation [18]. In each simulation round, we simulated the 1000 normally incident Fe ions with energies of 100 MeV, 1 GeV and 10 GeV, respectively. As is shown in Figure 6, the target material is silicon and the ionization energy deposition is counted in the cylindrical coordinate system, because the radial distribution is approximately axisymmetric about the Z axis. Due to the relatively large feature size, the radial spacing and the axial spacing are set to 10 nm and 1 μm, respectively. Furthermore, the calculation accuracy can be further improved by reducing these spacings.

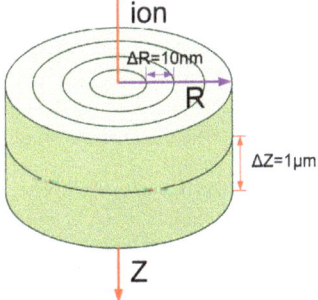

Figure 6. The schematic diagram of the cylindrical coordinate system used in the Geant4 simulation.

We could then obtain the e-h pairs' distribution (shown in Figure 7) by considering the average ionization energy 3.6 eV in silicon. From Figure 7, one could better visualize the differences in the EHPs' distribution generated by ions with different energies. As we can see, the 100 MeV Fe ion's incident depth is about 20 μm and its energy loss is limited in a relatively narrow radial distance. With the increase in ion energy, the ion's incident depth becomes larger and the EHPs' distribution can reach further radial distances. The EHPs' peak densities induced by Fe ions with energies 100 MeV, 1 GeV and 10 GeV are about 5.25×10^{21} cm^{-3}, 2.25×10^{21} cm^{-3} and 3.59×10^{20} cm^{-3}, respectively.

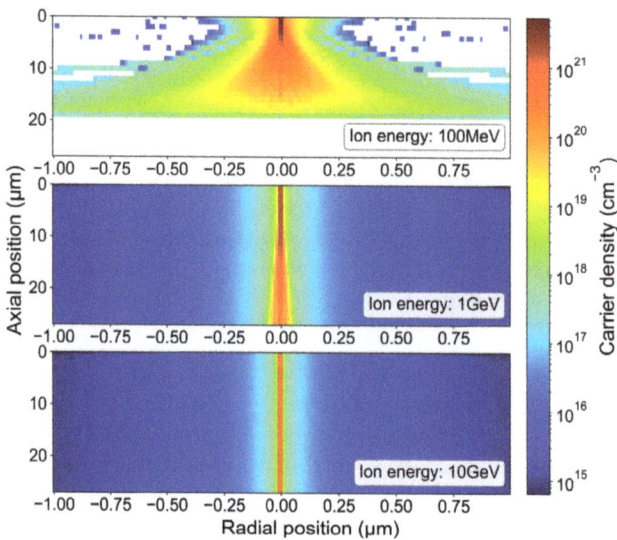

Figure 7. The 2-D profile of EHPs' distribution when Fe ion strike in silicon. (Note: the ion energies are 100 MeV, 1 GeV and 10 GeV, respectively.)

According to our previous study [14], the effective charge collection depth of the DUT is 20 µm; therefore, we should pay more attention to this distance. When the ion energy increases from 100 MeV to 10 GeV, the surface LET value decreases from about 0.31 pC/µm to 0.02 pC/µm.

3.2. Embedding Ion Track to TCAD Simulation

The most popular method to simulate the heavy ion-induced EHPs' distribution is the Gaussian distribution function in TCAD simulation. In general, the characteristic distance is a constant value and the LET value could be constant or be a function of incident depth. However, this is a simplified empirical model, and some details will be lost.

In the literature [19], the double Gaussian-fitted model is proposed to simulate the heavy ion-induced SEE in the TCAD toolkit, while the accuracy of the simulation is better than the simplistic Gaussian model as mentioned earlier. However, we need to spend a considerable amount of time manually fitting the necessary parameters at different incident depths to make the carrier density distribution as close as possible to the results of the Monte Carlo calculation.

In this paper, we chose a more direct method by defining the spatial distribution function (SDF) in TCAD simulation. As is shown in Figure 8, we should first define the heavy ion and target material in the Geant4 project. We then need to code the C++ script to read the energy deposition results from the Geant4 calculation and build the SDF to extract the corresponding carrier densities according to the different spatial locations, and then call this SDF function in the TCAD command file. Using this method, we could introduce the EHPs' distribution generated by Geant4 to the TCAD simulation directly.

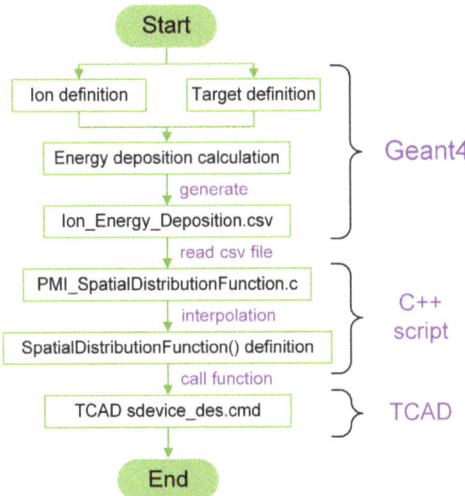

Figure 8. The workflow for introducing the EHPs' distribution generated by the Geant4 calculation to the TCAD simulation.

3.3. Simulation Results

According to the method in Sections 3.1 and 3.2 we introduced the EHPs' profiles of Fe ions with different energies to the 3-D TCAD process model, as in Figure 9. Crucially, we also need to optimize the meshing strategy to make the EHPs' profiles in the TCAD model and the Geant4 simulation results almost identical. Specifically, we need to use a tighter meshing (about 10 nm) in the central axis of the incident position, especially in the sensitive volumes such as the E/B/C stack structure and the junction regions.

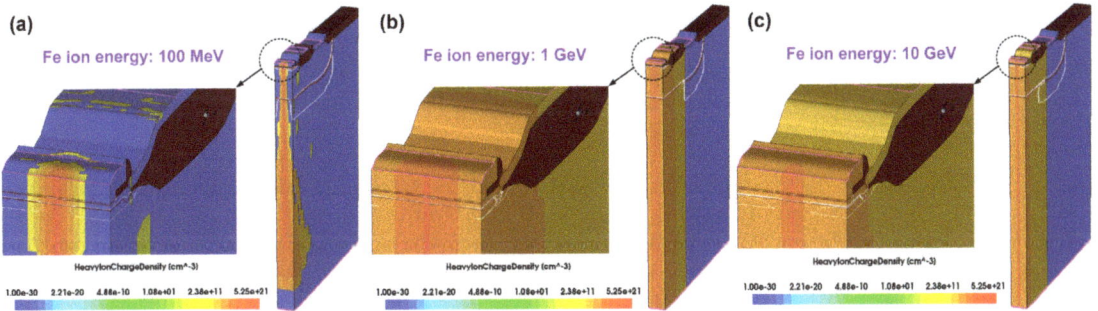

Figure 9. The EHPs' distribution of the TCAD simulation produced by incident Fe ions of different energies (**a**) 100 MeV, (**b**) 1 GeV and (**c**) 10 GeV. For a clearer comparison, the charge densities are adjusted to the same scale range.

We could then achieve the SET waveforms of the 3-D process simulation, as in Figure 10. To save simulation time, we have built half of the 3-D model and set the thickness in the z-direction to 1.5 μm. We have also chosen the emitter center normal incidence at room temperature. We could then obtain the information below.

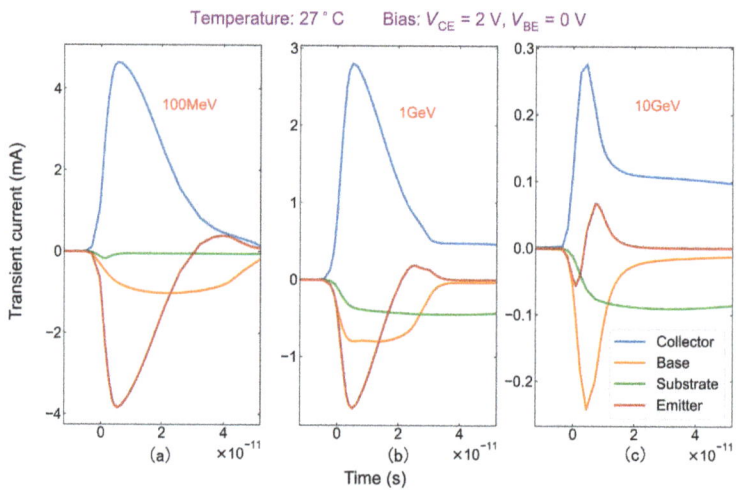

Figure 10. The transient waveforms of the TCAD simulation produced by incident Fe ions of different energies (**a**) 100 MeV, (**b**) 1 GeV and (**c**) 10 GeV at room temperature.

First, we are referring to the 100 MeV Fe ion incidence in Figures 9a and 10a. At this time, the Fe-induced EHPs have a peak density about 5.25×10^{21} cm^{-3} which is very high to cause the strongest emitter transient current and finally the strongest collector transient current (the transient peak could exceed 4 mA). However, the EHPs' lateral distribution distance is the smallest, and the total EHPs decrease rapidly with the depth of incidence. That is exactly why the substrate transient is much weaker than the emitter transient. Therefore, the collector transient's inflection point is dominated by the emitter transient.

Second, we will focus on the 1 GeV Fe ion incidence in Figures 9b and 10b. At this point, the EHPs' peak density is about 2.25×10^{21} cm^{-3} and the emitter transient is weaker than the 100 MeV Fe ion's case. However, the EHPs' lateral distribution distance is much larger, and the total EHPs stay almost constant along the incident depth. The ion shunt effect is still turned on and the share of the substrate transient peak increases significantly. The collector transient is then dominated by both the emitter transient and the substrate transient.

Third, when it comes to the 10 GeV Fe ion incidence in Figures 9c and 10c, the collector transient is the weakest. At this point, the incident Fe ions could deposit energy to the deepest distance. However, the EHPs' peak density is only about 3.59×10^{20} cm^{-3} and the ion shunt effect is turned off. According to the results in 2.3, it is hard to see the collector's transient inflection point at a cryogenic temperature.

In summary, the heavy ion's parameters will also have a significant impact on the SET's inflection point at cryogenic temperatures. Added to which, the conventional simulations of heavy ion-induced SEE generally set some fixed model parameters, which can help us to qualitatively analyze the experimental phenomena. However, some EHPs' distribution details are lost and this can affect the accuracy of our analysis. In particular, as the device feature size continues to decrease, the single heavy ion's incidence can affect the operation of more than one device. In this case, our proposed method could help to preserve as much detail as possible about the distribution of the initial ionized EHPs and obtain more accurate simulation results.

4. Conclusions

This paper presented an investigation into the inflection point of the single-event transient in a SiGe HBT at a cryogenic temperature. We focused on the impact of the heavy ion-induced initial EHPs' distribution on the inflection point by TCAD simulation. The collector's transient inflection point is jointly determined by the transient current of the emitter, substrate, and base. Moreover, the characteristics of the transient peaks with the temperature vary greatly among electrodes. The ions with high LET values will trigger the ion shunt effect which can lead to an earlier inflection point (about $-160\ °C$). And when the incident ion's LET value is too low to trigger the ion shunt effect, we will not see an inflection point of the collector's transient inflection point even at $-200\ °C$.

In addition, we proposed a method to directly introduce the initial ionized EHPs' distribution of the Geant4 simulation to the TCAD simulation which could improve the simulation accuracy and efficiency of the heavy ion-induced SEE.

However, the problem of how to improve the convergence of the TCAD 3-D simulation at cryogenic temperatures still remains to be solved and more efforts will be needed in the future.

Author Contributions: Conceptualization, X.P. and Y.L.; methodology, X.P. and C.L.; validation, H.Z., C.L. and X.P.; formal analysis, X.P.; investigation, X.P.; resources, X.P.; data curation, H.G.; writing—original draft preparation, X.P.; writing—review and editing, X.P. All authors have read and agreed to the published version of the manuscript.

Funding: This research was funded by the National Natural Science Foundation of China (Grant Nos. 61704127, 12005159, and 11775167).

Acknowledgments: The authors would like to thank Wen Zhao, Xinshuai Jiang and Yuanyuan Xue for their valuable discussion on Geant4 simulation.

Conflicts of Interest: The authors declare no conflict of interest.

References

1. Cressler, J.D. Radiation Effects in SiGe Technology. *IEEE Trans. Nucl. Sci.* **2013**, *60*, 1992–2014. [CrossRef]
2. Cressler, J.D. Silicon-Germanium as an Enabling Technology for Extreme Environment Electronics. *IEEE Trans. Device Mater. Reliab.* **2010**, *10*, 437–448. [CrossRef]
3. Rieh, J.S.; Jagannathan, B.; Greenberg, D.R.; Meghelli, M.; Rylyakov, A.; Guarin, F.; Yang, Z.; Ahlgren, D.C.; Freeman, G.; Cottrell, P.; et al. SiGe Heterojunction Bipolar Transistors and Circuits Toward Terahertz Communication Applications. *IEEE Trans. Microw. Theory Tech.* **2004**, *52*, 2390–2408. [CrossRef]
4. Ildefonso, A.; Lourenco, N.E.; Fleetwood, Z.E.; Wachter, M.T.; Tzintzarov, G.N.; Cardoso, A.S.; Roche, N.J.H.; Khachatrian, A.; McMorrow, D.; Buchner, S.P.; et al. Single-Event Transient Response of Comparator Pre-Amplifiers in a Complementary SiGe Technology. *IEEE Trans. Nucl. Sci.* **2017**, *64*, 89–96. [CrossRef]
5. Lourenco, N.E.; Fleetwood, Z.E.; Ildefonso, A.; Wachter, M.T.; Roche, N.J.H.; Khachatrian, A.; McMorrow, D.; Buchner, S.P.; Warner, J.H.; Itsuji, H.; et al. The Impact of Technology Scaling on the Single-Event Transient Response of SiGe HBTs. *IEEE Trans. Nucl. Sci.* **2017**, *64*, 406–414. [CrossRef]
6. Reed, R.A.; Marshall, P.W.; Pickel, J.C.; Carts, M.A.; Fodness, B.; Guofu, N.; Fritz, K.; Vizkelethy, G.; Dodd, P.E.; Irwin, T.; et al. Heavy-ion broad-beam and microprobe studies of single-event upsets in 0.20-μm SiGe heterojunction bipolar transistors and circuits. *IEEE Trans. Nucl. Sci.* **2003**, *50*, 2184–2190. [CrossRef]
7. Guofu, N.; Hua, Y.; Varadharajaperumal, M.; Yun, S.; Cressler, J.D.; Krithivasan, R.; Marshall, P.W.; Reed, R. Simulation of a new back junction approach for reducing charge collection in 200 GHz SiGe HBTs. *IEEE Trans. Nucl. Sci.* **2005**, *52*, 2153–2157. [CrossRef]
8. Vizkelethy, G.; Reed, R.A.; Marshall, P.W.; Pellish, J.A. Ion beam induced charge (IBIC) studies of silicon germanium heterojunction bipolar transistors (HBTs). *Nucl. Instrum. Methods Phys. Res. Sect. B Beam Interact. Mater. At.* **2007**, *260*, 264–269. [CrossRef]
9. Sutton, A.; Moen, K.; Cressler, J.D.; Carts, M.; Marshall, P.; Pellish, J.; Ramachandran, V.; Reed, R.; Alles, M.; Niu, G. Proton-induced SEU in SiGe digital logic at cryogenic temperatures. *Solid-State Electron.* **2008**, *52*, 1652–1659. [CrossRef]
10. Xu, Z.; Niu, G.; Luo, L.; Cressler, J.D.; Alles, M.L.; Reed, R.; Mantooth, H.A.; Holmes, J.; Marshall, P.W. Charge Collection and SEU in SiGe HBT Current Mode Logic Operating at Cryogenic Temperatures. *IEEE Trans. Nucl. Sci.* **2010**, *57*, 2085050. [CrossRef]
11. Luo, L.; Niu, G.; Moen, K.A.; Cressler, J.D. Compact Modeling of the Temperature Dependence of Parasitic Resistances in SiGe HBTs Down to 30 K. *IEEE Trans. Electron Devices* **2009**, *56*, 2169–2177. [CrossRef]
12. Knudson, A.R.; Campbell, A.B.; Hauser, J.R.; Jessee, M.; Stapor, W.J.; Shapiro, P. Charge Transport by the Ion Shunt Effect. *IEEE Trans. Nucl. Sci.* **1986**, *33*, 1560–1564. [CrossRef]
13. Pan, X.; Guo, H.; Feng, Y.; Liu, Y.; Zhang, J.; Fu, J.; Yu, G. Temperature dependence of single event transient in SiGe HBT for cryogenic application. *Chin. Phys. B* **2022**. [CrossRef]

14. Pan, X.; Guo, H.; Feng, Y.; Liu, Y.; Zhang, J.; Li, Z.; Luo, Y.; Zhang, F.; Wang, T.; Zhao, W.; et al. Single event transient characterization of SiGe HBT by SPA experiment and 3-D process simulation. *Sci. China Technol. Sci.* **2022**, *65*, 1193–1205. [CrossRef]
15. Niu, G.; Krithivasan, R.; Cressler, J.D.; Riggs, P.A.; Randall, B.A.; Marshall, P.W.; Reed, R.A.; Gilbert, B. A comparison of SEU tolerance in high-speed SiGe HBT digital logic designed with multiple circuit architectures. *IEEE Trans. Nucl. Sci.* **2002**, *49*, 3107–3114. [CrossRef]
16. Dodd, P.E.; Musseau, O.; Shaneyfelt, M.R.; Sexton, F.W.; D'Hose, C.; Hash, G.L.; Martinez, M.; Loemker, R.A.; Leray, J.L.; Winokur, P.S. Impact of ion energy on single-event upset. *IEEE Trans. Nucl. Sci.* **1998**, *45*, 2483–2491. [CrossRef]
17. Raine, M.; Gaillardin, M.; Sauvestre, J.-E.; Flament, O.; Bournel, A.; Aubry-Fortuna, V. Effect of the Ion Mass and Energy on the Response of 70-nm SOI Transistors to the Ion Deposited Charge by Direct Ionization. *IEEE Trans. Nucl. Sci.* **2010**, *57*, 1892–1899. [CrossRef]
18. Weller, R.A.; Mendenhall, M.H.; Reed, R.A.; Schrimpf, R.D.; Warren, K.M.; Sierawski, B.D.; Massengill, L.W. Monte Carlo Simulation of Single Event Effects. *IEEE Trans. Nucl. Sci.* **2010**, *57*, 1726–1746. [CrossRef]
19. McPherson, J.A.; Kowal, P.J.; Pandey, G.K.; Chow, T.P.; Ji, W.; Woodworth, A.A. Heavy Ion Transport Modeling for Single-Event Burnout in SiC-Based Power Devices. *IEEE Trans. Nucl. Sci.* **2019**, *66*, 474–481. [CrossRef]

Disclaimer/Publisher's Note: The statements, opinions and data contained in all publications are solely those of the individual author(s) and contributor(s) and not of MDPI and/or the editor(s). MDPI and/or the editor(s) disclaim responsibility for any injury to people or property resulting from any ideas, methods, instructions or products referred to in the content.

Communication

Study of Single Event Latch-Up Hardness for CMOS Devices with a Resistor in Front of DC-DC Converter

Jindou Xin [1,2], Xiang Zhu [1,2,*], Yingqi Ma [1,2,*] and Jianwei Han [1,2]

[1] State Key Laboratory of Space Weather, National Space Science Center, Chinese Academy of Sciences, Beijing 100190, China
[2] School of Astronomy and Space Science, University of Chinese Academy of Sciences, Beijing 100049, China
* Correspondence: zhuxiang@nssc.ac.cn (X.Z.); myq@nssc.ac.cn (Y.M.)

Abstract: Bulk silicon Complementary Metal Oxide Semiconductor (CMOS) devices have distinct single event latch-up (SEL) problems in aerospace. Therefore, it is essential that CMOS devices are designed with appropriate circuit-level methods. Traditional resistor hardness satisfies the current aerospace trend of low cost, high performance, and miniaturization. Therefore conventional resistor hardness is often applied in circuit-level designs due to the reduction of latch-up current. In circuits containing a DC-DC buck converter, the resistor is connected to the back of the converter in the traditional method. However, the traditional method is unable to take devices out of the latch-up owing to the small resistance range. To solve this problem, the paper proposes an improved design for the resistor in front of the DC-DC buck converter. The proposed method enables the devices to exit the latch-up by increasing the resistance range according to the input characteristic of the DC-DC buck converter. The paper quantifies the range of the resistor through the parametric model containing the resistor and the DC-DC buck converter. Two CMOS devices are chosen for pulsed laser experiments, verifying that the proposed method increases the resistance ranges by 300% to 400% compared to the conventional method. It is also demonstrated that the proposed method exits the devices from latch-up within the resistor ranges. That is, the resistance ranges of 34 Ω~41 Ω and 51 Ω~56 Ω reduce the latch-up currents of the devices to below holding currents of 72.1 mA and 24.2 mA, respectively.

Keywords: CMOS devices; single event latch-up (SEL); single event effect (SEE); resistor; pulsed laser

1. Introduction

Bulk silicon Complementary Metal Oxide Semiconductor (CMOS) devices are widely applied in satellite electronic systems owing to their low power consumption, high integration, and low production cost [1,2]. However, CMOS devices are often subject to collisions with high-energy protons and heavy ions from the cosmic space environment. Therefore, CMOS devices are susceptible to Single Event Effect (SEE) [3–6]. In particular, Single Event Latch-up (SEL), a special SEE, can alter devices' currents and even cause devices to burn up in severe cases [7–10]. From a circuit-level hardness perspective, SEL is generated by the conduction of parasitic PNP and NPN transistors inside the devices, creating low resistance paths between the devices' power supplies and grounds with resulting devices' current rise when the devices are exposed to the space radiation [11–14]. The hazard of SEL to CMOS devices is gradually increasing as commercial aerospace applications become more widespread [15,16]. Consequently, SEL hardness assurance has developed into an extremely significant challenge for CMOS devices in aerospace applications [17].

From a circuit-level hardness perspective, the devices will exit the SEL when the latch-up currents or latch-up voltages fall below the holding currents or voltages. To improve the SEL immunity of CMOS devices, three dominant research directions are proposed, respectively process-level hardness, layout-level hardness, and circuit-level

hardness [18–20]. Both process-level and layout-level designs enable the devices to be protected from SEL, while neither is applicable to commercial devices that have already been designed for production [21–23]. For the SEL problem in commercial CMOS devices, circuit-level designs are primarily adopted [24,25], which include power off-restart, constant current source, and cold backup. The power off-restart [26–28] adopts power disconnection to eliminate the latch-up of the devices. However, the approach will result in a functional interruption of the devices during power loss. The constant current source [29] keeps the devices' currents below the latch-up holding currents through a constant current source. The method is effective in increasing the latch-up hardness assurance of the devices, but it will limit the dynamic currents and affect the dynamic functionalities of the devices. The cold backup [30] uses a cold backup to set up multiple identical SEL-sensitive devices. If the current device occurs the latch-up, it will switch to the backup device to complete task requirements. The cold backup approach can effectively mitigate the latch-up hazard. However, it leads to the problem of manufacturing complex circuit structures and increased power consumption.

To address the above issues, a resistor in series behind a DC-DC buck converter (front of CMOS devices) is usually applied in conventional circuit-level hardness. In aerospace circuits, the satellite power supplies are 28 V and above, with the devices' voltages often at 5.5 V, 3.3 V and below. Therefore, to ensure that the devices are safely connected to the satellite power supplies, the DC-DC buck converter should be connected between the power supplies and the devices. In circuits containing a buck converter, a resistor is connected in series at the output of the DC-DC buck converter (i.e., the input of the devices). The resistor can effectively reduce latch-up current and latch-up harm by dividing voltages and limiting currents. Nevertheless, the dividing voltages of the resistor cannot exceed the normal operating voltage ranges of the devices, which will result in the resistor taking small ranges of values. The drawback will further cause the devices to fail to exit the latch-up. The more detailed deficiencies of the conventional resistor are described in depth in Section 2.1.

To overcome the limitation of the conventional resistor, the paper proposes an improved design with a resistor placed in front of the DC-DC buck converter. The design allows the devices to exit the latch-up by increasing the resistance range combined with the wider input voltage range of the DC-DC buck converter. The larger voltage input range of the converter indicates a wider resistance range. Since the latch-up current decreases as the resistance increases, therefore, the latch-up hazard of the devices becomes smaller with the higher resistance. When the latch-up current is reduced below the holding current, the devices will exit the latch-up. In order to quantify the range of the resistor that brings the devices out of the latch-up, the paper investigates the resistance calculation method by building a parametric model containing the resistor and the converter. After that, pulsed laser experiments will be implemented using two CMOS devices. As well, it is verified that the proposed method enables the devices to exit the latch-up within the resistance range.

The primary contributions of the paper are as follows:

(1) The proposed method addresses the prominent limitation of the traditional method. Conventional resistor hardness design only acts as a current limit for the latch-up, and does not allow the devices to exit the latch-up. The method proposed in the paper enables the device to exit the latch-up by combining the resistor in concert with the DC-DC buck converter.

(2) The method of taking the resistance is studied to improve the lack of mathematical analysis of the resistor in traditional latch-up hardness. The paper systematically analyzes the operating principle of the resistor in front of the DC-DC buck converter, establishes the corresponding parametric model, and proposes the method of taking the resistance. It has extremely valuable guidance for the proposed method in practical hardness assurance applications.

(3) The proposed method has the advantages of continuous operation with power, maintaining the dynamic functions of the devices, and occupying a smaller circuit design

area compared to power-off restart, constant current source, and cold backup in the circuit-level hardness methods. Furthermore, the proposed method is compatible with the current trend of low cost, high performance, and miniaturization in aerospace.

The paper is organized as follows: In Section 2, comparing the conventional method with the proposed method, it is demonstrated that the proposed method enables the device to exit the latch-up within the range of the resistor. As well, the range of resistance is quantified. In Section 3, to verify the SEL hardness performance of the proposed method, laser experiments are carried out. In Section 4, the paper discusses the resistive power consumption in the proposed method. Finally, a conclusion is given in Section 5.

2. Method

2.1. Inadequacy of Conventional Method

The section will provide comprehensive descriptions including the conventional resistor's connection, the principle of resistor operation, and the constraints of the traditional method. It focuses on the problem of the traditional method by elaborating on the latch-up hardness principle. Figure 1 is a schematic diagram of the circuit for a conventional resistor hardness design.

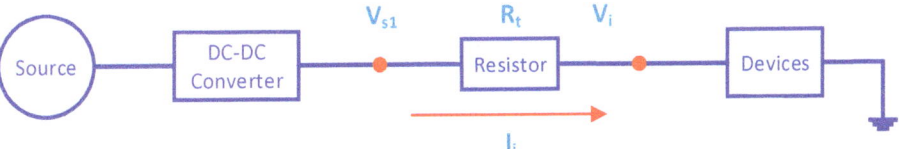

Figure 1. Schematic diagram of the circuit for a conventional resistor hardness design.

The R_t represents the resistor in the conventional method, which is connected at the output of the DC-DC buck converter. The V_{s1} denotes the output voltage of the converter. The V_i and I_i represent the voltage and current of the device, respectively. Thus, the is expressed as:

$$(V_{s1} - V_i) \cdot \frac{1}{R_t} = I_i \tag{1}$$

According to Equation (1), the resistor changes the device current by limiting the circuit current. The device current decreases as the resistance increases. It indicates that the higher the resistance, the lower the device's latch-up current. However, the increase in resistance is limited, since it will result in a reduced voltage of the device. When the device's voltage is below the normal voltage tolerance range, it will prevent the device from operating properly. Therefore, the resistor is subject to certain constraints in the actual latch-up hardness.

Based on the above resistive hardness principle in combination with the latch-up property, two constraints are derived [31–33]:

- Condition 1: The resistor does not affect the normal operation of the device. The operating voltage of the device should not exceed the normal voltage tolerance range. Otherwise the device cannot operate successfully. The voltage tolerance range is typical −10% to 10% of the rated voltage.
- Condition 2: Reduce the latch-up current to below the SEL holding current when the device is experiencing the latch-up [34–36]. According to the latch-up criterion, when the latch-up current drops below the latch-up maintenance point, the device will exit the latch-up state because the latch-up current cannot be maintained.

The traditional method of taking the resistance will be studied with respect to the constraints. As well, the range of resistance is researched to illustrate the latch-up hardness problem that exists with the conventional method. The 3.3 V CMOS process device is selected as the object of the study, i.e., V_{s1} = 3.3 V, then the voltage tolerance is −0.3 V~0.3 V.

The range of R_{t1} satisfying condition 1 in the conventional method is represented as:

$$0 \leq R_{t1} \leq \frac{V_{s1} - V_i}{I_i} \quad (2)$$

where $V_{s1} - V_i$ denotes the voltage tolerance. From Equation (2), it is known that R_{t1} has a harsh range of $0 \sim \frac{0.3}{I_i}$ Ω owing to its small voltage tolerance range.

The range of resistance R_{t2} fulfilling condition 2 in the conventional method is expressed as:

$$R_{t2} \geq \frac{V_{s1} - V_h}{I_l} \quad (3)$$

where V_h and I_l represent the latch-up holding voltage and latch-up current, respectively. As the resistor is required to meet both the normal operation and to make the device exit the latch-up, it is obtained that $R_{t2} \leq R \leq R_{t1}$. According to the test data of several devices, V_h is about 1.32 V~2.45 V, which means that $V_{s1} - V_h > V_{s1} - V_i$. Usually I_l is 2~3 times and more than I_i, as well as combined with the actual data, it is evident that $R_{t1} < R_{t2}$. It indicates that the conventional method does not allow the device to exit the latch-up.

To address the limitation of the conventional method, the paper proposes a latch-up hardness design with a resistor placed in front of the DC-DC buck converter. The details of the proposed method will be described in the next section.

2.2. The Proposed Method

The section describes in detail the connection method, operating principle, design advantages, parameter model and resistance-taking the method of the proposed method. Emphasis will be placed on the design advantages of the proposed method to allow the device to exit the latch-up and the discussion of the resistor-taking method by building a parametric model.

2.2.1. Take the Device out of the Latch-Up

Figure 2 depicts the schematic circuit diagram of the proposed hardness method. The resistor employed in the proposed method is named R_p, which is connected to the input of the DC-DC converter. V_{s2} indicates the supply voltage to which the converter is attached. V_d and I_d separately represent the input voltage and input current of the converter. Thus, R_p is given as:

$$(V_{s2} - V_d) \cdot \frac{1}{R_p} = I_d \quad (4)$$

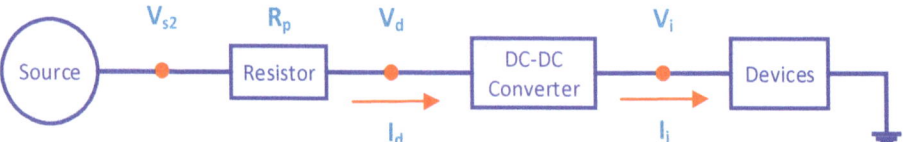

Figure 2. Schematic diagram of the connection of the resistor in the proposed method.

From Equation (4), it is obvious that the proposed method operates by varying the input current of the DC-DC converter to regulate the current of the device. The hardness mechanism of R_p is that R_p reduces the device current by limiting the input current of the converter. As well, the device current decreases as the input current is reduced. However, compared to conventional design, the advantage of the proposed approach is that the objective of exiting the device from latch-up will be achieved by increasing the voltage tolerance based on the wide input range of the DC-DC converter. The input voltage range of the DC-DC buck converter is more extensive than the device's voltage tolerance range, for example, the input range of LTM4644 converter is 2.4 V~14 V, which is much higher than the 0.3 V voltage tolerance of the 3.3 V device. The effect of a larger voltage tolerance is to make a larger range of resistance that satisfies condition 1. Following the operating principle of R_p, it is known that a larger resistor makes the device latch-up current lower by further reducing the input current of the converter. When the device latch current falls below the latch-up holding current, the device will exit the latch-up.

2.2.2. Method of Taking the Resistance

To further investigate the proposed hardness method of taking the resistance that simultaneously meets conditions 1 and 2, a parametric model of the resistor placed at the input of the DC-DC buck converter is developed, as shown in Figure 3. The parametric model consists of supply voltage V_{s2}, resistor R_p, switch S, inductor L, capacitor C, diode D_i and feedback network. The feedback network is composed of resistors R_1 and R_2, an error amplifier and a duty ratio modulator. The function of the feedback network is to generate the duty cycle signal and control the state of the switch.

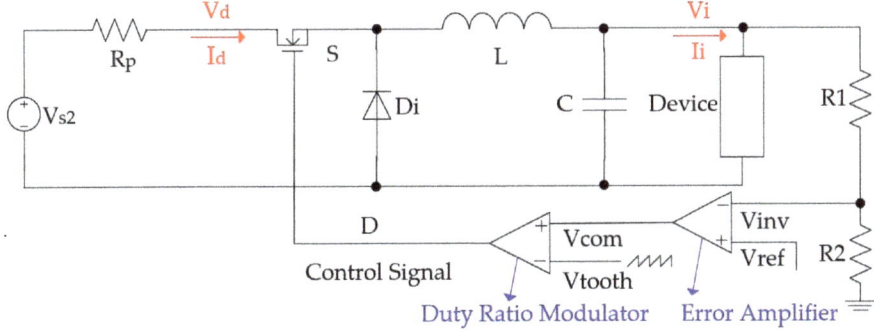

Figure 3. Parametric model of the resistor placed at the input of the DC-DC buck converter.

According to the I-V characteristic of R_p, R_p is represented as:

$$\frac{V_{s2} - V_d}{R_p} = I_d \tag{5}$$

where V_d and I_d denote the input voltage and input current of the DC-DC buck converter, respectively.

Following the DC-DC buck converter power conservation principle and duty cycle equation, it is known that:

$$V_d \cdot I_d = \frac{1}{\mu} \cdot V_i \cdot I_i \tag{6}$$

$$V_i = D V_d \tag{7}$$

where μ and D respectively denote the conversion efficiency and duty cycle of the converter. μ and D are related to the operating voltage and current of the converter which are available according to the datasheet or actual test values. Substituting Equations (6) and (7) into Equation (5) gives that:

$$- D^2 R_p I_i + D\mu V_{s2} - \mu V_i = 0 \tag{8}$$

From condition 1, the range of R_{p1} that satisfies the normal operation of the device is derived as:

$$R_{p1} \leq \frac{R_{eqn}(D\mu V_{s2} - \mu V_{ii})}{D^2 V_{ii}} \tag{9}$$

where R_{eqn} represents the equivalent resistance of the device in the normal state. V_{ii} indicates the minimum value of the device voltage V_i in the voltage tolerance range. For 3.3 V devices, V_{ii} is typically 3.0 V.

The boundary formula for the R_{p2} meeting condition 2 is given by:

$$R_{p2} \geq \frac{R_{eql}(D\mu V_{s2} - \mu V_h)}{D^2 V_i} \qquad (10)$$

where R_{eql} signifies the equivalent resistance of the device in the latch-up state. In summary, the range of R_p that simultaneously satisfies the normal operation of the device and enables the device to exit the latch-up is:

$$R_{p1} \leq R_p \leq R_{p2} \qquad (11)$$

3. Pulsed Laser Experiments

3.1. Experimental Setup and Devices Selection

To verify the latch-up hardness performance of the proposed method and the method of taking the resistance, pulsed laser experiments are carried out. The mechanism of SEL induced by pulsed laser experiments in CMOS devices is approximately the same as that of heavy ion experiments, both of which induce latch-up in CMOS devices by ionization of electron-hole pairs. However, the primary differences between pulsed laser experiments and heavy ion experiments are the small spot diameter and high resolution of the pulsed laser, which allows accurate simulation of SEE caused by individual high-energy particles in space [37,38]. In addition, the irradiation intensity and irradiation time of CMOS devices by the pulsed laser are precisely controllable [39]. The pulsed laser test setup adopts the self-researched equipment of the National Space Science Centre of the Chinese Academy of Sciences. Figure 4 shows a schematic diagram of the pulsed laser unit. The laser setup consists of the component laser generator, the optical path system, the 3D mobile table, the synchronization control system and the host computer [40–42]. Table 1 shows the main parameters of the laser equipment.

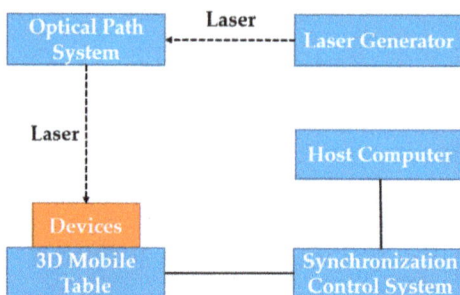

Figure 4. A schematic diagram of the pulsed laser equipment.

Table 1. The main parameters of the laser equipment.

Machine Type	Wavelength	Wideband	Frequency	Energy
Nd:YAG	1064 nm	25 ps	1~1k Hz	1.5 nJ

Based on the operating principle and parametric model analysis of the proposed method, it is shown that the method in the paper is generally applicable to latch-up sensitive devices. To verify the applicability of the proposed method, two CMOS chips, A3PE1500 and AD7472, are selected as the test objects for laser experiments. Table 2 summarizes the key parameters of the two subjects. The normal operating and latch-up holding currents for device 1 are 0.072 A and 0.088 A respectively; for device 2 the normal operating and latch-up holding currents are 0.022 A and 0.031 A accordingly.

Table 2. The key parameters of the two test subjects.

Device Number	Model	Operating Voltage	Operating Current
Device 1	A3PE1500	3.3 V	72 mA
Device 2	AD7472	3.3 V	22 mA
Device Number	**SEL Current**	**Holding Voltage**	**Holding Current**
Device 1	356.6 mA	2.1 V	88 mA
Device 2	97.3 mA	1.7 V	31 mA

The devices will be triggered to produce the SEL when a pulsed laser is an incident on the active regions inside the devices. To ensure that the laser energy is effectively injected into the active areas, the devices must be back-opened before laser experiments. Figure 5 illustrates the practical picture of the devices in the pulsed laser experiments. The diagram contains mainly the pulsed laser, the devices and the DC-DC buck converter. The input of the DC-DC buck converter is connected to the supply voltage, and the output is attached to the power supply of the devices.

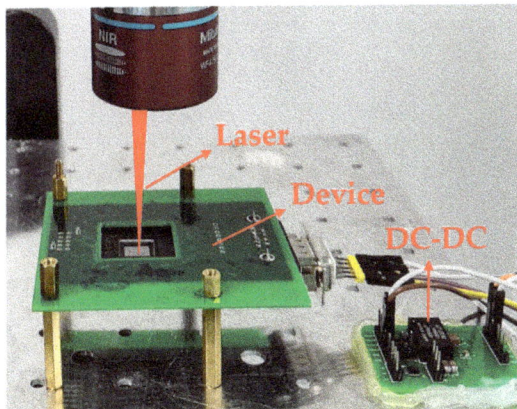

Figure 5. The practical picture of the devices in the pulsed laser experiments.

3.2. Experimental Method

The purposes of the experiments are to verify that the proposed method enables the devices to exit the latch-up as well as the resistor-taking method. To better illustrate the experimental results, comparative tests are designed in the paper for the conventional and experimental groups respectively. The resistor of the conventional group is connected to the output of the DC-DC buck converter, while the resistor of the experimental group is attached to the input of the DC-DC buck converter. Besides, the experimental manipulation is the same for both. The experimental operation is divided into three major steps, which are adjusting the position of the devices to be tested, testing the functions of the devices and changing the resistance.

- First, the devices are adjusted to a horizontal state by adjusting the 3D moving table to ensure that the laser energy is injected into the devices at the same depth.
- Then, with the circuit connected correctly, the power is turned on to test the functions of the devices. The voltages and currents of the devices in the initial state and the latch-up state in the two sets of experiments are detected and recorded respectively.
- Finally, by connecting different resistors, the electrical parameters of the devices in the initial state and in the latch-up state are recorded in both sets of experiments.

When the initial voltages of the devices exceed the voltage tolerance range, it means that the resistance is already the maximum value, and the experiment will end at this time. The following experimental results will be obtained by collating the relevant test data.

3.3. Experimental Results

3.3.1. Exiting the Devices from the Latch-Up by the Proposed Method

To demonstrate that the proposed method enables the devices to exit the latch by increasing the resistance range, the section first investigates the resistance range of the proposed method compared to the conventional design. Next, a comparative analysis is performed on the latch-up currents variation over the range of resistance values.

Figure 6 depicts the ranges of the resistance under the conventional method and the proposed design, respectively. Figure 6a shows a resistance range of 0 to 8.2 Ω in the conventional method within the normal operating voltage range of device 1. The proposed method, however, has a resistance range of 0 to 41 Ω. In comparison to the conventional method, the proposed method increases the resistance range by up to 400%. Figure 6b depicts the resistance ranges of 0~14 Ω and 0~56 Ω for the conventional method and the proposed method, respectively, in the operating voltage range of device 2. A 300% increase in resistance range can be achieved with the proposed design. It is concluded that the proposed design improves the resistance range by 300% to 400%.

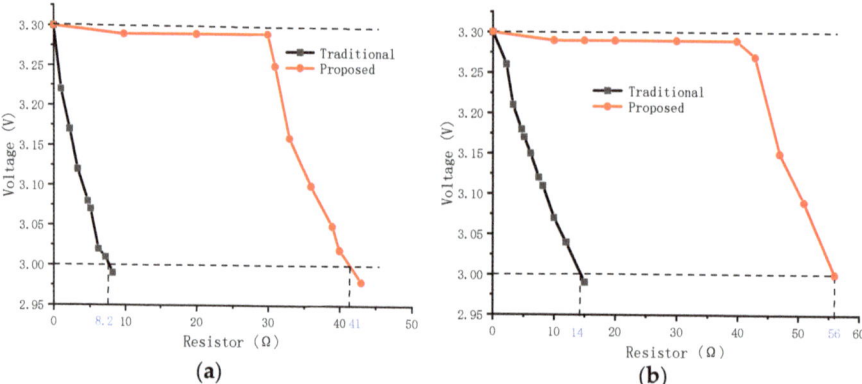

Figure 6. The ranges of the resistance under the conventional method and the proposed design, respectively. (**a**) Description of the ranges of the resistance in device 1; (**b**) Description of the ranges of the resistance in device 2.

The increase in the resistance range will further improve the latch-up hardness of devices. Under certain conditions, the resistor will make the devices drop out of the latch-up. Figure 7 shows the relationships between the resistor and the latch-up current of the two devices under two different methods. Figure 7a describes the conventional method of reducing the device 1 latch-up current to 134.7 mA at resistor maximum. Even so, the device 1 remains in an abnormal latch-up state. Nevertheless, the proposed method reduces the device current to below the latch-up holding current of 72.1 mA at a resistance of 34 Ω. Consequently, the proposed hardness design with resistances of 34 Ω and above will keep device 1 from latch-up. Figure 7b illustrates that the conventional method and the proposed design respectively reduce the device 2 latch-up current to 50.3 mA and 24.2 mA (below the holding current). However the proposed method improves the latch-up hardness of the device by making it latch-up-free under certain conditions because of the large resistance range.

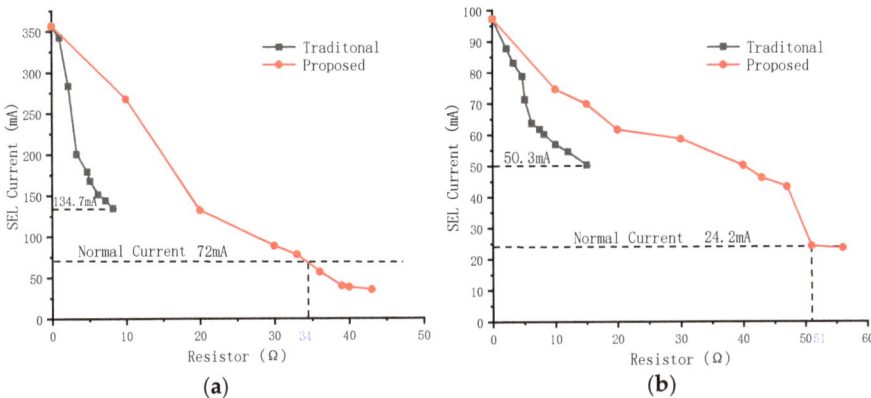

Figure 7. Graph of resistance versus latch-up current for two devices under two different methods. (**a**) Relationship between resistance and latch-up current of device 1; (**b**) Relationship between resistance and latch-up current of device 2.

3.3.2. Verification of the Resistance-Taking Method

To verify the resistance range of the proposed method, the main parameters of devices in Table 2 are substituted into Equations (4) and (11) respectively. The theoretical range of resistance is calculated to be 37 Ω~48 Ω and 53 Ω~62 Ω for device 1 and device 2 separately. Table 3 indicates the theoretical and actual ranges of resistance. According to the experimental results in Figure 6, it can be seen that R_{p1} of the two devices are 0 Ω~41 Ω and 0 Ω~56 Ω, respectively. According to the data in Figure 7, it is evident that R_{p2} of the two devices are 34 Ω and above, and 51 Ω and above, accordingly.

Table 3. The theoretical and actual ranges of resistance.

Devices	Type	R_{p1} (Ω)	R_{p2} (Ω)	$R_{p1} \cap R_{p2}$ (Ω)
Device 1	Theory Value	0~48	≥37	37~48
	Test Value	0~41	≥34	34~41
Device 2	Theory Value	0~62	≥53	53~62
	Test Value	0~56	≥51	51~56

It is noticed that the actual results of the resistance are smaller than the theoretical results. This phenomenon may be due to the capacitor and inductor of the converter having parasitic resistance in the actual circuit, resulting in an actual low resistance.

4. Discussion of Resistor Power Consumption

The issue with the proposed method is that it will cause an increase in the power consumption of the circuits, due to the increased resistance range compared to the traditional design. Figure 8 summarizes the power consumption data of the proposed resistor versus the conventional resistor in the devices. The power consumption of the resistor in the proposed method is about 0.11 W to 0.19 W, which is more than 50% higher than that of the conventional design.

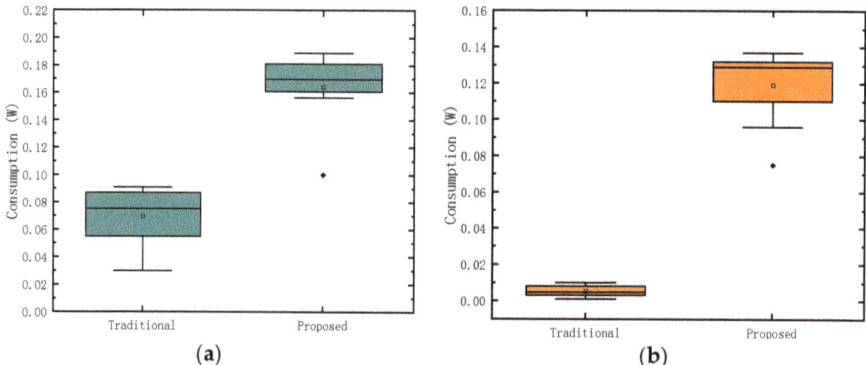

Figure 8. Description of the power consumption generated by the resistor in the two devices with two different methods. (**a**) Displays the power consumption generated by the resistor in device 1; (**b**) Introduction of the power consumption generated by the resistor in device 2.

To reduce the power consumption problem of the proposed method, the approach of reducing the converter input voltage is proposed. The minimum input voltage of the DC-DC buck converter is above the device voltage, i.e., $V_{s2} > 3.3$ V. Therefore, 6 V, 5 V, and 4.5 V supply voltages are chosen to explore the effect of voltage reduction on resistor power consumption. Figure 9 shows the resistive power consumption for the two devices with supply voltages of 6 V, 5 V, and 4.5 V, correspondingly. It is observed that the resistive power consumption reduces with decreasing supply voltage. Compared to the resistor power consumption with a supply voltage of 6 V, the resistor power consumption with a supply voltage of 4.5 V is reduced by more than 87% to about 0.06 W~0.08 W. Power consumption is acceptable in engineering.

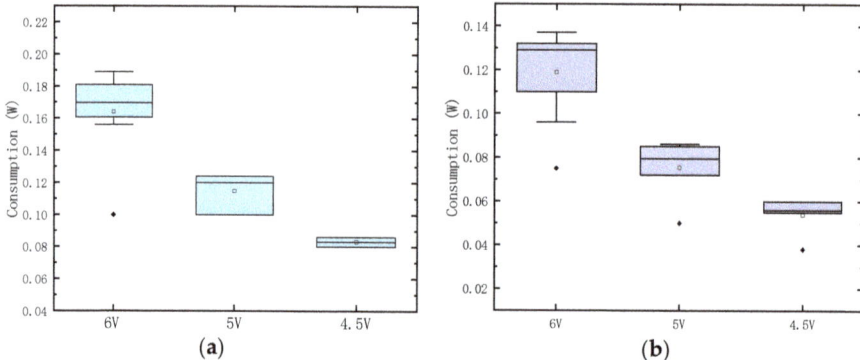

Figure 9. Power consumption of the resistor at 6 V, 5 V, and 4.5 V for the two devices separately. (**a**) The power consumption generated by the resistor in device 1; (**b**) The power consumption generated by the resistor in device 2.

5. Conclusions

The paper proposes a circuit-level SEL hardness design for a resistor in front of a DC-DC buck converter. The method improves the latch-up hardness performance by improving the resistance range compared to the conventional hardness design. The proposed method enables devices to exit the latch-up when the resistance takes the value of $R_{p2} \sim R_{p1}$. The proposed method is validated with the A3PE1500 and AD7472 CMOS devices to effectively increase the resistance range by 300% to 400%. It is also demonstrated that the resistor enables devices to operate normally and exit the latch-up within the boundary range when devices are suffering from SEL.

Author Contributions: Conceptualization, J.X. and J.H.; methodology, J.X., X.Z. and Y.M.; validation, J.X., Y.M., J.H. and X.Z.; formal analysis, J.X.; investigation, J.X.; resources Y.M. and X.Z.; data curation, J.X.; writing—original draft preparation, J.X.; writing—review and editing, J.X., Y.M. and X.Z.; visualization, J.X.; supervision, J.H., Y.M. and X.Z.; project administration, Y.M. and J.H.; funding acquisition, X.Z. All authors have read and agreed to the published version of the manuscript.

Funding: This research was funded by the National Natural Science Foundation of China (No. U2241280).

Data Availability Statement: The data used to support the findings of this study are available from the corresponding author upon request.

Acknowledgments: This work was supported by the Effects Centre of State Key Laboratory of Space Weather of the National Space Science Center, and research group members. The authors offer deep appreciation for their kind support.

Conflicts of Interest: The authors declare no conflict of interest.

References

1. Guo, J.; Zhu, L.; Sun, Y.; Cao, H.; Huang, H.; Wang, T.; Qi, C.; Zhang, R.; Cao, X.; Xiao, L.; et al. Design of Area-Efficient and Highly Reliable RHBD 10T Memory Cell for Aerospace Applications. *IEEE Trans. VLSI Syst.* **2018**, *26*, 991–994. [CrossRef]
2. Li, Z.; Berti, L.; Wouters, J.; Wang, J.; Leroux, P. Characterization of the Total Charge and Time Duration for Single-Event Transient Voltage Pulses in a 65-Nm CMOS Technology. *IEEE Trans. Nucl. Sci.* **2022**, *69*, 1593–1601. [CrossRef]
3. Liu, J.; Zhou, Z.; Wang, D.; Zhou, S.-Q.; Sun, X.-M.; Ren, W.-P.; You, B.-H.; Gao, C.-S.; Xiao, L.; Yang, P.; et al. Prototype of Single-Event Effect Localization System with CMOS Pixel Sensor. *Nucl. Sci. Tech.* **2022**, *33*, 136. [CrossRef]
4. Prasad, G.; Mandi, B.C.; Ali, M. Double-Node-Upset Aware SRAM Bit-Cell for Aerospace Applications. *Microelectron. Reliab.* **2022**, *133*, 114526. [CrossRef]
5. Autran, J.L.; Moindjie, S.; Munteanu, D.; Dentan, M.; Moreau, P.; Pellissier, F.P.; Bucalossi, J.; Borgese, G.; Malherbe, V.; Thery, T.; et al. Real-Time Characterization of Neutron-Induced SEUs in Fusion Experiments at WEST Tokamak During D-D Plasma Operation. *IEEE Trans. Nucl. Sci.* **2022**, *69*, 501–511. [CrossRef]
6. Roffe, S.; Akolkar, H.; George, A.D.; Linares-Barranco, B.; Benosman, R.B. Neutron-Induced, Single-Event Effects on Neuromorphic Event-Based Vision Sensor: A First Step and Tools to Space Applications. *IEEE Access* **2021**, *9*, 85748–85763. [CrossRef]
7. Biereigel, S.; Kulis, S.; Leroux, P.; Moreira, P.; Kolpin, A.; Prinzie, J. Single-Event Effect Responses of Integrated Planar Inductors in 65-Nm CMOS. *IEEE Trans. Nucl. Sci.* **2021**, *68*, 2587–2597. [CrossRef]
8. Biereigel, S.; Kulis, S.; Moreira, P.; Kölpin, A.; Leroux, P.; Prinzie, J. Radiation-Tolerant All-Digital PLL/CDR with Varactorless LC DCO in 65 Nm CMOS. *Electronics* **2021**, *10*, 2741. [CrossRef]
9. Monda, D.; Ciarpi, G.; Saponara, S. Diagnosis of Faults Induced by Radiation and Circuit-Level Design Mitigation Techniques: Experience from VCO and High-Speed Driver CMOS ICs Case Studies. *Electronics* **2021**, *10*, 2144. [CrossRef]
10. Wang, Y.; Liu, F.; Li, B.; Li, B.; Huang, Y.; Yang, C.; Zhang, J.; Wang, G.; Luo, J.; Han, Z.; et al. Dependence of Temperature and Back-Gate Bias on Single-Event Upset Induced by Heavy Ion in 0.2-Mm DSOI CMOS Technology. *IEEE Trans. Nucl. Sci.* **2021**, *68*, 1660–1667. [CrossRef]
11. Ladbury, R.; Bay, M.; Zinchuk, J. Threats to Resiliency of Redundant Systems Due to Destructive SEEs. *IEEE Trans. Nucl. Sci.* **2021**, *68*, 970–979. [CrossRef]
12. Gardenghi, R.A.; Houlne, R.C. Power Supply Considerations for Pulsed Solid-State Radar. In Proceedings of the Nineteenth IEEE Symposium on Power Modulators, San Diego, CA, USA, 26–28 June 1990; IEEE: San Diego, CA, USA, 1990; pp. 146–152.
13. Kohinata, H.; Arai, M.; Fukumoto, S. An Experimental Study on Latch Up Failure of CMOS LSI. In Proceedings of the 2008 Second International Conference on Secure System Integration and Reliability Improvement, Yokohama, Japan, 14–17 July 2008; IEEE: Yokohama, Japan, 2008; pp. 215–216.
14. Yu, Y.; Biess, J.J. Some Design Aspects Concerning Input Filters for DC-DC Converters. In Proceedings of the 1971 IEEE Power Electronics Specialists Conference, Pasadena, CA, USA, 19–20 April 1971; IEEE: Pasadena, CA, USA, 1971; pp. 66–76.
15. Liang, S.; Hou, L.Z.; Gu, Q.T.; Salama, C.A.T. Latch-up Modeling of BiCMOS Merged Bipolar-MOS Structures. *Solid State Electron.* **1992**, *35*, 1461–1469. [CrossRef]
16. Duzellier, S. Radiation Effects on Electronic Devices in Space. *Aerosp. Sci. Technol.* **2005**, *9*, 93–99. [CrossRef]
17. Cecchetto, M.; García Alía, R.; Wrobel, F. Impact of Energy Dependence on Ground Level and Avionic SEE Rate Prediction When Applying Standard Test Procedures. *Aerospace* **2019**, *6*, 119. [CrossRef]
18. Hiblot, G.; Serbulova, K.; Hellings, G.; Chen, S.-H. TCAD Study of Latch-up Sensitivity to Wafer Thinning below 500 Nm. In Proceedings of the 2021 International Semiconductor Conference (CAS), Sinaia, Romania, 6–8 October 2021; IEEE: Sinaia, Romania, 2021; pp. 121–124.

19. Osipenko, P.N.; Antonov, A.A.; Klishin, A.V.; Vasilegin, B.V.; Gorbunov, M.S.; Dolotov, P.S.; Zebrev, G.I.; Anashin, V.S.; Emeliyanov, V.V.; Ozerov, A.I.; et al. Fault-Tolerant SOI Microprocessor for Space Applications. *IEEE Trans. Nucl. Sci.* **2013**, *60*, 2762–2767. [CrossRef]
20. Dodds, N.A.; Hooten, N.C.; Reed, R.A.; Schrimpf, R.D.; Warner, J.H.; Roche, N.J.; McMorrow, D.; Wen, S.; Wong, R.; Salzman, J.F.; et al. Effectiveness of SEL Hardening Strategies and the Latchup Domino Effect. *IEEE Trans. Nucl. Sci.* **2012**, *59*, 2642–2650. [CrossRef]
21. Jiang, Z.-H.; Ker, M.-D. Schottky-Embedded Isolation Ring to Improve Latch-Up Immunity Between HV and LV Circuits in a 0.18 Mm BCD Technology. *IEEE J. Electron Devices Soc.* **2022**, *10*, 516–524. [CrossRef]
22. Tsai, H.-W.; Ker, M.-D. Active Guard Ring to Improve Latch-Up Immunity. *IEEE Trans. Electron Devices* **2014**, *61*, 4145–4152. [CrossRef]
23. Chen, C.-C.; Ker, M.-D. Optimization Design on Active Guard Ring to Improve Latch-Up Immunity of CMOS Integrated Circuits. *IEEE Trans. Electron Devices* **2019**, *66*, 1648–1655. [CrossRef]
24. Jiang, Z.-H.; Ker, M.-D. Latch-Up Prevention with Autodetector Circuit to Stop Latch-Up Occurrence in CMOS-Integrated Circuits. *IEEE Trans. Electromagn. Compat.* **2022**, *64*, 1785–1792. [CrossRef]
25. Tsai, H.-W.; Ker, M.-D. Latch-Up Protection Design with Corresponding Complementary Current to Suppress the Effect of External Current Triggers. *IEEE Trans. Device Mater. Relib.* **2015**, *15*, 242–249. [CrossRef]
26. Becker, H.N.; Miyahira, T.F.; Johnston, A.H. Latent Damage in CMOS Devices from Single-Event Latchup. *IEEE Trans. Nucl. Sci.* **2002**, *49*, 3009–3015. [CrossRef]
27. Jeong, S.-H.; Lee, N.-H.; Cho, S.-I. A Design of High-speed Power-off Circuit and Analysis. *Trans. Korean Inst. Electr. Eng.* **2014**, *63*, 490–494. [CrossRef]
28. Andjelković, M.S.; Petrović, V.; Stamenković, Z.; Ristić, G.S.; Jovanović, G.S. Circuit-Level Simulation of the Single Event Transients in an On-Chip Single Event Latchup Protection Switch. *J. Electron. Test.* **2015**, *31*, 275–289. [CrossRef]
29. Rui, C.; Ying, F.; Yongtao, Y.; Shipeng, S.G.; Guoqiang, F.; Xiang, Z.; Yingqi, M.; Jianwei, H. Mitigation Technique and Experimental Verification of Single Event Latch-up Effect in Circuit Level for CMOS Device. *At. Energy Sci. Technol.* **2014**, *48*, 721–726.
30. Niaraki Asli, R.; Shirinzadeh, S. High Efficiency Time Redundant Hardened Latch for Reliable Circuit Design. *J. Electron. Test.* **2013**, *29*, 537–544. [CrossRef]
31. Ball, D.R.; Sheets, C.B.; Xu, L.; Cao, J.; Wen, S.-J.; Fung, R.; Cazzaniga, C.; Kauppila, J.S.; Massengill, L.W.; Bhuva, B.L. Single-Event Latchup in a 7-Nm Bulk FinFET Technology. *IEEE Trans. Nucl. Sci.* **2021**, *68*, 830–834. [CrossRef]
32. Li, D.; Liu, T.; Wu, Z.; Cai, J.; Zhao, P.; He, Z.; Liu, J. An Investigation of FinFET Single-Event Latch-up Characteristic and Mitigation Method. *Microelectron. Reliab.* **2020**, *114*, 113901. [CrossRef]
33. Yongtao, Y.; Guoqiang, F.; Rui, C.; Shipeng, S.G.; Jianwei, H. Experimental Study on Single Event Latchup of SRAM K6R4016V1D and Its Protection. *At. Energy Sci. Technol.* **2012**, *46*, 587–591.
34. Matino, H. Analysis of the holding current in CMOS latch-up. *IBM J. Res. Dev.* **1985**, *29*, 588–592. [CrossRef]
35. Mergens, M.P.J.; Russ, C.C.; Verhaege, K.G.; Armer, J.; Jozwiak, P.C.; Mohn, R. High Holding Current SCRs (HHI-SCR) for ESD Protection and Latch-up Immune IC Operation. *Microelectron. Reliab.* **2003**, *43*, 993–1000. [CrossRef]
36. Chen, M.-J.; Lee, H.-S.; Chen, J.-H.; Hou, C.-S.; Lin, C.-S.; Jou, Y.-N. A Physical Model for the Correlation between Holding Voltage and Holding Current in Epitaxial CMOS Latch-Up. *IEEE Electron Device Lett.* **1998**, *19*, 276–278. [CrossRef]
37. Burnell, A.J.; Chugg, A.M.; Harboe-Sørensen, R. Laser SEL Sensitivity Mapping of SRAM Cells. *IEEE Trans. Nucl. Sci.* **2010**, *57*, 1973–1977. [CrossRef]
38. Faraud, E.; Pouget, V.; Shao, K.; Larue, C.; Darracq, F.; Lewis, D.; Samaras, A.; Bezerra, F.; Lorfevre, E.; Ecoffet, R. Investigation on the SEL Sensitive Depth of an SRAM Using Linear and Two-Photon Absorption Laser Testing. *IEEE Trans. Nucl. Sci.* **2011**, *58*, 2637–2643. [CrossRef]
39. Jones, R.; Chugg, A.M.; Jones, C.M.S.; Duncan, P.H.; Dyer, C.S.; Sanderson, C. Comparison between SRAM SEE Cross-Sections from Ion Beam Testing with Those Obtained Using a New Picosecond Pulsed Laser Facility. *IEEE Trans. Nucl. Sci.* **2000**, *47*, 539–544. [CrossRef]
40. Feng, G.; Shangguan, S.; Ma, Y.; Han, J. SEE Characteristics of Small Feature Size Devices by Using Laser Backside Testing. *J. Semicond.* **2012**, *33*, 014008. [CrossRef]
41. Chen, R.; Han, J.-W.; Zheng, H.-S.; Yu, Y.-T.; Shangguang, S.-P.; Feng, G.-Q.; Ma, Y.-Q. Comparative Research on "High Currents" Induced by Single Event Latch-up and Transient-Induced Latch-Up. *Chin. Phys. B* **2015**, *24*, 046103. [CrossRef]
42. Yingqi, M.; Jianwei, H.; ShiPeng, S.; Rui, C.; Xiang, Z.; Yue, L.; Yueying, Z. SEE Characteristics of COTS Devices by 1064 nm Pulsed Laser Backside Testing. In Proceedings of the 2018 IEEE Nuclear & Space Radiation Effects Conference (NSREC 2018), Waikoloa Village, HI, USA, 16–20 July 2018; IEEE: Waikoloa Village, HI, USA, 2018; pp. 1–4.

Disclaimer/Publisher's Note: The statements, opinions and data contained in all publications are solely those of the individual author(s) and contributor(s) and not of MDPI and/or the editor(s). MDPI and/or the editor(s) disclaim responsibility for any injury to people or property resulting from any ideas, methods, instructions or products referred to in the content.

Communication

The Effects of Total Ionizing Dose on the SEU Cross-Section of SOI SRAMs

Peixiong Zhao [1,*], Bo Li [2], Hainan Liu [2], Jinhu Yang [1,3], Yang Jiao [1,3], Qiyu Chen [1,3], Youmei Sun [1] and Jie Liu [1,*]

1. Institute of Modern Physics, Chinese Academy of Sciences, Lanzhou 730000, China
2. Institute of Microelectronics, Chinese Academy of Sciences, Beijing 100029, China
3. School of Nuclear Science and Technology, University of Chinese Academy of Sciences, Beijing 100049, China
* Correspondence: zhaopeixiong@impcas.ac.cn (P.Z.); j.liu@impcas.ac.cn (J.L.)

Abstract: The total ionizing dose (TID) effects on single-event upset (SEU) hardness are investigated for two silicon-on-insulator (SOI) static random access memories (SRAMs) with different layout structures in this paper. The contrary changing trends of TID on SEU sensitivity for 6T and 7T SOI SRAMs are observed in our experiment. After 800 krad(Si) irradiation, the SEU cross-sections of 6T SRAMs increases by 15%, while 7T SRAMs decreases by 60%. Experimental results show that the SEU cross-sections are not only affected by TID irradiation, but also strongly correlate with the layout structure of the memory cells. Theoretical analysis shows that the decrease of SEU cross-section of 7T SRAM is caused by a raised OFF-state equivalent resistance of the delay transistor N5 after TID exposure, which is because the radiation-induced charges are trapped in the shallow trench, and isolation oxide (STI) and buried oxide (BOX) enhance the carrier scattering rate of delay transistor N5.

Keywords: single event upset (SEU); total ionizing dose (TID); silicon-on-insulator (SOI); synergistic effect; radiation-hardened by design (RHBD)

1. Introduction

In the space environment, there are many high-energy radiation particles, such as electrons, protons, heavy ions, and so on [1,2]. These high-energy radiation particles will cause the macroscopic electrical properties of devices to change, degrade, or even fail. The single-event effect (SEE) and effect of total ionizing dose (TID) are the main causes of the failure of spacecrafts and satellites [3–11]. The radiation hardening technology of integrated circuits mainly include radiation-hardened by design (RHBD) storage cells (heavy ion tolerant cell, dual interlocked storage cell, etc.) [12–14] and radiation-hardened-by-process (RHBP) front-end-of-line (guard-band, silicon-on-insulation, etc.) [15–17]. The silicon-on-insulation (SOI) process, realized physically, isolates the channel region from the substrate region, which not only significantly reduces the effective collection region, but also eliminates the single event latch-up (SEL) [18,19] and tunneling [20] effects commonly found in bulk silicon devices. Therefore, the SOI process is naturally resistant to irradiation [21–23] and has important applications in the field of radiation-hardened integrated circuits (ICs).

The SEE, TID effect, and synergy between the TID and SEE in electronic devices have been extensively studied. The effect of ion parameters on the multi-bit upset effect in 45 and 28 nm SOI static random access memories (SRAMs) has been investigated by Raine et al., in which the 4-bit upsets phenomenon was observed under oblique incidence conditions [24], and the multi-bit upsets phenomenon caused by the non-charge sharing effect was observed under positive incidence conditions [25]. Moreover, Liu et al. have carried out proton and heavy ion irradiation experiments on radiation-hardened SOI SRAMs. The results showed that single-event upset (SEU) can be triggered only when secondary ions hit both the delay transistor and OFF-state NMOS transistor [26]. Schwank et al. have irradiated many kinds of SRAM devices with various radiation sources, such as γ, X-ray, and proton, and then measured the SEU cross-section of the devices. The experimental results showed that the TID irradiation has a significant effect on the SEU cross-section of the device, and the SEU cross-section increases with increasing irradiation dose. Meanwhile, the SEU cross-section had a certain dependence on parameters such as test data patterns, irradiation test temperature, etc. [27–32].

Previous studies have shown that TID significantly affects the SEU cross-section of the electron device, and numerous studies have been conducted on the irradiation doses, data patterns, and experimental temperature. However, the effect of the layout structure on the SEU cross-section after TID irradiation has had relatively few studies, and the physical mechanism is not yet fully understood. Therefore, in this paper, we design two SOI SRAM devices with different layout structures and investigate the mechanism of the effect of the layout structure on SEU cross-section of SOI SRAMs. This paper is organized as follows. In Section 2, the test circuits and experimental methods are presented. The experimental results for TID effects on SEU hardness of SOI SRAMs with 6T and 7T cell designs are described in Section 3. In Section 4, the experimental results are discussed. In Section 5, conclusions are drawn.

2. Test Circuit and Experimental Setup

2.1. Test Circuit

We design two SOI SRAMs with different layout structures, based on the 130 nm SOI CMOS process. The memory capacity is 64 kbit and organized by 8 k × 8 bits. Device operates using a dual power supply for the input–output (I/O) circuitry (higher voltage) and memory array (lower voltage). The nominal supply voltages (Vdd) are 1.5 and 3.3 V for the core blocks and I/O, respectively. Figure 1a shows a schematic diagram of the layout structure and size of the 6T SRAM cell with dimensions 3.7 × 3.2 µm. The access transistors N3 and N4 share drain electrodes with the pull-down transistors N1 and N2, respectively. Figure 1b shows the layout structure and size of the 7T SRAM memory cell after hardened design by the delay transistor N5. The cell size of the 7T SRAM is 3.9 × 3.4 µm. The gate electrode of the delay transistor N5 is connected to the gate electrode of the access transistors N3 and N4. During the read/write operation, the delay transistor N5 will be in the ON-state, which has a very low resistance. While the delay transistor N5 will be in the OFF-state when the data hold state is entered, and the resistance of OFF-state N5 is very high, which can effectively suppress the single-event transient disturbance and significantly improve the stability of the 7T memory cell. As shown in Figure 1c, the structures of N1, N2, N3, N4, P1, and P2 were designed via body under source FET (BUSFET) [33]. Figure 1d shows a schematic diagram of the device structure of the delay transistor N5, which is equivalent to a resistor and transistor in parallel.

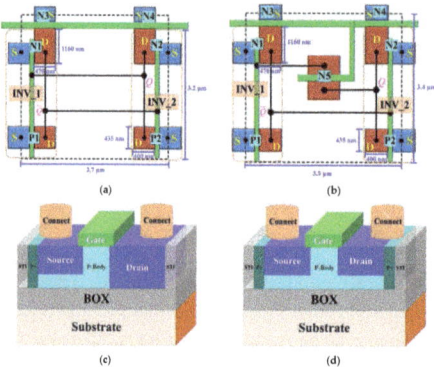

Figure 1. Schematics of memory cell structure of (**a**) 6T SRAM, (**b**) 7T SRAM, (**c**) BUSFET (body under source FET) structure of N1, N2, N3, N4, P1, P2, and (**d**) delay transistor N5 structure.

2.2. TID Experiment

As shown in Figure 2, TID exposures were carried out with 60 Co-γ ray at The Xinjiang Technical Institute of Physics and Chemistry, Chinese Academy of Sciences, with dose rate of 200 rad(Si)/s. A data pattern of 55 h was written into SRAM before irradiation, and it was then set to data hold operation during irradiation. Different devices from the same wafer were chosen for two times of SEU tests. For Kr ion, we chose nine devices and divided them into three groups, with one group irradiated to 200 krad(Si), another group irradiated to 400 krad(Si), and the last group being the reference sample without TID irradiation. Additionally, six devices were divided into two groups for Bi ion SEU test: 800 krad(Si) and reference sample without TID irradiation.

Figure 2. DUT in the terminal of the ^{60}Co irradiation.

2.3. Heavy Ion Irradiation

As shown in Figure 3, the heavy ion irradiation experiments were performed at the heavy ion research facility in Lanzhou (HIRFL) in the Institute of Modern Physics, Chinese Academy of Sciences. The ion species, energy, LET, and range are shown in Table 1. The LET values calculated by SRIM2013 [34] varied from 20.5 to 99.8 MeV·cm^2/mg. In the following experiments, the LET values were at the device surface. The ion ranges in silicon were always greater than 50 μm. Three levels of metal were applied to the SOI SRAM studied in this paper, and the thickness of the overlayer was measured at 7.2 μm. The range of Kr and Bi ion was enough to punch through the silicon film of our 130-nm SOI SRAM because the thickness of silicon film was only 260 nm. The SEU cross-sections of SRAMs were characterized in a dynamic mode, i.e., the SRAMs were written with a specific pattern to the memory array, and the read repeatedly and errors were counted until 200 errors were recorded. To evaluate the effect of the data pattern applied during TID exposure on SEU hardness, the SEU characterizations were performed with TID data pattern 55h and its complement data pattern AAh.

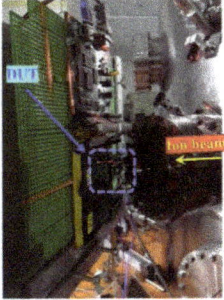

Figure 3. DUT in the terminal of the heavy-ion beamline.

Table 1. Ion spices, energy, LET, and range in silicon.

Ion Species	Air/Al-Foil (mm)/(μm)	Energy at Device Surface (MeV)	LET at Device Surface (MeV·cm^2/mg)	Ion Range (μm)
^{86}Kr	30/0	1841	20.5	274
	50/100	1154	27.2	150
	50/180	480	37.6	59
^{209}Bi	30/0	923	99.8	54

3. Experimental Results

3.1. Effect of TID on the 6T SRAM SEU cross-section

Figure 4 provides the results for the SEU cross-sections of 6T SRAM characterized by Kr ion of devices in three different groups: (1) fresh; (2) after deposition of 200 krad(Si); and (3) 400 krad(Si). The mean value of SEU data with an error bar at each dose level is depicted in Figure 4. As shown in Figure 4, SEU cross-section of 6T SRAM increased by a factor of 0.1% (20.5 MeV·cm^2/mg), 12.9% (27.2 MeV·cm^2/mg), and 5.2% (37.6 MeV·cm^2/mg) of after deposition of 200 krad(Si), with respect to the fresh condition, and increased by a factor of 3.7% (20.5 MeV·cm^2/mg), 4.0% (27.2 MeV·cm^2/mg), and 13.7% (37.6 MeV·cm^2/mg) of after deposition 400 krad(Si). It can be clearly observed that the SEU cross-section of 6T SRAMs shows an increasing trend after TID irradiation, and the maximum increase is 13.7%.

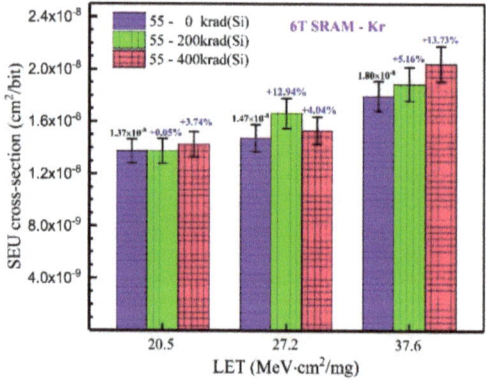

Figure 4. SEU cross-sections versus LET for 6T SRAMs. The purple horizontal column represents data for TID = 0 rad(Si), the green vertical column represents data for TID = 200 krad(Si), and the pink checkerboard represents data for TID = 400 krad(Si).

Zheng et al. investigated the effect of the total dose effect on the SEU cross-section of the SRAMs. The experimental results show that the SEU cross-section of the SRAMs gradually increases after the TID irradiation, and the main oxide trap charge regions of the nanoscale feature device are the buried oxide and shallow trench isolation oxide regions [27,31,35–37]. For the test chip we designed, the SEU cross-section of the 6T SRAM increases slightly after TID irradiation, due to two main reasons: (1) the gate oxide layer thickness of our test chip is only 1.5 nm, so the gate oxide layer cannot trap enough trap charges; (2) the transistor used in the test chip adopts the structure of body under source FET (BUSFET), which eliminates the formation of parasitic leakage channels between the source and drain electrodes caused by the radiation-induced charges trapped in buried oxide (BOX). Therefore, the radiation-induced charges trapped in the shallow trench isolation oxide (STI), rather than in the BOX, were responsible for the increase of the SEU cross-section of the 6T SRAM.

3.2. Effect of Data Pattern on the 6T SRAM SEU cross-section

Figure 5 provides the results for the SEU cross-sections characterized by the Kr ion (20.5 MeV·cm^2/mg) of devices in three different groups: (1) fresh; (2) after deposition of 200 krad(Si); and (3) 400 krad(Si). The mean value of SEU data with error bar at each dose level is depicted in Figure 5. As shown in Figure 5, the mean SEU cross-section of 6T SRAM increases by a factor of 0.1% (55 h), 2.8% (AAh) after deposition 200 krad(Si), with respect to the fresh condition, and it increases by a factor of 3.7% (55h), 6.8% (AAh) after deposition 400 krad(Si). It was observed that the data patterns have little effect on the SEU cross-section of the 6T SRAM after TID irradiation. There are two main reasons for these experimental results. First, the degree of ionization damage of the ultrathin gate oxide layer at 1.5 V was basically the same as the case without voltage addition; second, the main sensitive area of TID of the nanodevice shifted from the gate oxide region to the STI and BOX regions. Therefore, the SEU cross-section of the 6T SRAM after TID irradiation dose had no dependence on the data pattern that was applied during TID exposure.

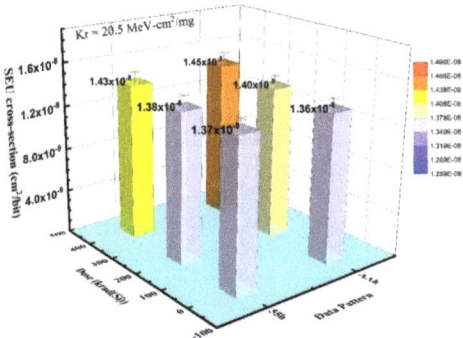

Figure 5. SEU cross-sections characterized by Kr ion versus TID for 6T SRAMs under different data patterns.

3.3. Effect of TID on the 7T SRAM SEU cross-section

Figure 6 shows the SEU cross-sections of 6T SRAM and 7T SRAM characterized by the Bi (99.8 MeV·cm^2/mg) ion as a function of TID. The SEU cross-sections of 6T SOI SRAM characterized by Bi ion are also increased by TID. As shown in Figure 6, the SEU cross-section of 6T SRAM increased by a factor 9.1% (55 h) and 4.0% (AAh) of after deposition 800 krad(Si), with respect to the fresh condition. Similarly, we did not observe a significant correlation between the SEU cross-section and data pattern applied during TID exposure for 7T SRAM. However, it is interesting to note that the SEU cross-section of the 7T SRAM showed an opposite changing trend to the 6T SRAM. As shown in Figure 6, SEU cross-section of 7T SRAM decreased by factors of 42.9% (55 h) and 56.6% (AAh) after deposition 800 krad(Si), with respect to the fresh condition. Because the 6T SRAM and 7T SRAM were fabricated in the same wafer, the physical dimensions and electrical characteristics of all transistors (N1, N2, N3, N4, P1, and P2) in the memory cell are very similar, except for the delay-hardened transistor, N5. Therefore, we can conclude that the change in the electrical characteristics of the N5 was responsible for the reduction in the SEU cross-section of the 7T SRAM after TID irradiation.

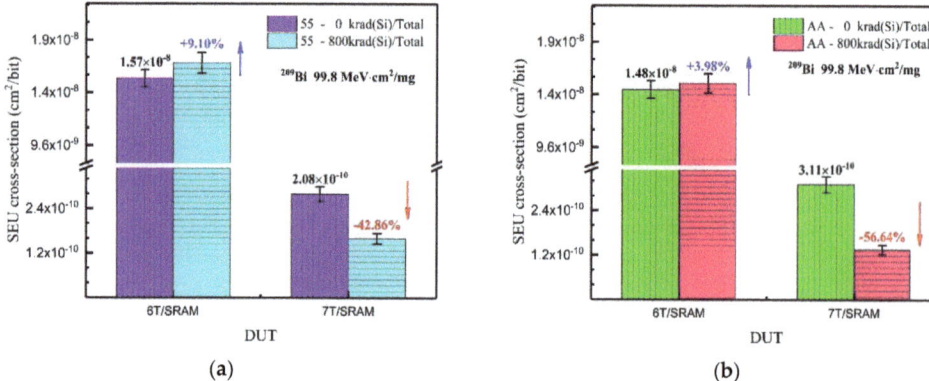

Figure 6. SEU cross-sections characterized by Bi ion versus layout structure for SOI SRAMs. The vertical column represents data for TID = 0 rad(Si), and the horizontal column represents data for TID = 800 krad(Si). The data pattern applied during SEU testing was (**a**) 55h and (**b**) AAh.

Furthermore, we investigated the TID effect on the "1→0" upset and "0→1" upset for 7T SRAM SEU types. As shown in Figure 7, the mean "1→0" upset cross-section of 7T SRAM decreased by a factor of 3.1% (55 h), 37.9% (AAh) of after deposition 800 krad(Si), with respect to the fresh condition, and the mean "0→1" upset cross-section decreased by a factor of 37.9% (55 h), 66.7% (AAh) after deposition 800 krad(Si). Therefore, the decrease of the cross-section of "0→1" upset was mainly responsible for the decrease of 7T SRAM SEU cross-section.

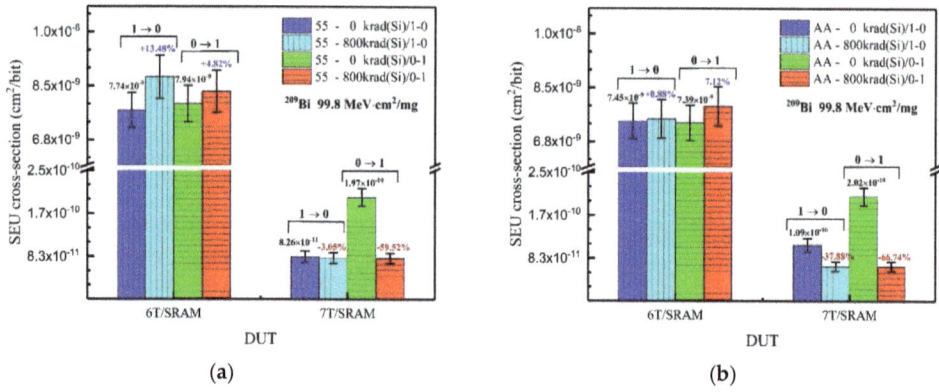

Figure 7. SEU cross-sections characterized by Bi ion versus layout structure for SOI SRAMs after deposition 800 krad(Si). The vertical column represents data for the "1→0" upset, the horizontal column represents data for the "0→1" upset. The data pattern applied during SEU testing is (**a**) 55h and (**b**) AAh.

4. Discussion

4.1. Transient Propagation Circuit Analysis for 7T SRAM

It is generally believed that, in silicon, electrons have much higher mobility than holes, resulting in the electrons are quickly collected at the drain contacts. Thus, the pull-down nMOSFET biased OFF-state determines the SEU resistance of the 7T SRAM. The equivalent circuits of transient pulse propagation, corresponding to two different SEU types in the 7T SRAM, are shown in Figure 8.

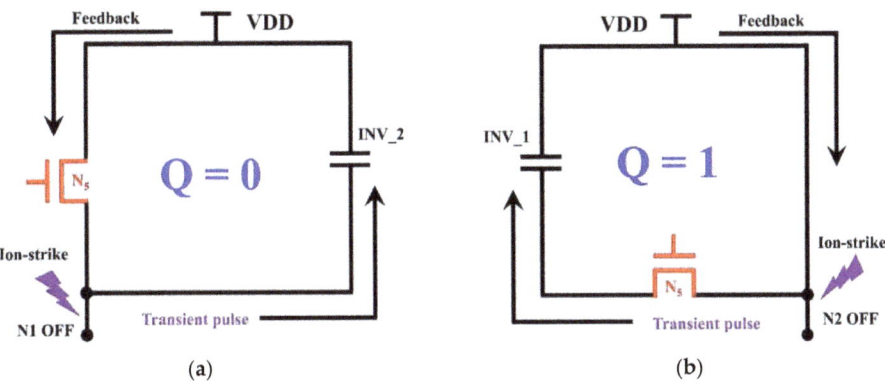

Figure 8. The equivalent circuits of transient pulse propagation corresponding to the (**a**) "0→1" upset type and (**b**) "1→0" upset type in the 7T SRAM.

As shown in Figure 8a, when the Q node is set to low potential, the pull-down nMOSFET N1 in the inverter 1 is turned OFF. Normally, incident heavy ions hitting the N1 will produce a transient pulse, caused by charge collection in the drain. After that, the transient pulse acts on the gate of inverter 2 to gradually increase the potential of the Q node. The potential perturbation in the Q node well further feeds back to the gate of inverter 1 through the delay transistor N5 to gradually decrease the potential of the $\bar{Q}$ node. Finally, the single event transient pulse signal is latched to the memory cell of the 7T SRAM. In this case, the delay transistor N5 indirectly delayed and suppressed the feedback signal of the single event transient pulse, so that the delay-efficiency was lower; thus, the cross-section of "0→1" upset was higher.

As shown in Figure 8b, when the Q node was set to the high potential, the pull-down nMOSFET N2 in the inverter 2 was turned OFF, when the incident heavy ions hitting the N2 produced a transient pulse, caused by charge collection in the drain. After that, the transient pulse through the delay transistor N5 acted on the gate of inverter 1 to gradually increase the potential of the $\bar{Q}$ node. The potential perturbation in the $\bar{Q}$ node well further fed back to the gate of inverter 1 to gradually decrease the potential of the Q node. Finally, the single event transient pulse signal was latched to the memory cell of the 7T SRAM. In this case, the delay transistor N5 directly delayed and suppressed the single event transient pulse, so that the delay-efficiency was higher; thus, the cross-section of "1→0" upset was lower.

4.2. Effect of TID on the OFF-State Equivalent Resistance of Delay Transistor N5

According to previous studies, TID irradiation significantly affects the device's carrier mobility [37–40]. As shown in the equation 1, the carrier mobility of the transistors was mainly affected by three scatterings: phonon scattering, surface scattering, and charged impurity scattering. It has been found that the scattering rate of charged impurities is mainly determined by the semiconductor process, and it is rarely affected by TID. However, phonon and surface scattering are proportional to the electric field intensity perpendicular to the channel direction; the higher the density of oxide trap charge is, the stronger the vertical electric field component is. Therefore, the effect of the TID on the carrier scattering rate was mainly to increase the phonon and surface scattering rates.

$$\frac{1}{\mu_n} = \frac{1}{\mu_1}\left(E_{eff}\right)^{|\alpha_1|} + \frac{1}{\mu_2}\left(E_{eff}\right)^{|\alpha_2|} + \frac{1}{\mu_3}Q_{ot}\left(\frac{1}{N_i}\right)^{\alpha_3} \tag{1}$$

where α_i and μ_i are the fitting parameters, N_i is the charge density of inversion layer, E_{eff} is effective vertical electric field intensity, and Q_{ot} is the oxide trap charge density.

As shown in Figure 9, a substantial amount of radiation-induced charge was trapped in the STI and BOX of the delay transistor N5 after TID radiation. As a result, the oxide trapped charge generated a vertical electric field in the channel region of the transistor N. This results in an increase in phonon and surface scattering. Finally, the carrier mobility rate of the delay transistor N5 decreased, and the equivalent OFF-state resistance increased, thus leading to a decrease in the SEU cross-section of the 7T SRAM.

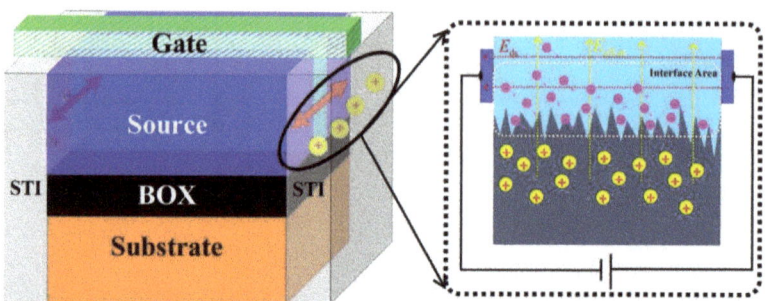

Figure 9. Schematic illustration of the physical mechanism of radiation-induced carrier scattering rate increases in the delay transistor N5 of the 7T SRAM.

4.3. The Advantages of Suppressing SEU with TID

In the natural space radiation environment, there are protons and electrons that can cause TID effects, as well as heavy ions that can cause transient SEE. When the integrated circuit is on-orbit, it will be affected by a variety of radiation effects; that is, there is electrical performance degradation caused by TID, and there is also the transient voltage pulse caused by SEE. Hence, TIDs and SEEs have a natural synergy in the space radiation environment. We found that, when delay transistors are used for hardened circuit design, the TID can cause degradation in the performance of delay transistors, thus suppressing the SEU. Therefore, not only does using delay-hardened transistors for hardened circuit design not affect the operating speed of nano-devices, it also improves the stability of memory cells. Furthermore, the TID effect can be used to suppress the transient SEE and achieve self-optimizing design in natural radiation space radiation environments. Our study provides new insight into radiation-hardened by design (RHBD) technology for nano-integrated circuits.

5. Conclusions

The total ionizing dose (TID) effects on single-event upset (SEU) hardness of silicon-on-insulator (SOI) static random access memories (SRAMs) with 6T and 7T cell design were explored in this paper. Experimental results show that the SEU cross-section of 6T SOI SRAM is increased by TID and has no dependence on the data pattern applied during TID exposure. However, it is interesting to note that the SEU cross-section of 7T SRAM decreases significantly after TID exposure. Furthermore, in our experiment, opposite changes intendencies of SEU cross-section for 6T and 7T SOI SRAMs were observed after TID irradiation. The mechanism behind the experimental results of 6T SRAM is that OFF-state leakage of pull-down nMOSEFTs increases after TID irradiation, since the parasitic transistor is turned ON by radiation in the shallow trench isolation oxide (STI) region. However, the radiation-induced decrease in carrier mobility in delay transistor N5 of the 7T SRAM is responsible for the decrease of the SEU cross-section. Because radiation-induced charges trapped in the STI and buried oxide (BOX) improve the carrier scattering rate, the OFF-state equivalent resistance of delay transistor N5 increases, causing the stronger suppression of transient pulses and feedback signals, ultimately leading the SEU cross-section decreases. Our experimental results provide a new insight into the radiation-hardened by design (RHBD) used in nano ICs.

Author Contributions: Conceptualization, P.Z.; data curation, P.Z.; formal analysis, P.Z., B.L., and H.L.; funding acquisition, P.Z. and J.L.; investigation, P.Z., B.L., and H.L.; project administration, P.Z. and J.L.; supervision, J.L.; validation, J.Y., Y.J., Q.C., and Y.S.; writing—original draft, P.Z.; writing—review and editing, J.L. All authors have read and agreed to the published version of the manuscript.

Funding: This research was funded by the National Natural Science Foundation of China (Nos. 12105341, 61874135, and 12035019).

Data Availability Statement: Data available on request due to restrictions eg privacy. The data presented in this study are available on request from the corresponding author. The data are not publicly available due to the research results having potential commercial value.

Acknowledgments: We acknowledge the support of the HIRFL in heavy-ion irradiation and for the TID irradiation equipment in the Xinjiang Technical Institute of Physics and Chemistry, Chinese Academy of Sciences.

Conflicts of Interest: The authors declare no conflict of interest.

References

1. Barth, J.L.; Dyer, C.S.; Stassinopoulos, E.G. Space, atmospheric, and terrestrial radiation environments. *IEEE Trans. Nucl. Sci.* **2003**, *50*, 466–482. [CrossRef]
2. Xapsos, M. A Brief History of Space Climatology: From the Big Bang to the Present. *IEEE Trans. Nucl. Sci.* **2019**, *66*, 17–37. [CrossRef]
3. Dodd, P.E.; Massengill, L.W. Basic mechanisms and modeling of single-event upset in digital microelectronics. *IEEE Trans. Nucl. Sci.* **2003**, *50*, 583–602. [CrossRef]
4. Baumann, R.C. Radiation-induced soft errors in advanced semiconductor technologies. *IEEE Trans. Device Mater. Reliab.* **2005**, *5*, 305–316. [CrossRef]
5. Dodd, P.E.; Shaneyfelt, M.R.; Schwank, J.R.; Felix, J.A. Current and Future Challenges in Radiation Effects on CMOS Electronics. *IEEE Trans. Nucl. Sci.* **2010**, *57*, 1747–1763. [CrossRef]
6. Kobayashi, D. Scaling Trends of Digital Single-Event Effects: A Survey of SEU and SET Parameters and Comparison With Transistor Performance. *IEEE Trans. Nucl. Sci.* **2021**, *68*, 124–148. [CrossRef]
7. Binder, D.; Smith, E.C.; Holman, A.B. Satellite Anomalies from Galactic Cosmic Rays. *IEEE Trans. Nucl. Sci.* **1975**, *22*, 2675–2680. [CrossRef]
8. Barillot, C.; Calvel, P. Review of commercial spacecraft anomalies and single-event-effect occurrences. *IEEE Trans. Nucl. Sci.* **1996**, *43*, 453–460. [CrossRef]
9. Normand, E. Single-event effects in avionics. *IEEE Trans. Nucl. Sci.* **1996**, *43*, 461–474. [CrossRef]
10. Harboe-Sorensen, R.; Poivey, C.; Zadeh, A.; Keating, A.; Fleurinck, N.; Puimege, K.; Guerre, F.X.; Lochon, F.; Kaddour, M.; Li, L.; et al. PROBA-II Technology Demonstration Module In-Flight Data Analysis. *IEEE Trans. Nucl. Sci.* **2012**, *59*, 1086–1091. [CrossRef]
11. Ecoffet, R. Overview of In-Orbit Radiation Induced Spacecraft Anomalies. *IEEE Trans. Nucl. Sci.* **2013**, *60*, 1791–1815. [CrossRef]
12. Bessot, D.; Velazco, R. Design of SEU-hardened CMOS memory cells: The HIT cell. In Proceedings of the RADECS 93—Second European Conference on Radiation and its Effects on Components and Systems, St. Malo, France, 13–16 September 1993; pp. 563–570. [CrossRef]
13. Calin, T.; Nicolaidis, M.; Velazco, R. Upset hardened memory design for submicron CMOS technology. *IEEE Trans. Nucl. Sci.* **1996**, *43*, 2874–2878. [CrossRef]
14. Schwank, J.R.; Shaneyfelt, M.R.; Dodd, P.E. Radiation Hardness Assurance Testing of Microelectronic Devices and Integrated Circuits: Radiation Environments, Physical Mechanisms, and Foundations for Hardness Assurance. *IEEE Trans. Nucl. Sci.* **2013**, *60*, 2074–2100. [CrossRef]
15. Hughes, H.L.; Benedetto, J.M. Radiation effects and hardening of MOS technology: Devices and circuits. *IEEE Trans. Nucl. Sci.* **2003**, *50*, 500–521. [CrossRef]
16. Narasimham, B.; Gambles, J.W.; Shuler, R.L.; Bhuva, B.L.; Massengill, L.W. Quantifying the Effect of Guard Rings and Guard Drains in Mitigating Charge Collection and Charge Spread. *IEEE Trans. Nucl. Sci.* **2008**, *55*, 3456–3460. [CrossRef]
17. Liu, X.; Cai, L.; Liu, B.; Yang, X.; Cui, H.; Li, C. Total Ionizing Dose Hardening of 45 nm FD-SOI MOSFETs Using Body-Tie Biasing. *IEEE Access* **2019**, *7*, 51276–51283. [CrossRef]
18. Soliman, K.; Nichols, D.K. Latchup in CMOS Devices from Heavy Ions. *IEEE Trans. Nucl. Sci.* **1983**, *30*, 4514–4519. [CrossRef]
19. Bruguier, G.; Palau, J.M. Single particle-induced latchup. *IEEE Trans. Nucl. Sci.* **1996**, *43*, 522–532. [CrossRef]
20. Hsieh, C.M.; Murley, P.C.; Brien, R.R.O. Dynamics of Charge Collection from Alpha-Particle Tracks in Integrated Circuits. In Proceedings of the 1981 International Reliability Physics Symposium, Orlando, FL, USA, 7–9 April 1981; pp. 38–42.
21. Schwank, J.R.; Ferlet-Cavrois, V.; Shaneyfelt, M.R.; Paillet, P.; Dodd, P.E. Radiation effects in SOI technologies. *IEEE Trans. Nucl. Sci.* **2003**, *50*, 522–538. [CrossRef]
22. Barnaby, H.J. Total-Ionizing-Dose Effects in Modern CMOS Technologies. *IEEE Trans. Nucl. Sci.* **2006**, *53*, 3103–3121. [CrossRef]

23. Schwank, J.R.; Shaneyfelt, M.R.; Fleetwood, D.M.; Felix, J.A.; Dodd, P.E.; Paillet, P.; Ferlet-Cavrois, V. Radiation Effects in MOS Oxides. *IEEE Trans. Nucl. Sci.* **2008**, *55*, 1833–1853. [CrossRef]
24. Raine, M.; Hubert, G.; Gaillardin, M.; Paillet, P.; Bournel, A. Monte Carlo Prediction of Heavy Ion Induced MBU Sensitivity for SOI SRAMs Using Radial Ionization Profile. *IEEE Trans. Nucl. Sci.* **2011**, *58*, 2607–2613. [CrossRef]
25. Raine, M.; Gaillardin, M.; Lagutere, T.; Duhamel, O.; Paillet, P. Estimation of the Single-Event Upset Sensitivity of Advanced SOI SRAMs. *IEEE Trans. Nucl. Sci.* **2018**, *65*, 339–345. [CrossRef]
26. Liu, M.S.; Liu, H.Y.; Brewster, N.; Nelson, D.; Golke, K.W.; Kirchner, G.; Hughes, H.L.; Campbell, A.; Ziegler, J.F. Limiting Upset Cross Sections of SEU Hardened SOI SRAMs. *IEEE Trans. Nucl. Sci.* **2006**, *53*, 3487–3493. [CrossRef]
27. Schwank, J.R.; Shaneyfelt, M.R.; Felix, J.A.; Dodd, P.E.; Baggio, J.; Ferlet-Cavrois, V.; Paillet, P.; Hash, G.L.; Flores, R.S.; Massengill, L.W.; et al. Effects of Total Dose Irradiation on Single-Event Upset Hardness. *IEEE Trans. Nucl. Sci.* **2006**, *53*, 1772–1778. [CrossRef]
28. Pereira, E.C.F.; Gonçalez, O.L.; Vaz, R.G.; Federico, C.A.; Both, T.H.; Wirth, G.I. The effects of total ionizing dose on the neutron SEU cross section of a 130 nm 4 Mb SRAM memory. In Proceedings of the 2014 15th Latin American Test Workshop—LATW, Fortaleza, Brazil, 12–15 March 2014; pp. 1–4.
29. Xiao, Y.; Guo, H.-X.; Zhang, F.-Q.; Zhao, W.; Wang, Y.-P.; Zhang, K.-Y.; Ding, L.-L.; Fan, X.; Luo, Y.-H.; Wang, Y.-M. Synergistic effects of total ionizing dose on single event upset sensitivity in static random access memory under proton irradiation. *Chin. Phys. B* **2014**, *23*, 118503. [CrossRef]
30. Artola, L.; Gaillardin, M.; Hubert, G.; Raine, M.; Paillet, P. Modeling Single Event Transients in Advanced Devices and ICs. *IEEE Trans. Nucl. Sci.* **2015**, *62*, 1528–1539. [CrossRef]
31. Zheng, Q.; Cui, J.; Liu, M.; Zhou, H.; Liu, Y.; Wei, Y.; Su, D.; Ma, T.; Lu, W.; Yu, X.; et al. Total Ionizing Dose Influence on the Single-Event Upset Sensitivity of 130-nm PD SOI SRAMs. *IEEE Trans. Nucl. Sci.* **2017**, *64*, 1897–1904. [CrossRef]
32. Zheng, Q.; Cui, J.; Lu, W.; Guo, H.; Liu, J.; Yu, X.; Wei, Y.; Wang, L.; Liu, J.; He, C.; et al. The Increased Single-Event Upset Sensitivity of 65-nm DICE SRAM Induced by Total Ionizing Dose. *IEEE Trans. Nucl. Sci.* **2018**, *65*, 1920–1927. [CrossRef]
33. Schwank, J.R.; Shaneyfelt, M.R.; Draper, B.L.; Dodd, P.E. BUSFET-a radiation-hardened SOI transistor. *IEEE Trans. Nucl. Sci.* **1999**, *46*, 1809–1816. [CrossRef]
34. Ziegler, J.F.; Ziegler, M.D.; Biersack, J.P. SRIM—The stopping and range of ions in matter. *Nucl. Instrum. Methods Phys. Res. Sect. B Beam Interact. Mater. At.* **2010**, *268*, 1818–1823. [CrossRef]
35. Schwank, J.R.; Dodd, P.E.; Shaneyfelt, M.R.; Feli, J.A.; Hash, G.L.; Ferlet-Cavrois, V.; Paillet, P.; Baggio, J.; Tangyunyong, P.; Blackmore, E. Issues for single-event proton testing of SRAMs. *IEEE Trans. Nucl. Sci.* **2004**, *51*, 3692–3700. [CrossRef]
36. Faccio, F.; Cervelli, G. Radiation-induced edge effects in deep submicron CMOS transistors. *IEEE Trans. Nucl. Sci.* **2005**, *52*, 2413–2420. [CrossRef]
37. Faccio, F.; Michelis, S.; Cornale, D.; Paccagnella, A.; Gerardin, S. Radiation-Induced Short Channel (RISCE) and Narrow Channel (RINCE) Effects in 65 and 130 nm MOSFETs. *IEEE Trans. Nucl. Sci.* **2015**, *62*, 2933–2940. [CrossRef]
38. Wu, X.; Lu, W.; Wang, X.; Xi, S.-B.; Guo, Q.; Li, Y.-D. Total ionizing dose effect on 0.18 μm narrow-channel NMOS transistors. *Acta Phys. Sin.* **2013**, *62*, 136101. [CrossRef]
39. Zhou, H.; Cui, J.-W.; Zheng, Q.-W.; Guo, Q.; Ren, D.-Y.; Yu, X.-F. Reliability of partially-depleted silicon-on-insulator n-channel metal-oxide-semiconductor field-effect transistor under the ionizing radiation environment. *Acta Phys. Sin.* **2015**, *64*, 086101. [CrossRef]
40. Zhang, H.; Bi, J.; Wang, H.; Hu, H.; Li, J.; Ji, L.; Liu, M. Study of total ionizing dose induced read bit errors in magneto-resistive random access memory. *Microelectr. Reliab.* **2016**, *67*, 104–110. [CrossRef]

MDPI
St. Alban-Anlage 66
4052 Basel
Switzerland
www.mdpi.com

Electronics Editorial Office
E-mail: electronics@mdpi.com
www.mdpi.com/journal/electronics

Disclaimer/Publisher's Note: The statements, opinions and data contained in all publications are solely those of the individual author(s) and contributor(s) and not of MDPI and/or the editor(s). MDPI and/or the editor(s) disclaim responsibility for any injury to people or property resulting from any ideas, methods, instructions or products referred to in the content.

www.ingramcontent.com/pod-product-compliance
Lightning Source LLC
LaVergne TN
LVHW070443100526
838202LV00014B/1653